Rare Earth Elements: Theory, Properties and Applications

Rare Earth Elements: Theory, Properties and Applications

Edited by Blossom Jenkins

SYRAWOOD PUBLISHING HOUSE
New York

Published by Syrawood Publishing House,
750 Third Avenue, 9th Floor,
New York, NY 10017, USA
www.syrawoodpublishinghouse.com

Rare Earth Elements: Theory, Properties and Applications
Edited by Blossom Jenkins

International Standard Book Number: 978-1-64740-402-4 (Hardback)

Cataloging-in-publication Data

Rare earth elements : theory, properties and applications / edited by Blossom Jenkins.
p. cm.
Includes bibliographical references and index.
ISBN 978-1-64740-402-4
1. Rare earths. 2. Rare earth metals. 3. Nonferrous metals. I. Jenkins, Blossom.
QE390.2.R37 R37 2023
338.274 94--dc23

TABLE OF CONTENTS

PREFACE

This book was inspired by the evolution of our times; to answer the curiosity of inquisitive minds. Many developments have occurred across the globe in the recent past which has transformed the progress in the field.

The rare-earth elements (REEs) or lanthanides refer to a group of 17 nearly identical lustrous silvery-white soft heavy metals. These comprise the 15 lanthanides listed on the periodic table as well as yttrium and scandium. REEs are lustrous but they tarnish in air. The rare earth metals have high electrical conductivity. There are extremely minute differences in the solubility and complex formation between these elements. Lanthanum, cerium, neodymium and yttrium are the most abundant rare earth elements. The primary economic sources of REEs are the minerals loparite, bastnasite, monazite, and lateritic ion-adsorption clays. They react with other metallic and non-metallic elements to form compounds having distinct chemical properties. REEs are utilized in the production of phosphors, catalysts and polishing compounds. The compounds that comprise rare earth elements have a wide range of uses including lasers, magnetic materials, electrical and electronic components, glass and industrial processes. They are also used for defense applications such as radar and sonar systems, electronic displays, and guidance systems. This book contains some path-breaking studies related to the rare earth elements. Researchers and students studying these elements will be greatly assisted by it.

This book was developed from a mere concept to drafts to chapters and finally compiled together as a complete text to benefit the readers across all nations. To ensure the quality of the content we instilled two significant steps in our procedure. The first was to appoint an editorial team that would verify the data and statistics provided in the book and also select the most appropriate and valuable contributions from the plentiful contributions we received from authors worldwide. The next step was to appoint an expert of the topic as the Editor-in-Chief, who would head the project and finally make the necessary amendments and modifications to make the text reader-friendly. I was then commissioned to examine all the material to present the topics in the most comprehensible and productive format.

I would like to take this opportunity to thank all the contributing authors who were supportive enough to contribute their time and knowledge to this project. I also wish to convey my regards to my family who have been extremely supportive during the entire project.

Editor

1

Electrochemical Recovery and Behaviors of Rare Earth (La, Ce, Pr, Nd, Sm, Eu, Gd, Tb, Dy, Ho, Er, Tm and Yb) Ions on Ni Sheets

Min Hee Joo [1,2], So Jeong Park [1], Sung Min Hong [1,2], Choong Kyun Rhee [1] and Youngku Sohn [1,2,*]

1 Department of Chemistry, Chungnam National University, Daejeon 34134, Korea; wnalsgml4803@naver.com (M.H.J.); jsjs5921@naver.com (S.J.P.); qwqe212@naver.com (S.M.H.); ckrhee@cnu.ac.kr (C.K.R.)

2 Department of Chemical Engineering and Applied Chemistry, Chungnam National University, Daejeon 34134, Korea

* Correspondence: youngkusohn@cnu.ac.kr

Abstract: The electrochemical behaviors of rare earth (RE) ions have extensively been studied because of their high potential applications to the reprocessing of used nuclear fuels and RE-containing materials. In the present study, we fully investigated the electrochemical behaviors of RE(III) (La, Ce, Pr, Nd, Sm, Eu, Gd, Tb, Dy, Ho, Er, Tm, and Yb) ions over a Ni sheet electrode in 0.1 M $NaClO_4$ electrolyte solution by cyclic voltammetry between +0.5 and −1.5 V (vs. Ag/AgCl). Amperometry electrodeposition experiments were performed between −1.2 and −0.9 V to recover RE elements over the Ni sheet. The successfully RE-recovered Ni sheets were fully characterized by scanning electron microscopy, energy dispersive X-ray spectroscopy, Fourier transform infrared spectroscopy, X-ray photoelectron spectroscopy, and photoluminescence spectroscopy. The newly reported recovery data for RE(III) ions over a metal electrode provide valuable information on the development of the treatment methods of RE elements.

Keywords: rare earth ions; electrochemical recovery; amperometry electrodeposition; cyclic voltammetry; actinide ions

---

1. Introduction

Rare earth (RE) elements have widely been used in diverse industry materials such as magnets, fluorescent lamps, catalysts, and batteries [1–5]. Lanthanide (Ln) elements are commonly co-present with actinide (An) elements in nuclear oxide fuels. Therefore, the treatment and the recycling of used RE-containing industry materials and oxide fuels are of very attractive research projects [6–10]. In the reprocessing of spent nuclear oxide fuels, the pyroprocessing technology has been employed to reprocess Ln and An ions present in a pre-prepared electrolyte condition [11]. For this treatment, electrochemistry is an indispensable technique, and the understanding electrochemical behaviors (e.g., electrochemical reduction-oxidation reactions) of the ions over electrodes is very important. For the recovery of RE elements from used industry materials [12,13], the electrochemical recovery method has been a useful method to selectively recover a desired element in an electrolyte with mixed RE ions [12–14]. To achieve these goals, electrochemical behaviors of RE ions in an electrolyte have fundamentally been studied by electrochemistry [15–19]. Yang et al. studied electrochemical behaviors and electrodeposition of Eu (III) ions over a Al-Ga cathode, and co-reduction behaviors of Eu(III), Al(III) and Ga(III) over a W electrode [16]. They showed efficient reduction and co-reduction of Eu(III) ions over the electrodes. Liu et al. studied direct separation of U from RE (La, Nd, Ce, and Sm)-containing LiCl-KCl eutectic and showed that U was efficiently electrodeposited as Al-U alloys (Al_3U and Al_4U) on the Al electrode [17]. Kim and Lee focused on obtaining fundamental electrochemical properties of both Nd^{3+}/Nd^{2+} and Ce^{3+}/Ce in a molten LiCl-KCl eutectic salt at a high temperature of 773 K using a glassy carbon counter electrode, a Ag/Ag^+ reference electrode, and a W working electrode, and reported the measured diffusion constant, the standard reduction potential, and the standard-state Gibb's free energy of the RE ions [18]. Separation of a neighboring RE pair (e.g., Sm^{3+}/Sm^{2+} and Eu^{3+}/Eu^{2+}) is another goal to be achieved. Ge et al. used a reactive Cu electrode for selective recovery as Cu-Sm intermetallic compounds [19]. Al-Eu compound was reported to be selectively recovered over an Al electrode at a negative potential of −2.225 V (vs. Ag/Ag^+), higher than that of Al-Sm compound recovery on the electrode [19]. The molten-salt condition has commonly been performed at a high temperature [16–19] while the electrodeposition in ionic liquid conditions has been introduced for a mild temperature condition [20–22]. Xu et al. efficiently electrodeposited a rare earth iron alloy of Nd-Fe film on a Cu electrode in an ionic liquid condition containing Nd(III) and Fe(II) ions [20]. They proposed a deposition mechanism that Fe was initially deposited to subsequently catalyze the reduction of Nd(III) to Nd(0), and consequently Nd-Fe was co-deposited.

Although there are many literatures for the recovery of RE elements under various conditions, few studies have been systematically reported for the recovery of all the RE elements over a metal electrode in a $NaClO_4$ condition at room temperature. Because the electrochemical behaviors of RE and An elements are very similar, the selective recovery and the treatment are challenging and needed to be continuously developed. The novelty of this study is that all the RE elements were introduced to show that they had a systematic relationship on redox potentials with atomic number. Therefore, this data set is very useful for better understanding the redox behaviors of rare earth ions and the development of Ln and An treatment methods as well as other heavy metal ions present in water.

2. Results and Discussion

To examine the electrochemical behaviors of RE(III) ions, Figure 1 displays cyclic voltammetry (CV) curves for various 10 mM RE(III) ions in 0.1 M $NaClO_4$ electrolyte over bare Ni sheets. In a blank 0.1 M $NaClO_4$ electrolyte, there was no discernible reduction–oxidation (redox) peaks except for the current increases below −1.0 V (vs. Ag/AgCl) and above 0.0 V. Upon addition of 10 mM RE(III) ions, a strong negative current increase was commonly observed, starting from −0.5 V. There was no significant current increase between −0.5 and +0.5 V. The negative current increase was due to both the hydrogen evolution reaction and the RE reduction/complexation. Interestingly, as the atomic number was increased (or as the atomic size was decreased) from #57 (La) to #70 (Yb), the reduction peak became distinctly appeared. This indicates that the reduction potentials for La (#57), Ce (#58), and Pr (#59) are close to the hydrogen reduction potential, and the reduction potentials for Er (#68), Tm (#69), and Yb (#70) are far from the hydrogen reduction potential. The reduction potential for La was barely detected at −1.2 V (vs. Ag/AgCl) and that for Yb was clearly seen at −0.95 V (vs. Ag/AgCl). The reduction potentials for Nd, Sm, and Eu were observed to be between −1.11 and −0.92 V (vs. Ag/AgCl). Xu et al. reported that the equilibrium potentials of RE elements were in a relationship with atomic size [23]. Herein, the present redox potentials of RE(III) ions appear to be also influenced by other factors such as electronegativity and complex formation degree.

To examine the electrochemical recovery of RE(III) ions over Ni sheets and the corresponding morphology, Figure 2 shows scanning electron microscope (SEM) and optical microscope images before and after electrodeposition. On the basis of the images, all the RE elements were efficiently deposited on Ni sheets. The deposited areas were clearly discriminated from the undeposited areas by optical microscope images. For the SEM images of La, Ce, and Pr, the surface morphology showed a uniform film state, although some cracks were present in the deposited areas. For the SEM images of Nd, Sm, Eu, Gd, and Tb, the morphology appeared to be a film state aggregated by ultrafine particles. For the SEM images of Dy, Ho, Er, Tm, and Yb, the particle sizes were appeared to be much bigger and become nanoparticle structures. Overall, it can be concluded that the morphology was changed from uniform thin film state to nanoparticle structures as the atomic number was increased. The present data can provide valuable information on the development of thin film fabrication by electrodeposition.

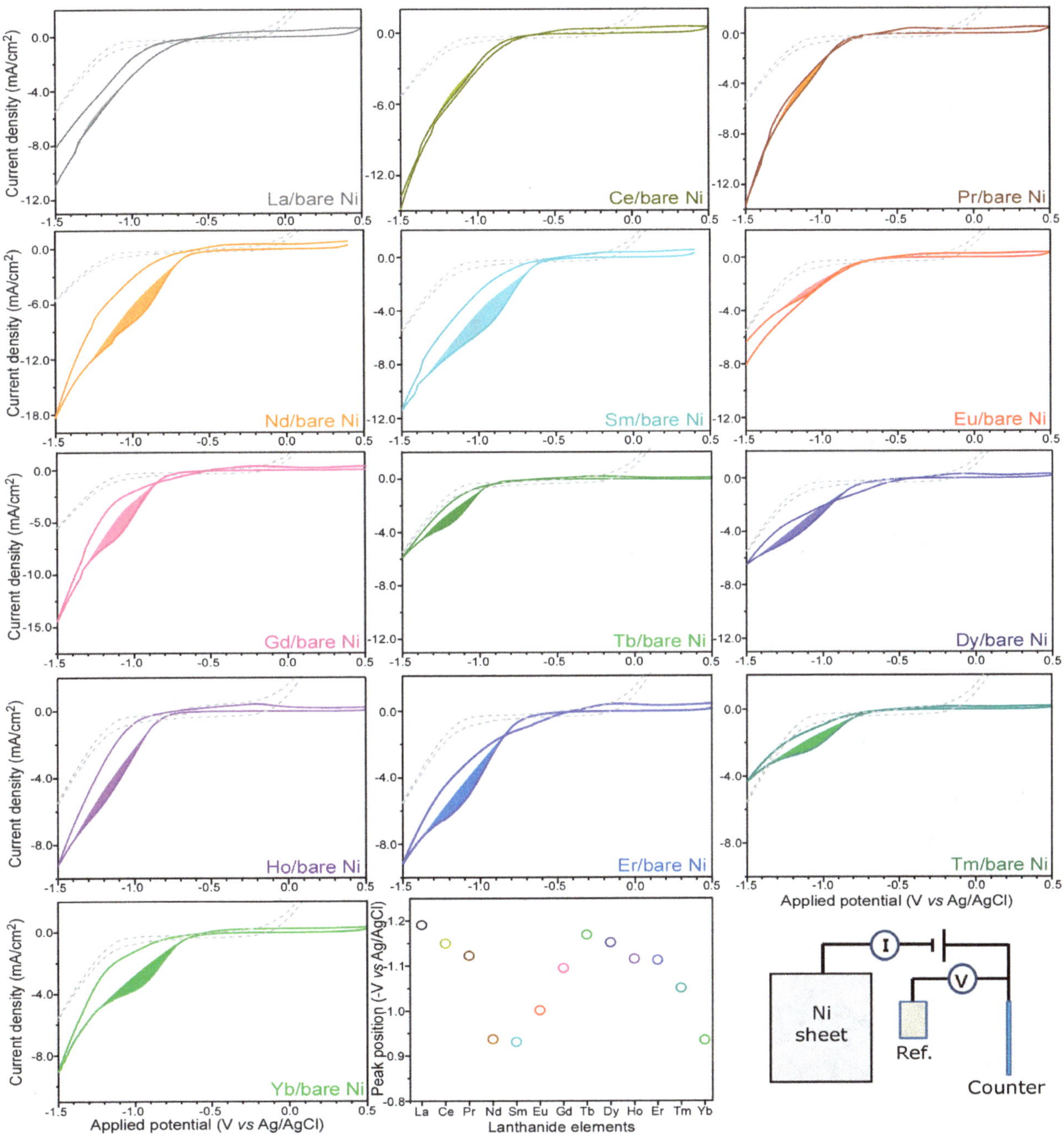

Figure 1. Cyclic voltammetry curves at a scan rate of 0.2 V/s for various 10 mM RE(III) ions in 0.1 M $NaClO_4$ electrolyte over bare Ni sheets. The reduction peak positions (V vs. Ag/AgCl) are shown with RE elements. A schematic of the three-electrode system used in this study is shown.

To confirm the elements of the recovered RE elements, Figure 3 shows the energy-dispersive X-ray spectroscopy (EDXS) data for the samples shown in Figure 2. For the EDX spectrum (not shown here) of a bare Ni sheet, the detected signals were of Ni (major), C (minor), and O (minor) elements: Ni L (0.84 KeV), C K (0.26 KeV), and O K (0.52 KeV) [24]. Upon electrodeposition, the signals of RE, C, and O were strongly increased. The strong O signal was plausibly due to RE–O, –OH, $-ClO_4$, and $-CO_3$ species, confirmed by Fourier transform infrared spectroscopy (FT-IR) and X-ray photoelectron spectroscopy (XPS) shown below. The EDXS signals of RE and Cl (Cl K = 2.62 KeV) elements were newly appeared.

The Cl K signal was due to ClO_4 species, further discussed in the FT-IR spectra below. The Ni L signals were observed to be dependent on the amounts of RE deposited on a Ni sheet. The Gd/Ni sample showed the strongest Ni signal while Dy/Ni showed the weakest signal. This indicates that Dy element was the most recovered by electrodeposition after a given time for 5 min. For the EDXS signals (Ln M) of RE elements, the peak position was linearly increased to a higher energy position with increasing the atomic number. The La M (#57 for La element) and the Yb M (#70 for Yb element) signals were positioned at 0.83 and 1.52 KeV, respectively [24]. The peak separation between the two peaks was estimated to be 0.69 KeV. For the radioactive promethium, Pm M can be expected to be observed around 1.1 KeV. On the basis of inductively coupled plasma elemental analysis of Ce and the Ln M EDXS signal intensities, the recovery percentages (%) were roughly estimated to be La (15.6%), Ce (3%), Pr (3.2%), Nd (5.7%), Sm (5.9%), Eu (4.3%), Gd (2.7%), Tb (3.3%), Dy (12.6%), Ho (11.5%), Er (12.7%), Tm (15.3%), and Yb (18.7%) over a 5 mm × 10 mm size electrode in 5 min.

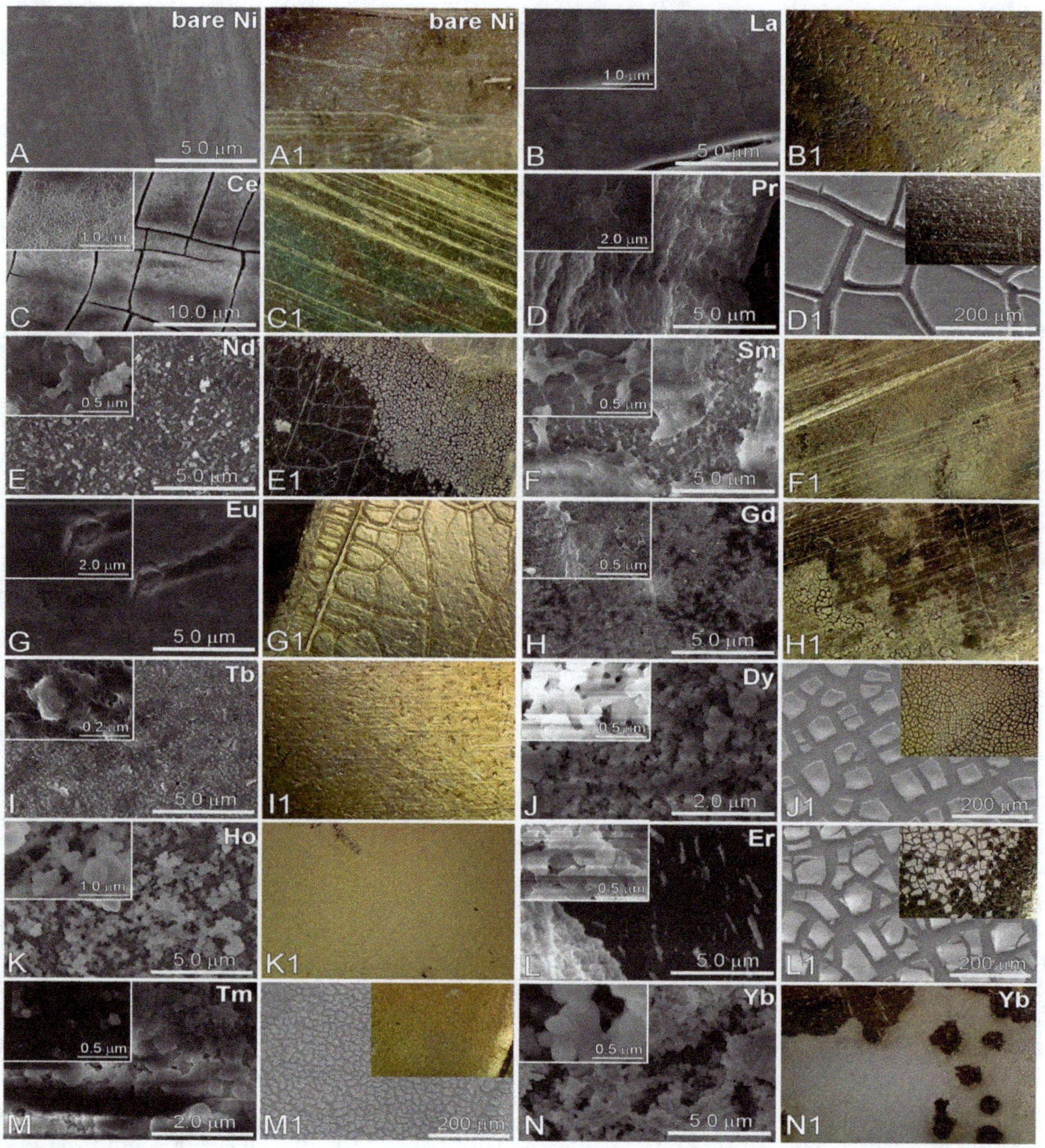

Figure 2. SEM images (**A**–**N**) and optical microscope (300×) images (**A1**–**N1**) are shown for a bare Ni sheet (**A**) and electrodeposited RE elements on a Ni sheet (from **B**–**N**).

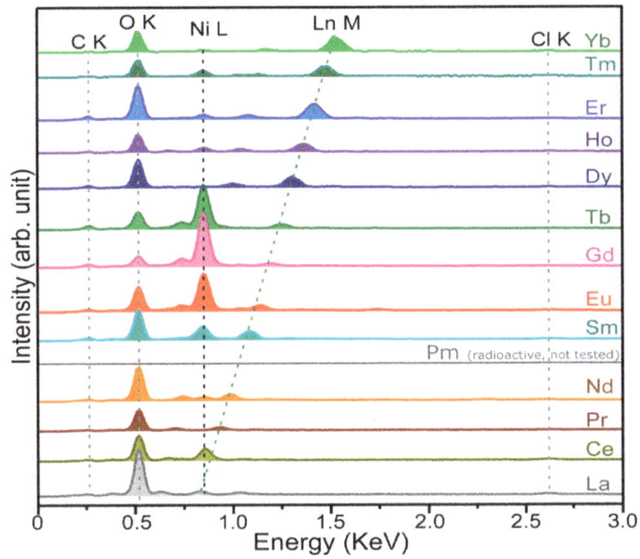

Figure 3. Energy-dispersive X-ray spectroscopy (EDXS) data for electrodeposited RE elements over a Ni sheet.

To deduce the chemical structures of the electrodeposited materials, Figure 4 displays the FT-IR spectra for all the electrodeposited samples. Very interestingly, all the FT-IR spectra were found to be similar, except for the signal intensity, which was determined by the amounts of recovered materials. The FT-IR intensity was in good consistent with the EDXS signal intensity. The Gd/Ni sample showed the weakest IR intensity. As discussed above, this sample showed the strongest EDXS Ni signal (from the Ni support). This indicates that the electrodeposited Gd was the thinnest. In the FT-IR spectra, a broad peak around 3600 cm^{-1} was distinctly observed, attributed to RE–OH stretching vibrations [25–28]. The corresponding O–H bending vibrational mode was observed at 1625 cm^{-1}. The commonly appeared peak at 620 cm^{-1} was attributed to a RE–O vibrational mode [26]. A broad peak at 1080 cm^{-1} could be related with a vibration of ClO_4^- group [29]. Two strong peaks were also observed at 1350 and 1403 cm^{-1}, the most plausibly attributed to the stretching vibration modes of CO_3^{2-} group [25–28]. On the basis of the FT-IR data, it could be concluded that the electrodeposited samples were complexes of RE, OH (and/or H_2O), CO_3^{2-}, and ClO_4^- groups, further discussed below.

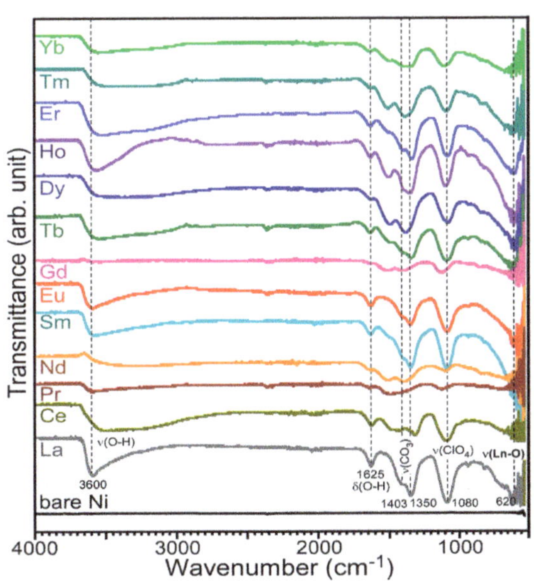

Figure 4. Transmittance FT-IR spectra for electrodeposited RE elements over a Ni sheet.

To further examine the crystal structures of the electrodeposited samples, the X-ray diffraction (XRD) patterns of a selected Eu sample and a bare Ni were displayed in Figure 5. The XRD signals were barely detected because the film was ultrathin (and low crystallinity), and the electrode was too small to be properly aligned. This needs further investigation to increase a potential applicability. For bare Ni sheet, the XRD signals were very strong and there were no significant impurity signals. The strong peaks at 2θ = 45° and 52° were attributed to the (111) and (200) crystal planes of metallic Ni [30]. The other (but clearly discriminated from the background signal) weak XRD signals were observed around 2θ = 10°, 20°, 28°, and 50°. These XRD peak positions are similar to those of $Ln_2(OH)_x(NO_3)_y(SO_4)_z{\cdot}nH_2O$ complex [25] and metal carbonate hydroxide structures [25,27,28,31,32]. Assuming that anions of SO_4^{2-} and NO_3^- are exchanged by CO_3^{2-} and ClO_4^- but the crystal phase is not changed [25], the corresponding crystal planes are assigned on the four broad peaks in Figure 5. Based on the present XRD information, the FT-IR data, and the literature information, the recovered RE elements were proposed to be presented as $RE_2(OH)_x(CO_3)_{y-z}(ClO_4)_z{\cdot}nH_2O$ complex, where O and RE were the most abundantly present in the complex [25]. The EDXS data commonly showed RE, C, O, and Cl elements, and the atomic compositions of RE and O were much higher than those of C and Cl.

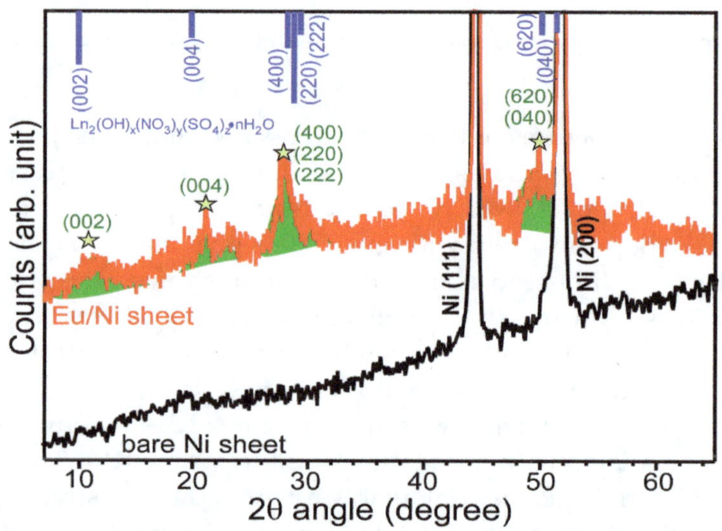

Figure 5. XRD data for bare Ni and selected electrodeposited Eu on a Ni sheet. The literature XRD patterns for $Ln_2(OH)_x(NO_3)_y(SO_4)_z{\cdot}nH_2O$ complex are shown for comparison.

Surface chemical states of a selected electrodeposited sample were examined by X-ray photoelectron spectroscopy (XPS). Figure 6 displays survey, high-resolution C 1s, Ni 2p, and Eu 3d XPS spectra before and after Eu electrodeposition over a Ni sheet. All the binding energies (BEs) here were not calibrated using an internal standard, but the XPS spectrometer showed an Au $4f_{7/2}$ XPS peak at 84.0 eV for a cleaned Au film (as an external standard). For the survey XPS scan of a bare Ni sheet, Ni, C and O elements were only observed. An O 1s XPS signal was observed at a BE of 530.2 eV for a bare Ni sheet, due to oxidation of Ni surface. The surface oxide was ultrathin and could not be detected by the bulk XRD technique, shown above. For the survey scan of an electrodeposited Eu on Ni, Eu 3d and Eu 4d signals were strongly and newly appeared. In addition, N 1s and Cl 2p XPS signals were also newly appeared. The N 1s peak position was observed at a BE of 406.0 eV [33,34], plausibly due to nitrates trapped in the surface. The Cl 2p BE was observed at 207.0 eV, due to ClO_4 species [33,35]. The Cl element was also detected by EDXS, shown above. In the C 1s XPS for bare Ni, two peaks were observed at 283.8 eV (major) and 287.2 eV (minor), attributed to adventitious C–C and C–O species, respectively. In the C 1s XPS for electrodeposited Eu on Ni, a C 1 XPS peak at a BE of 288.5 eV was newly observed, attributed to O–C=O species such as carbonates [33]. In the Ni 2p XPS for bare Ni, several peaks were observed. Two shaper peaks at 868.6 and 851.4 eV with a spin orbit splitting of 17.2 eV were assigned to the Ni $2p_{1/2}$ and Ni $2p_{3/2}$ XPS signals of metallic Ni [30,33,36], respectively.

Two broader peaks at 872.6 and 854.6 eV with a spin orbit splitting of 17.7 eV were assigned to the Ni $2p_{1/2}$ and Ni $2p_{3/2}$ XPS signals of Ni(II) state (e.g., NiO), respectively [30,36]. The corresponding satellite peaks of Ni(II) oxidation state were observed at 879.3 and 861.0 eV, respectively [33,36]. On the basis of the Ni 2p XPS, the surface of Ni sheet was oxidized to Ni(II) state. Upon Eu deposition, the Ni 2p signal was not observed, an indication that the electrodeposited overlayer was much thicker than the XPS probe depth. In the Eu 3d XPS, two major signals were observed at 1163.3 and 1133.5 eV with a spin-orbit splitting energy of 29.8 eV, attributed to Eu $3d_{3/2}$ and Eu $3d_{5/2}$ XPS signals of Eu(III) state, respectively [37]. A weaker Eu $3d_{3/2}$ peak around 1156.0 eV could be due to some Eu(II) states [37].

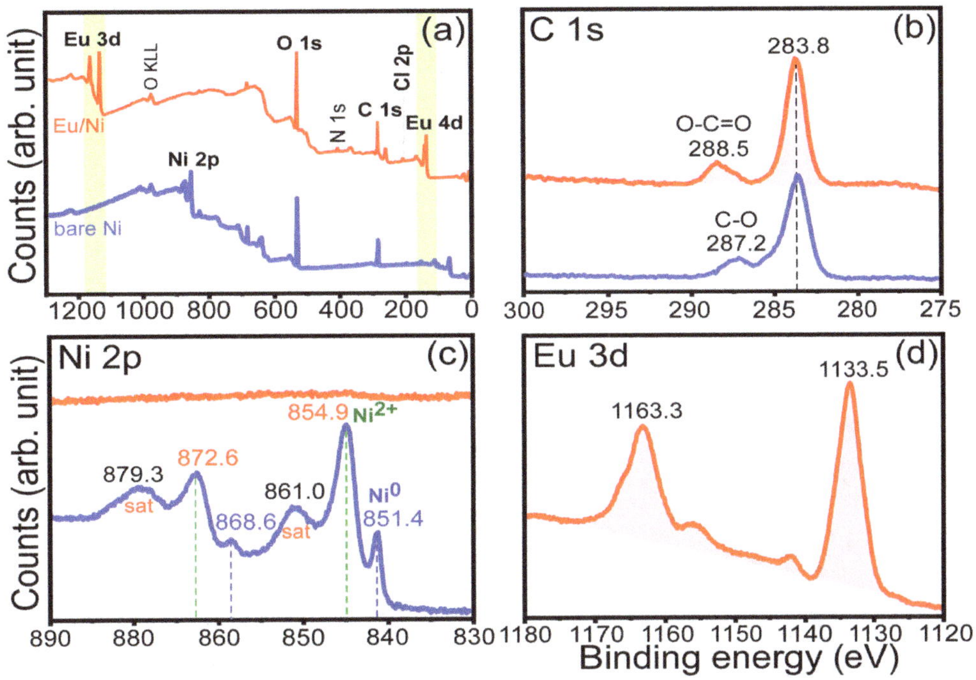

Figure 6. Survey (**a**), high-resolution C 1s (**b**), Ni 2p (**c**), and Eu 3d (**d**) XPS spectra before and after Eu electrodeposition over a bare Ni sheet.

To further confirm the oxidation state of Eu elements, photoluminescence spectroscopy was employed for the Eu electrodeposited sample because Eu(III) ion has unique $^5D_0 \rightarrow {}^7F_{0,1,2,3,4}$ transitions in the visible region between 550 and 720 nm [37–40], while Eu(II) shows no visible light emission under an excitation of UV light. Figure 7 shows excitation and emission spectra for a thick electrodeposited Eu sample and the corresponding 2D- and 3D-PL contour mapping images. For the excitation spectra setting at emission wavelengths (λ_{em}) of 613 and 590 nm, the PL profiles were found to be very similar, but the intensity was stronger for the spectra at λ_{em} = 613 nm. This indicates that the emission under 613 nm light was stronger than the emission under 590 nm light. Various peaks were observed at 299, 319, 363, 383, 395, 416, and 466 nm, assigned to the $^5F_4$, $^5H_5$, $^5D_4$, $^5G_J/^5L_7$, $^5L_6$, $^5D_3$, and $^5D_2$ transitions from the ground $^7F_0$ state of Eu(III), respectively [37]. For the emission spectra taken at excitation wavelengths (λ_{ex}) of 280 nm (indirect excitation), 320 nm (the $^5F_4 \leftarrow {}^7F_0$ direct excitation), and 395 nm (the $^5L_6 \leftarrow {}^7F_0$ direct excitation), the two emission profiles at direct excitations of λ_{ex} = 320 and 395 nm are similar, but the emission profile at an indirect excitation of λ_{ex} = 280 nm is dissimilar to the other two. The PL intensity of the direct transition to an Eu(III) excited energy level was found to be stronger than that of the indirect transition. Several PL peak positions were commonly observed at 578, 590, 613, 647, and 698 nm, commonly associated to the $^5D_0 \rightarrow {}^7F_0$, $^5D_0 \rightarrow {}^7F_1$, $^5D_0 \rightarrow {}^7F_2$, $^5D_0 \rightarrow {}^7F_3$, and $^5D_0 \rightarrow {}^7F_4$, transitions, respectively [37–40]. The $^5D_0 \rightarrow {}^7F_2$ transition was observed to be the most intense. Because of the strong emission, the sample appeared to be red for the electrodeposited Eu (inset photo in Figure 7A). The $^5D_0 \rightarrow {}^7F_2$ transition is known to be electric dipole transition and hypersensitive to Eu(III) local environment. The transition becomes dominant when Eu(III) ion is located at an asymmetric site. On the other hand, the $^5D_0 \rightarrow {}^7F_1$ transition is insensitive to the local environment. For this reason, ($^5D_0 \rightarrow {}^7F_2$)/($^5D_0 \rightarrow {}^7F_1$) intensity ratio is regarded as an asymmetric ratio for Eu(III) location site. The ratio was estimated to be 2.3 for the emission spectra at direction excitations. The ratio was 1.2 for the emission at indirect excitation. The corresponding 2D and 3D contour PL mapping profiles show densely spaced regions, an indication of strong emission signals. The densely spaced regions are mainly localized at the upper energy levels ($^5F_4$, $^5H_5$, $^5D_4$, ^5G_J/$^5L_7$, $^5L_6$, $^5D_3$, and $^5D_2$) $\leftarrow {}^7F_0$ excitation transitions and $^5D_0 \rightarrow {}^7F_{0,1,2,3,4}$ emission transitions.

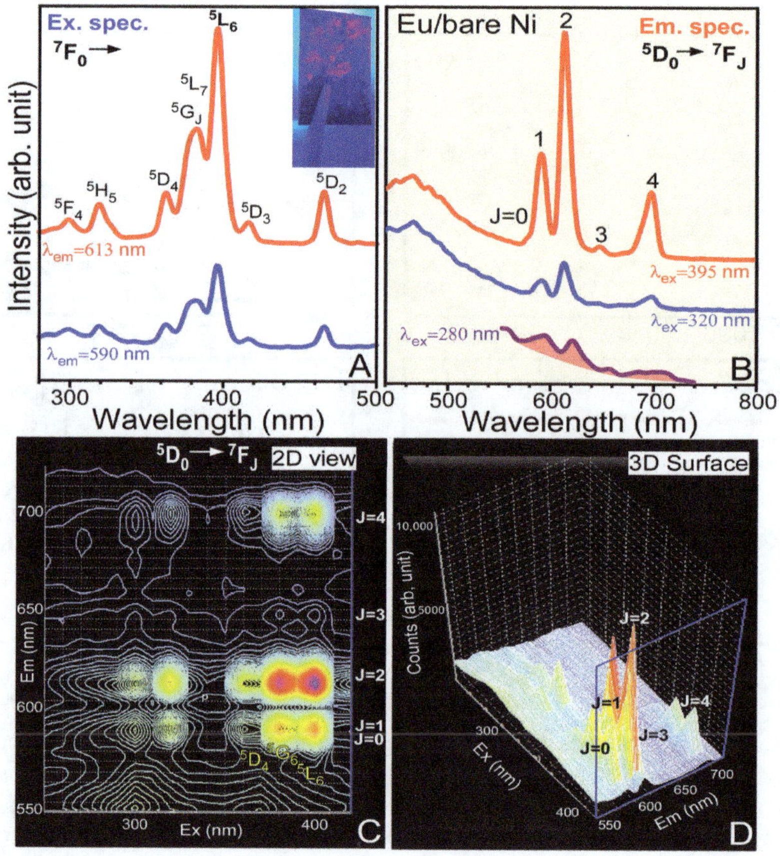

Figure 7. Excitation (**A**) and emission (**B**) spectra for a thick electrodeposited Eu on a Ni sheet, and the corresponding 2D- and 3D-PL contour mapping images (**C**, **D**).

3. Experimental Section

3.1. Sample Preparations

Ni sheets (99.96%) were cut with a size of 5 mm × 20 mm; cleaned by ultrasonication in acetone, isopropyl alcohol, and water, repeatedly; and dried under an infrared lamp. All the purchased RE(III) ions were nitrate forms and used as received without any further purification: La(III) nitrate hexahydrate (99.999%, Sigma-Aldrich, Saint Louis, MO, USA), Ce(III) nitrate hexahydrate (99%, Sigma-Aldrich), Pr(III) nitrate pentahydrate (99.9%, Alfa Aesar, Ward Hill, MA, USA), Nd(III) nitrate hexahydrate (99.9%, Alfa Aesar), Sm(III) nitrate hexahydrate (99.9%, Alfa Aesar), Eu(III) nitrate hexahydrate (99.9%, Alfa Aesar), Gd(III) nitrate hexahydrate (99.9%, Sigma-Aldrich), Tb(III) nitrate hydrate (99.9%, Alfa Aesar), Dy(III) nitrate pentahydrate (99.9%, Sigma-Aldrich), Ho(III) nitrate pentahydrate (99.99%, Alfa Aesar), Er(III) nitrate hydrate (99.9%, Sigma-Aldrich), Tm(III) nitrate hydrate (99.9%, Alfa Aesar), and Yb(III) nitrate hydrate (99.9%, Sigma Aldrich). Sodium perchlorate ($NaClO_4 \geq 98.0\%$, Sigma-Aldrich) was used as received and made to an aqueous 0.1 M solution as a supporting electrolyte. RE(III) ions were 10 mM concentration in the supporting electrolyte. Cyclic voltammetry tests were conducted using a WPG100 Potentiostat/Galvanostat (WonATech Co., Ltd., Seoul, Korea) electrochemical workstation. with a conventional three-electrode arrangement: a Ag/AgCl (3.5 M KCl) reference electrode, a Pt wire (0.5 mm) counter electrode, and a Ni sheet electrode (5 mm × 20 mm) in a 0.1 M $NaClO_4$ electrolyte with and without RE(III) ions (10 mM). For the recovery of the RE elements, the electrodeposition over a Ni sheet was performed by amperometry for 5 min setting at the reduction potential observed in the CV data. After the electrodeposition, the electrode was gently washed with deionized water and dried under an infrared lamp before further characterization.

3.2. Characterization of the RE Recovered Materials

The morphology of the RE-recovered samples was examined using a field-emission Hitachi S-4800 SEM (FE-SEM, Hitach Ltd., Tokyo, Japan). The crystal phases of the electrodeposited sample were identified using a MiniFlex II X-ray diffractometer (Rigaku Corp., Tokyo, Japan) equipped with a Cu K_α radiation source. Energy-dispersive X-ray spectroscopy (EDXS) was performed using a JSM 7000 F scanning electron microscope (JEOL Ltd., Tokyo, Japan) at an acceleration voltage of 20 kV. The FT-IR spectra were recorded using a Nicolet iS 10 FT-IR spectrometer (Thermo Scientific Korea, Seoul, Korea) with an attenuated total reflection mode. Photoluminescence (PL) emission and excitation spectra were obtained using a Sinco FS-2 fluorescence spectrometer (Sinco, Seoul, Korea) setting at scan speed = 300 nm/min, slit width = 5 nm, PMT voltage = 700 V, and integration time = 20 ms. The emission profiles were recorded at various excitation wavelengths to plot 2D/3D PL contour profiles. X-ray photoelectron spectroscopy was employed using Thermo-VG Scientific K-alpha$^+$ spectrometer (Thermo VG Scientific, Waltham, MA, USA) with a monochromatic Al K_α X-ray source (spot size = 400 μm) and a hemispherical energy analyzer (CAE mode, pass energy = 50.0 eV, and step size = 0.100 eV). An Avio 500 inductively coupled plasma optical emission spectrometer (Perkin Elmer, Waltham, MA, USA) was used to examine the amount of Ce ions present in the electrolyte after the amperometry test for 5 min.

4. Conclusions

In this work, we first showed a big data set for the recovery of rare earth (La, Ce, Pr, Nd, Sm, Eu, Gd, Tb, Dy, Ho, Er, Tm, and Yb) ions over a Ni sheet in a 0.1 M $NaClO_4$ electrolyte. Cyclic voltammetry tests showed the reduction potentials between −1.2 and −0.8 V (vs. Ag/AgCl). The observed reduction potentials were reported for all the RE elements with increasing the atomic number. SEM images showed that the electrodeposition method was useful to recover the RE elements as thin film or nanoparticle structures on a Ni sheet. As the atomic number was increased the morphology was changed from uniform thin film state to nanoparticle structures. The electrodeposited La (#59) on Ni showed a uniform film morphology while the electrodeposited Yb (#70) on a Ni sheet showed nanoparticle morphology. EDXS data confirmed the elements of RE, C, O, and Cl. The EDXS Ni L signal was used to estimate the relative amounts of electrodeposited RE materials. On the basis of the FT-IR and XRD data, it was tentatively concluded that the electrodeposited recovered materials were of a $RE_2(OH)_x(CO_3)_{y-z}(ClO_4)_z{\cdot}nH_2O$ complex. For a selected electrodeposited Eu on Ni, the oxidation state of 3+ was confirmed by Eu 3d XPS signals. The Eu(III) oxidation state was further confirmed by the PL spectra showing the $^5D_0 \rightarrow {}^7F_{0,1,2,3,4}$ emission transitions in the visible region between 570 and 720 nm. Future potential experiments may include a clear elucidation of the crystal structure, and electrodeposition over a porous Ni mesh for energy storage and energy production (e.g., CO_2 reduction and water splitting) electrochemical electrodes.

Overall, the present study contains very valuable electrochemical behaviors of all the RE(III) ions in a $NaClO_4$ electrolyte when a Ni sheet is used as a working electrode. The newly established big data set for RE elements could be used to predict the redox behaviors of Ln and An elements present in solutions. Furthermore, their morphologies, chemical states and the electrochemical recovery method could provide valuable information on the reprocessing of used nuclear oxide fuels, the recovery of RE-containing industry wastes, and fabrication of thin films.

Author Contributions: Conceptualization, C.K.R. and Y.S.; methodology, Y.S.; formal analysis, M.H.J., S.J.P. and S.M.H.; investigation, M.H.J., S.J.P. and S.M.H.; data curation, Y.S.; writing—original draft preparation, M.H.J. and Y.S.; writing—review and editing, C.K.R. and Y.S.; supervision, C.K.R. and Y.S.; funding acquisition, Y.S. All authors have read and agreed to the published version of the manuscript.

References

1. Cardoso, C.E.D.; Almeida, J.; Lopes, C.B.; Trindade, T.; Vale, C.; Pereira, M.E. Recovery of Rare Earth Elements by Carbon-Based Nanomaterials—A Review. *Nanomaterials* **2019**, *9*, 814. [CrossRef] [PubMed]
2. Wójcik, G. Sorption Behaviors of Light Lanthanides(III) (La(III), Ce(III), Pr(III), Nd(III) and Cr(III) Using Nitrolite. *Materials* **2020**, *13*, 2256. [CrossRef] [PubMed]
3. Dev, S.; Sachan, A.; Dehghani, F.; Ghosh, T.; Briggs, B.R.; Aggarwal, S. Mechanisms of biological recovery of rare-earth elements from industrial and electronic wastes: A review. *Chem. Eng. J.* **2020**, *397*, 124596. [CrossRef]
4. Wang, Y.; Guo, X.; Bai, Y.; Sun, X. Effective removal of calcium and magnesium sulfates from wastewater in the rare earth industry. *RSC Adv.* **2019**, *9*, 33922–33930. [CrossRef]
5. Edahbi, M.; Plante, B.; Benzaazoua, M. Environmental challenges and identification of the knowledge gaps associated with REE mine wastes management. *J. Clean. Prod.* **2019**, *212*, 1232–1241. [CrossRef]
6. Binnemans, K.; Jones, P.T.; Blanpain, B.; Van Gerven, T.; Yang, Y.; Walton, A.; Buchert, M. Recycling of rare earths: A critical review. *J. Clean. Prod.* **2013**, *51*, 1–22. [CrossRef]
7. Jha, M.K.; Kumari, A.; Panda, R.; Kumar, J.R.; Yoo, K.; Lee, J.Y. Review on hydrometallurgical recovery of rare earth metals. *Hydrometallurgy* **2016**, *165*, 2–26. [CrossRef]
8. Binnemans, K.; Jones, P. Perspectives for the recovery of rare earths from end-of-life fluorescent lamps. *J. Rare Earths* **2014**, *32*, 195–200. [CrossRef]
9. Yurramendi, L.; Gijsemans, L.; Forte, F.; Aldana, J.L.; Del Río, C.; Binnemans, K. Enhancing rare-earth recovery from lamp phosphor waste. *Hydrometallurgy* **2019**, *187*, 38–44. [CrossRef]
10. Liu, H.; Zhang, S.-G.; Pan, D.; Tian, J.; Yang, M.; Wu, M.; Volinsky, A.A. Rare earth elements recycling from waste phosphor by dual hydrochloric acid dissolution. *J. Hazard. Mater.* **2014**, *272*, 96–101. [CrossRef]
11. Choi, E.-Y.; Jeong, S.M. Electrochemical processing of spent nuclear fuels: An overview of oxide reduction in pyroprocessing technology. *Prog. Nat. Sci.* **2015**, *25*, 572–582. [CrossRef]
12. Abbasalizadeh, A.; Malfliet, A.; Seetharaman, S.; Sietsma, J.; Yang, Y. Electrochemical Recovery of Rare Earth Elements from Magnets: Conversion of Rare Earth Based Metals into Rare Earth Fluorides in Molten Salts. *Mater. Trans.* **2017**, *58*, 400–405. [CrossRef]
13. Tunsu, C.; Petranikova, M.; Ekberg, C.; Retegan, T. A hydrometallurgical process for the recovery of rare earth elements from fluorescent lamp waste fractions. *Sep. Purif. Technol.* **2016**, *161*, 172–186. [CrossRef]
14. Yang, D.-W.; Liu, Y.-L.; Yin, T.-Q.; Jiang, S.-L.; Zhong, Y.-K.; Wang, L.; Li, M.; Chai, Z.-F.; Shi, W. Application of Binary Ga–Al Alloy Cathode in U Separation from Ce: The Possibility in Pyroprocessing of Spent Nuclear Fuel. *Electrochim. Acta* **2020**, *353*, 136449. [CrossRef]
15. Yuan, L.; Yin, T.-Q.; Liu, Y.; Yang, D.; Yan, P.Y.; Wang, G.; Chai, Z. Thermodynamics and Kinetics Properties of Lanthanides (La, Ce, Pr, Nd) on the Liquid Bismuth Electrode in LiCl-KCl Molten Salt. *J. Electrochem. Soc.* **2020**, *167*, 122507. [CrossRef]
16. Yang, D.-W.; Chang, K.-K.; Hu, X.-F.; Liu, Y.-L.; Yin, T.-Q.; Jiang, S.-L.; Liu, J.; Han, W.; Chai, Z.-F.; Yuan, L. Electrochemical Behaviors of Eu (III) on the Liquid Binary Al-Ga Alloy Cathode. *J. Electrochem. Soc.* **2019**, *166*, D882–D889. [CrossRef]
17. Liu, Y.-L.; Liu, K.; Luo, L.; Yuan, L.; Chai, Z.; Shi, W. Direct separation of uranium from lanthanides (La, Nd, Ce, Sm) in oxide mixture in LiCl-KCl eutectic melt. *Electrochim. Acta* **2018**, *275*, 100–109. [CrossRef]
18. Kim, S.; Lee, S.-H. Electrochemical properties of $NdCl_3$ and $CeCl_3$ in molten LiCl-KCl eutectic salt. *Appl. Sci.* **2020**, *10*, 7252. [CrossRef]
19. Ge, J.; Yang, Q.; Wang, Y.; Zhuo, W.; Du, M.; Zhang, J. Selective Electrodeposition of Europium and Samarium in Molten LiCl-KCl with Copper and Aluminum Electrodes. *J. Electrochem. Soc.* **2020**, *167*, 022501. [CrossRef]
20. Xu, X.; Sturm, S.; Zavašnik, J.; Rozman, K.Z. Electrodeposition of a Rare-Earth Iron Alloy from an Ionic-Liquid Electrolyte. *ChemElectroChem* **2019**, *6*, 2860–2869. [CrossRef]

21. Hunter, J.P.; Dolezalova, S.; Ngwenya, B.T.; Morrison, C.A.; Love, J.B. Understanding the recovery of rare-earth elements by ammonium salts. *Metals* **2018**, *8*, 465. [CrossRef]
22. Krishna, G.M.; Rout, A.; Venkatesan, K. Voltammetric investigation of some lanthanides in neutral ligand-ionic liquid. *J. Electroanal. Chem.* **2020**, *856*, 113671. [CrossRef]
23. Zhang, M.; Zhang, M.; Yan, Y.; Qiu, M.; Chen, J. Theoretical investigation of lanthanide and transition metal on Al cathode: Equilibrium potential and atomic radii analysis by a mathematical equation. *Colloids Surf. A Physicochem. Eng. Asp.* **2020**, *590*, 124490. [CrossRef]
24. NIST DTSA-II Software. Available online: https://cstl.nist.gov/div837/837.02/epq/dtsa2/ (accessed on 30 September 2020).
25. Wu, X.; Li, J.; Zhu, Q.; Liu, W.; Sun, X.; Sakka, Y. One-step freezing temperature crystallization of layered rare-earth hydroxide ($Ln_2(OH)_5NO_3·nH_2O$) nanosheets for a wide spectrum of Ln (Ln = Pr–Er, and Y), anion exchange with fluorine and sulfate, and microscopic coordination probed via photoluminescence. *J. Mater. Chem. C* **2015**, *3*, 3428–3437. [CrossRef]
26. Kang, J.-G.; Min, B.-K.; Sohn, Y. Synthesis and characterization of $Gd(OH)_3$ and Gd_2O_3 nanorods. *Ceram. Int.* **2015**, *41*, 1243–1248. [CrossRef]
27. Sohn, Y. Yb_2O_3 nanowires, nanorods and nano-square plates. *Ceram. Int.* **2018**, *44*, 3341–3347. [CrossRef]
28. Kang, J.-G.; Gwag, J.S.; Sohn, Y. Synthesis and characterization of $Dy(OH)_3$ and Dy_2O_3 nanorods and nanosheets. *Ceram. Int.* **2015**, *41*, 3999–4006. [CrossRef]
29. Chen, Y.; Zhang, Y.; Zhao, L.-J. ATR-FTIR spectroscopic studies on aqueous $LiClO_4$, $NaClO_4$, and $Mg(ClO_4)_2$ solutions. *Phys. Chem. Chem. Phys.* **2004**, *6*, 537. [CrossRef]
30. Li, J.; Li, P.; Li, J.; Tian, Z.Q.; Yu, F. Highly-Dispersed Ni-NiO Nanoparticles Anchored on an SiO_2 Support for an Enhanced CO Methanation Performance. *Catalists* **2019**, *9*, 506. [CrossRef]
31. Zhang, Y.; Cao, H.; Zhang, J.; Xia, B. Synthesis of $LiNi_{0.6}Co_{0.2}Mn_{0.2}O_2$ cathode material by a carbonate co-precipitation method and its electrochemical characterization. *Solid State Ion.* **2006**, *177*, 3303–3307. [CrossRef]
32. Antoniou, S.; Pashalidis, I.; Gessner, A.; Kumke, M.U. Spectroscopic investigations on the effect of humic acid on the formation and solubility of secondary solid phases of $Ln_2(CO_3)_3$. *J. Rare Earths* **2011**, *29*, 516–521. [CrossRef]
33. NIST. *X-Ray Photoelectron Spectroscopy Database*; Version 4.1; National Institute of Standards and Technology: Gaithersburg, MD, USA, 2012. Available online: http://srdata.nist.gov/xps/ (accessed on 30 September 2020).
34. Miyakawa, M.; Nakata, M.; Tsuji, H.; Fujisaki, Y. Simple and reliable direct patterning method for carbon-free solution-processed metal oxide TFTs. *Sci. Rep.* **2018**, *8*, 12825. [CrossRef] [PubMed]
35. Sahadeo, E.; Song, J.H.; Gaskell, K.; Kim, N.; Rubloff, G.W.; Lee, S.B.; Gaskel, K. Investigation of the water-stimulated Mg^{2+} insertion mechanism in an electrodeposited MnO_2 cathode using X-ray photoelectron spectroscopy. *Phys. Chem. Chem. Phys.* **2018**, *20*, 2517–2526. [CrossRef] [PubMed]
36. Lee, S.; Kang, J.-S.; Leung, K.T.; Kim, S.K.; Sohn, Y. Magnetic Ni-Co alloys induced by water gas shift reaction, Ni-Co oxides by CO oxidation and their supercapacitor applications. *Appl. Surf. Sci.* **2016**, *386*, 393–404. [CrossRef]
37. Kang, J.-G.; Jung, Y.; Min, B.-K.; Sohn, Y. Full characterization of $Eu(OH)_3$ and Eu_2O_3 nanorods. *Appl. Surf. Sci.* **2014**, *314*, 158–165. [CrossRef]
38. Hong, S.; Rhee, C.K.; Sohn, Y. Photoluminescence, electro- and thermal catalytic properties of bare and Eu(III)-doped GaOOH, α- and β-Ga_2O_3 nanorods. *J. Alloy. Compd.* **2019**, *774*, 11–17. [CrossRef]
39. Pawlik, N.; Szpikowska-Sroka, B.; Pisarski, W.A. Energy transfer study on Tb^{3+}/Eu^{3+} co-activated sol-gel glass-ceramic materials containing MF_3 (M = Y, La) nanocrystals for NUV optoelectronic devices. *Materials* **2020**, *13*, 2522. [CrossRef]
40. Mangalam, V.; Pita, K. White Light Emission from Thin-Film Samples of ZnO Nanocrystals, Eu^{3+} and Tb^{3+} Ions Embedded in an SiO_2 Matrix. *Materials* **2019**, *12*, 1997. [CrossRef]

2

Biosorption of Rare Earth Elements by Different Microorganisms in Acidic Solutions

Anja Breuker, Simon F. Ritter and Axel Schippers *

Federal Institute for Geosciences and Natural Resources (BGR), Stilleweg 2, 30655 Hannover, Germany; anja.breuker@online.de (A.B.); sim.ritter@gmail.com (S.F.R.)

* Correspondence: axel.schippers@bgr.de

Abstract: Acidic solutions from metal bioleaching processes usually contain mixtures of metals in different concentrations which need to be separated and concentrated in downstream processing. Aim of this study was to explore and compare biosorption of rare earth elements (REE) by different microorganisms in acidic solutions. Biosorption of REE by bacteria and fungi showed element selective biosorption. The gram-positive bacterium *Bacillus subtilis* showed a higher selectivity to ytterbium (Yb) and lutetium (Lu) than the gram-negative bacteria *Leisingera methylohalidivorans* and *Phaeobacter inhibens*. In contrast, the tested fungi (*Catenulostroma chromoblastomyces*, *Pichia* sp.) showed a preference for the middle rare earth elements. Algae exhibited a low biosorption performance. Additionally, for *B. subtilis* and one yeast (*Pichia* sp.), better results were achieved with living than dead biomass. This study compares for the first time biosorption of different microorganisms at standardized conditions at low pH und application related conditions.

Keywords: biosorption; rare earth elements; acidic solution; *B. subtilis*; *Pichia* sp.

---

1. Introduction

The ability of microbial cells to sequester solutes selectively from aquatic solutions, via non-metabolically mediated pathways, has been termed biosorption [1]. With other words, biosorption can be defined as the removal of substances from solution by biological material [2,3]. Biosorption can be performed in a wide range of pH values 3–9 [4]. Nevertheless, the major part of the studies under acidic conditions were carried out for one or two elements in solution and mild acidic conditions (pH > 3) [5,6]. The pH is possibly the most important physico-chemical parameter for biosorption and competition between cations and protons for binding sites means that biosorption of metals like Cu, Cd, Ni, Co and Zn is often reduced at low pH values [3].

Bioleaching usually takes place in acidic solutions with low pH (<2–3). These solutions contain mixtures of heavy metals in different concentrations which need to be separated and concentrated in downstream processing. Here biosorption might be an option.

Due to high economy delivery risks, special attention has been paid to rare earth elements (REE) and their recovery from primary and secondary resources over the last years [7].

Biosorption of REE at low pH has been studied in some cases [8–16]. Comparing these studies, it is obvious that the conditions under which biosorption experiments took place are often not comparable due to different sorbent/metal ratios even if isotherm studies with increasing metal concentrations were carried out. Therefore, the results in the literature for various microorganisms often rather reflect the experimental conditions than comparable differences between the microorganisms. Aim of this study was to elucidate biosorption of REEs with different microorganisms at comparable acidic conditions. Biosorption experiments were carried out with a mixed solution containing eight REEs (lanthanum, cerium, neodymium, gadolinium, dysprosium, erbium, ytterbium and lutetium).

Additionally, to test the applicability, an experiment with a gadolinium-containing wastewater was performed.

2. Material and Methods

2.1. Microorganisms and Cultivation Conditions

Different Bacteria and Eukarya were selected and grown at specific cultivation conditions (Table 1) to obtain enough biomass for biosorption experiments. The organisms were mainly selected because REE biosorption was described for them in the literature, partly at low pH.

Two eukaryotic microorganisms from the BGR culture collection were selected because of their high biosorption performance at neutral pH studied in previous work.

Their phylogeny was investigated as described below.

The composition of the media is described as follows:

FP medium: meat peptone (Roth #2366.1) 8 g/L, casein peptone (Roth, #8986.1) 2 g/L, yeast extract 0.5 g/L, NaCl 2 g/L adjusted to pH 7.0, after autoclaving 0.2 µm filtered glucose (400 g/L) was added to a final concentration of 4 g/L.

NB medium: Difco™ Nutrient Broth.

Malt medium: biomalt 40 g/L, casein peptone 5 g/L, $(NH_4)_2HPO_4$ 1 g/L, adjusted to pH 6.

Cyanidium medium: $(NH_4)_2SO_4$ 1 g/L; K_2HPO_4 0.02 g/L; $MgSO_4{\cdot}7H_2O$ 0.02 g/L; soil extract 30 mL/L, adjusted to pH 4.0.

Preparation of soil extract for cyanidium medium: Fill a 6-L flask one third with garden or leaf soil of medium, but not too great humus content which does not contain fertilizers or plant protective agents. Success of soil extract depends on selection of suitable soils. Those with high clay content are usually less satisfactory. Add de-ionized water until it stands 5 cm above the soil and sterilize by heating in a steamer for one hour twice in a 24 h interval. Separate the decanted extract from particles by centrifugation. Fill into small containers of stock solution each of a size appropriate to making a batch of media, autoclave for 20 min at 121 °C and store in the refrigerator. (http://sagdb.uni-goettingen.de/culture_media/17%20Cyanidium%20Medium.pdf)

WC medium: Prepared as described by Guillard and Lorenzen 1972 [22].

2× Allens medium (1:1diluted): $(NH_4)_2SO_4$ 2.62 g/L; KH_2PO_4 0.54 g/L; MgSO4·7H2O 0.5 g/L; Ca $Cl_2{\cdot}2H_2O$ 0.14 g/L; $FeCl_3{\cdot}6H_2O$ 0.016 g/L

Aron's A6 trace element solution for Allens medium: $MnCl_2{\cdot}4H_2O$ 1.8 g/L; $ZnCl_2$ 0.105 g/L; $Na_2MoO_4{\cdot}2H_2O$ 0.39 g/L; $CoCl_2{\cdot}6H_2O$ 0.05 g/L; $CuCl_2{\cdot}2H_2O$ 0.06452 g/L; add 2% to medium.

Table 1. Cultivation conditions for microorganisms used for biosorption experiments.

Organism	Taxonomy	Medium	Tempe-Rature	Type of Flasks	rpm on the Shaker	Incuba-Tion Time (Days)	Reference
Leisingeria methylo-halidivorans DSM 14336	Bacteria (gram-negative, *Roseobacter* clade	Marine Broth medium (Difco, MB2216)	20 °C	Bullying flasks	80	2	[17]
Phaeobacter inhibens DSM 17395	Bacteria (gram-negative, *Roseobacter* clade	Marine Broth medium (Difco, MB2216)	20 °C	Bullying flasks	80	2	[17]
Bacillius subtilis ATCC 7003	Bacteria (gram-positive)	FP medium	30 °C	Erlen-meyer flasks	120	2–4	[18–20]
Pezicomy-cotina sp.	Eukarya (fungus)	FP medium	30 °C	Erlen-meyer flasks	120	~10	BGR culture M8
Fusarium sp.	Eukarya (fungus)	malt medium	30 °C	Erlen-meyer flasks	120	2–4	BGR culture
Catenulostro-ma chromo-blastomyces CBS 597.97	Eukarya (fungus)	malt medium	30 °C,	Erlen-meyer flasks	120	2–4	[16]
Pichia sp.	Eukarya (fungus, yeast)	malt medium	30 °C,	Erlen-meyer flasks	120	2–4	BGR culture
Pichia naganishii [MB#320511]	Eukarya (fungus, yeast)	NB medium	30 °C,	Erlen-meyer flasks	120	2–4	BGR culture
Saccharo-myces cerevisiae	Eukarya (fungus, yeast)	Sabouraud medium	30 °C,	Erlen-meyer flasks	120	2–4	[21]
Galderia sulphuraria	Eukarya, Rhodo-phyta (algae)	Cyanidium medium and 2× Allens medium	20 °C, 30 °C sunlight and dark	Baffled shake flasks	200	~14	[15]
Messastrum gracilis	Eukarya, Chloro-phyta (algae)	WC medium	20 °C, 30 °C, sunlight and dark	Baffled shake flasks	200	~14	[12]

2.2. *Biosorption Experiments with Suspended Biomass*

Cell biomass (BM) was harvested by centrifugation with the High Speed Centrifuge Sorvall RC6 plus (7000 g, 10–20 min; Thermo Electron, Langenselbold, Germany) of grown cultures, and washed twice with autoclaved Milli-Q® water at pH 2.5 (HCl). BM dry weight was measured with the HC 103 device (Mettler Toledo, Albstadt, Germany). BM was suspended in pH adjusted and autoclaved Milli-Q® water in measuring pistons at room temperature.

REE solutions (pH 2.5, HCl) were manually prepared from chloride salts of REEs (La, Ce, Nd, Gd, Dy, Er, Yb and Lu) and added to BM to the required final concentrations (BM 5 g/L, REEs 15 µmol/L each). REE/zirconium/iron solutions for competition experiments were prepared in the same way with addition of iron ($FeCl_2$) and zirconium ($ZrOCl_2$) to a final concentration of 264 µmol/L and 16 µmol/L, respectively. For tests with dead vs. living BM, the collected and centrifuged samples were autoclaved with a small amount of medium before experiments.

Experiments were done in duplicates for BM assays and controls (assay without BM) if not shown differently. REE concentrations were measured in samples of supernatants taken after different time intervals by inductively coupled plasma mass spectrometry (ICP-MS, Agilent 7700, Agilent Technologies, Waldbron, Germany) after centrifugation and filtration of samples (0.2 µm nylon membrane). The pH was measured with the pH-Meter 766 Calimatic device (Knick, Berlin, Germany, for medium) and the Five Easy Plus pH/mV device (Mettler Toledo, Albstadt, Germany, for samples). Calculations were carried out by subtraction of REE concentrations (c_e) from controls (c_i) taken at the same time to exclude possible precipitations. Relative biosorption was calculated as follows: $((c_i)-(c_e))/(c_i)*100$. Amount of REE uptake was calculated by dividing the difference of the mean values between control and sample assays through biomass.

2.3. *Biosorption Conditions Experiments with Immobilized Biomass*

For immobilization and recovery rate experiments with selected microorganisms, highly purified agarose was diluted in water and autoclaved (solution A). After autoclaving, solution A was acidified to pH 2.5 (HCl) and stored at 45 °C to prevent solidification. Suspended biomass was adjusted to pH 2.5 after washing twice with acidified autoclaved Milli-Q® water (pH 2.5, HCl), for the control samples acidified autoclaved Milli-Q® water was used (pH 2.5, HCl), (solution B). Solutions A and B were mixed and quickly poured into four forms and let solidify to get platelets. These platelets were put into Erlenmeyer flasks which were previously filled with the REE solution. Solid/solution ratio was 1:4, the experiments were done in duplicate with four 2.5 mL platelets (total 10 mL with acidified Milli-Q® water or acidified biomass) and 40 mL acidified REE solution. After completion of the experiment, the platelets were mixed with 238 µL concentrated nitric acid and incubated in a sandbathe over night at 80 °C. The dry remnants were diluted in 2% nitric acid and analyzed by ICP-MS.

2.4. *Phylogenetic Analysis*

High-molecular-weight DNA of the yeast strain (BGR culture collection) and the fungus M8 (BGR culture collection) was extracted from pure cultures following the FastDNA Spin Kit for Soil (Bio101) protocol [23].

Cloning and sequencing of the ITS (internal transcribed spacer) sequence and the EF1$^{\alpha}$ (translation elongation factor 1α) gene was done by the Microsynth AG (Switzerland). Sequences were manually edited and curated with BioEdit 7.2.5. Phylogenetic mapping was done at NCBI (www.ncbi.nlm.nih.gov) with BLAST [24]. Sequences were submitted to NCBI (accession numbers MN567305.1 and MN644911). For the fungus M8, similarly cloning and sequencing of the ITS sequence and 28 s Sequence was done by the Microsynth AG (Switzerland) and sequences were manually edited and curated with BioEdit 7.2.5 [25]. Phylogenetic mapping was done at NCBI (www.ncbi.nlm.nih.gov) with BLAST [24]. Sequences were submitted to NCBI (Accession numbers MN575689.1

and MN575688.1). The yeast strain was identified as *Pichia* sp., the fungal strain M8 was identified as *Pezicomycotina* sp.

3. Results and Discussion

3.1. Biosorption of REE by Bacteria

Initially biosorption tests with *B. subtilis* and four elements (lanthanum, neodymium, dysprosium and ytterbium, each 15 µmol/L, no parallels) showed a selective preference of *B. subtilis* for ytterbium at different amounts of biomass, and a reduction of uptake of REEs per biomass with increasing biomass (Figure S1, Table S1). Biosorption at pH 4 and pH 5 was also tested and resulted in more than 96% REE removal and did not show any difference concerning different amounts of biomass (data not shown). Uptake for lanthanum was 5.6, 5.5 and 5.8 µmol/g for two parallel assays at pH 4 and a biomass of 2.5 g/L, these data are in accordance with a previous study with a REE/BM ratio of 2.9 with lanthanum as single metal in solution [26].

Increasing the REE concentration with increasing REE/BM ratio decreased the relative biosorption with small influence on the preference for the heavy rare earth elements (HREEs) (Figure 1).

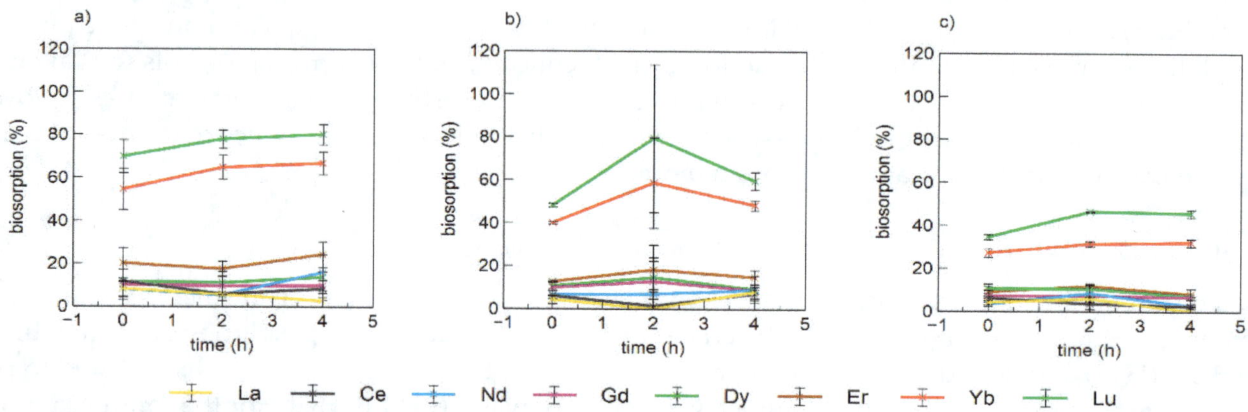

Figure 1. Biosorption of REEs by *B. subtilis* with increasing REE/BM ratio at pH 2.5, BM 4 g/L, (**a–c**): REE/BM ratio 3.75, 7.5 and 15, respectively; REE concentration of each element was 15, 30 and 60 µmol/L, respectively. Error bars show standard deviation of two parallels.

This aspect together with the result, that biosorption did not increase very much especially for ytterbium with a biomass of more than 5 g/L (Figure S1b,d,f,h) is essential for application of biosorption technologies because producing biomass is an energy and resource consuming process. In conclusion, for comparing different microorganisms in terms of possible applications, REE/BM ratio, pH, time and absolute REE and BM concentrations must be taken into consideration.

Therefore, to evaluate the biosorption pattern of the eight REEs in solution a biomass of 5 g/L for the gram-positive bacterium *B. subtilis* as well as for the gram-negative bacteria *L. methylohalidivorans* and *P. inhibens* was chosen for further experiments. The preference for the HREEs with different selectivity at pH 2.5 was confirmed (Figure 2). Additionally, for *B. subtilis*, a slight reduction of biosorption after four hours was found (Figure S1), for *L. methylohalidivorans* and *P. inhibens* a reduction of biosorption after 20 h was observed (Figure S2).

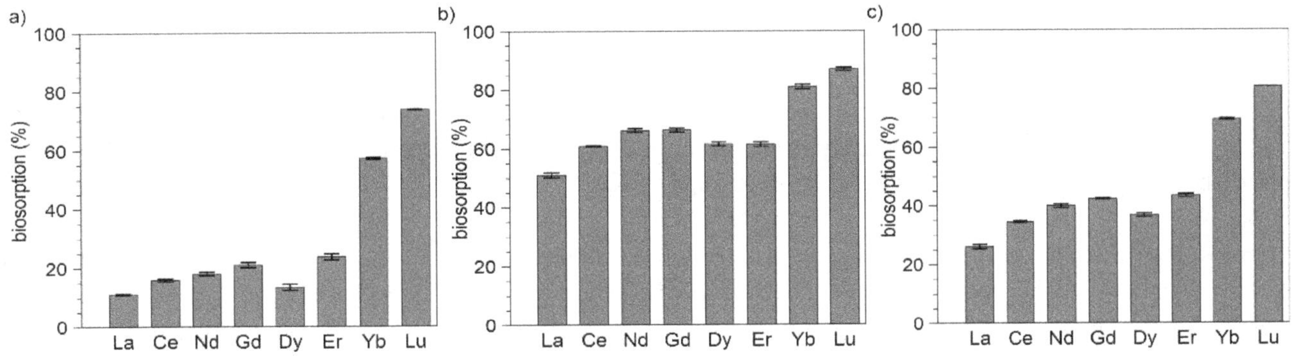

Figure 2. Biosorption of REEs by *B. subtilis* ((**a**) gram-positive), *L. methylohalidivorans* ((**b**) gram-negative) and *P. inhibens* ((**c**) gram-negative). BM 5 g/L, initial REE concentration was 15 μmol/L for each element, sampling took place after 2 h. Bars show mean of two parallel assays, error bars show standard deviation.

Takahashi et al. found that heavy REE (HREE, especially Tm, Yb, and Lu) are enriched on the cell surface of the gram-positive bacterium *B. subtilis* and the gram-negative bacterium *E. coli* compared to other REEs and concluded that the REE patterns suggest that there are at least two binding sites on the bacterial cell surface, i.e., carboxylate and phosphate groups [18]. Our results are in accordance with a later study of Takahashi et al. who furthermore found that HREEs form complexes with multiple phosphate sites (including phosphoester sites) with a larger coordination number (CN) at lower REE-bacteria ratios (REE/BM) [19]. In contrast, light and middle REEs form complexes with the phosphate sites at a lower CN. However, the fraction coordinated to carboxylate increased for all REEs with increasing REE/BM ratio, but the enrichment of HREE in the REE distribution patterns of the bacteria was less marked with an increasing REE/BM ratio [19]. Another previous study, which combined surface complexation modelling of macroscopic adsorption data with X-ray spectroscopic measurements to identify lanthanide sorption sites on the bacterial surface by the gram-negative *Pantoea agglomerans* suggested that there may be variations in the dominant sorption sites across the lanthanide series. The adsorption of both LREEs used (La and Nd) was best modelled assuming adsorption to phosphate sites, whereas Gd, Er and Yb could be modelled equally well with carboxyl or phosphate sites; samarium was better modelled with carboxyl relative to phosphate sites. However, one restriction of this study was that the observed differences in modelling were small [14] and it has to be mentioned, that the used REE/BM ratio was high, this could also contribute to a higher impact of the carboxyl sites. In contrast, concerning the europium/*B. subtilis* system it was found that carboxyl, phosphate, and hydroxyl (and/or amine) groups were present at the bacterial surface of *B. subtilis* [20]. At pH 5, the adsorption data could be well described by mainly carboxyl groups and with increasing pH, an increasingly deprotonated phosphate environment should be considered [20]. Concerning the data of the literature provided by Takashi et al. [18,19] the REE/BM ratios are similar to the here presented data.

Considering the obvious difference between gram-positive and gram-negative bacteria in cell wall structures, it was shown, that powder of a wild type vs. a lipoteichonic acid defective strain of *B. subtilis* showed a better biosorption performance, indicating that lipoteichonic acids contribute to the absorption of REEs and followed the order of effectiveness La < EU < TM [27].

Regarding the relevant chemical groups for lanthanide binding, Gadd et al. stated that peptidoglycan carboxyl sites are the main binding sites for metal cations in the gram-positive bacterial cell and that phosphate groups contribute significantly to biosorption in gram-negative species. Furthermore, other bacterial metal-binding components are proteinaceous S-layers, and sheaths largely composed of polymeric materials with proteins and polysaccharides [3]. Interestingly, a genetically engineered gram-negative bacterium *Caulobacter crescentus* with a high-density cell surface display of lanthanide binding tags (composed of small proteins on its S-layer) exhibited a preference for HREEs [28]. Considering literature results and the observed higher selectivity for ytterbium and lutetium of the gram-positive bacterium *B. subtilis* than of the tested gram-negative bacteria (at the same REE/BM ratio), the question remains open what kind of binding mainly contributes to the biosorption of REEs at acidic pH. A relative higher amount of phosphate groups contributing to binding of HREEs in gram-positive bacteria or a better accessibility and amount of carboxyl groups in the proteinaceous S-layer in gram-negative bacteria at low REE/BM ratios could explain the observed difference. This assumption together with results from Takahashi et al. [18] that HREEs form complexes with multiple phosphate sites (including phosphoester sites) with a larger coordination number at lower REE/BM ratios would also explain the diminishing relative difference between the absorption of LREEs and HREEs with increasing biomass (2.5 g/L vs. 5 g/L) and decreasing REE/BM ratio (6 µmol/g vs. 3 µmol/g). This also supports the results from Takahashi et al. [19] that, at low REE/BM ratios, HREEs are bound to phosphate sites with a higher coordination number (Figure 3).

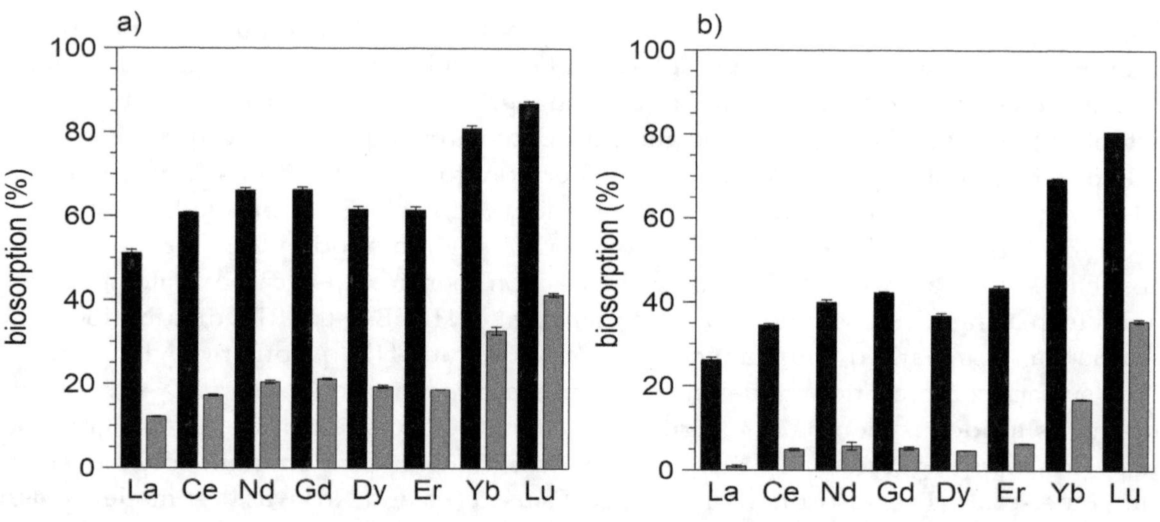

Figure 3. Biosorption of REEs by *L. methylohalidivorans* (**a**) and *P. inhibens* (**b**). BM 2.5 g/L (grey) and 5 g/L (black), initial REE concentration was 15 µmol/L for each element, sampling took place after 2 h. Bars show mean of two parallel assays, error bars show standard deviation.

Competition of metal cations for metal binding tags has been described in some studies [29]. For example, it has been shown that cations like Al^{3+} are strong inhibitors for biosorption of REEs by *Pseudomonas aeruginosa.* In the same study, lanthanum biosorption was strongly affected by the presence of europium and ytterbium, whereas the extent of removal of ytterbium did not decrease to the same extent than with the addition of lanthanum ions [11]. Comparison of the results of pretests with four elements (La, Nd, Dy, Yb) with those experiments with eight REEs (biomass 5 g/L, pH 2.5, REEs 15 µmol each) confirmed this observation (Table S1). To investigate the question, whether at acidic pH other cations like Fe(II) and Zr(IV) compete with the REEs for biosorption binding sites, the relative biosorption of the eight REEs in presence of iron ($FeCl_2$) and zirconium ($ZrOCl_2$) was tested, however, although biosorption and uptake of zirconium was high, competition was not found. The mean decrease of biosorption of all eight REEs after two hours was only 0.5% compared to the experiment without zirconium and iron (Figure 4). These data are in accordance with a previous study which pointed out that Tm was selectively removed by *B. subtilis* powders in the presence of Fe(II) [27].

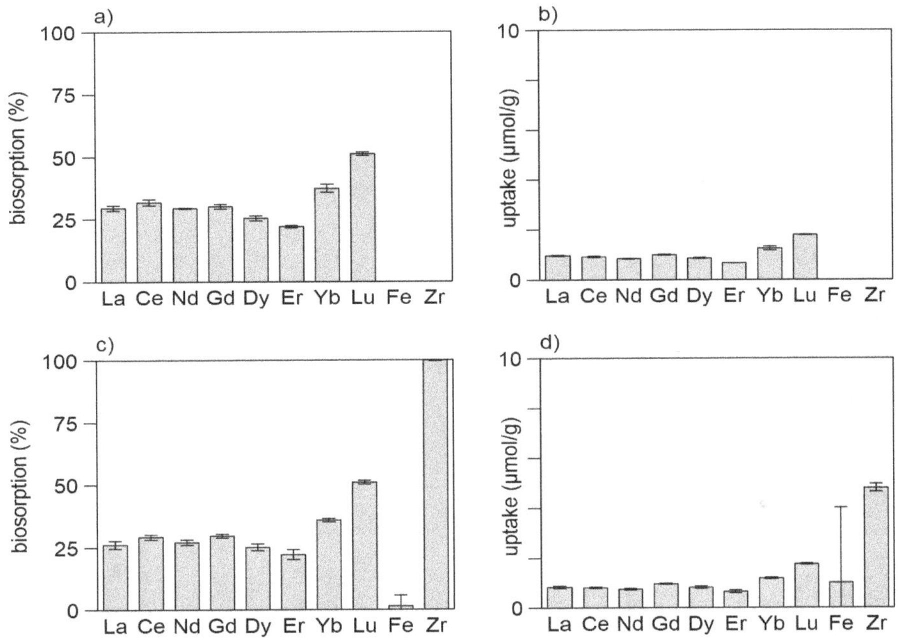

Figure 4. REE biosorption (**a**,**c**) and uptake (**b**,**d**) by *B. subtilis* in absence (**a**,**b**) and presence (**c**,**d**) of iron and zirconium (Fe(II): 264 µmol/L, Zr(IV) 16 µmol/L, BM 5g/L, pH 2.5). Bars show mean of two parallel assays, error bars show standard deviation.

In conclusion, at low pH, for selective removal of HREEs, biosorption by *B. subtilis* is a promising approach; for removal of all REEs, the marine bacterium *L. methylohalidivorans* seems suitable.

3.2. Biosorption of REE by Eukaryotes

Concerning the eukaryotes, several fungi and two algae were tested at low pH. Results are summarized in Table 2.

Table 2. Overview about biosorption capabilities of eukaryotes.

Organism	Result
Saccharomyces cerevisiae	Tested with BM of 5, 10 and 20 g/L, at a BM of 5 g/L biosorption was <50%, biosorption was 100% at a BM of 10 g/L, preference for Gd, Dy, Yb, Lu
Pichia naganishii [MB#320511]	Tested for La, Nd, Dy, Yb at a BM of 2.5, 5 and 10 g/L. At a BM >5 g/L, biosorption >80%.
Pichia sp.	Preference for Ce, Nd, Gd, Dy
Catenulostroma chromoblastomyces CBS 597.97	Preference for Ce, Nd, Gd, Dy, at a BM of 7.5 g/L ~30% biosorption
Pezicomycotina sp.	Tested with BM of 2, 5 and 10 g/L, biosorption <20%, preference for Gd, Yb, Lu
Fusarium sp.	Preference for Nd, Gd, Gy, Lu, tested for 4 elements (La, Nd, Dy, Yb), BM 4 g/L ~40%, BM 5 g/L biosorption ~100%
Messastrum gracilis	Literature data not confirmed [12]
Galdieria sulphuraria	Literature data not confirmed [15]

Concerning biosorption capacities of fungi, a preference for the MREEs (middle rare elements), partly together with a preference for ytterbium and lutetium was observed (Table 2). Interestingly, strain *Pezicomycotina* sp. which was isolated from an abandoned acid copper mine (Marsberg, Germany) showed the lowest biosorption (<20% at a BM of 10 g/L). This might be due to adaptation to the metal-rich environment inhibiting absorption of cations to the cell wall. Because the *Pichia* sp. strain was most promising for biosorption at low pH, the total biosorption at different pH values with two different amounts of biomass (2 and 4 g/L) was tested in order to estimate a suitable biomass for further work. Biosorption increased still after four hours, therefore it was tested up to 24 h. In a pretest, a compensation of pH effect on biosorption with increasing biomass but a decreasing biosorption per g/dry weight was observed (Figure S3). To discriminate a possible preference for elements, 2 g/L BM was chosen for further tests. Pretests with 1.5% agarose and 2% agarose as immobilization matrix carried out with *B. subtilis* showed a better biosorption with 1.5% agarose (Figure S4), therefore a recovery experiment with immobilized *Pichia* sp. in 1.5% agarose was conducted. The preference of certain REEs, specifically cerium, neodymium, gadolinium and dysprosium was confirmed by the data (Figure 5). For application, the metals could be recovered from the ashes.

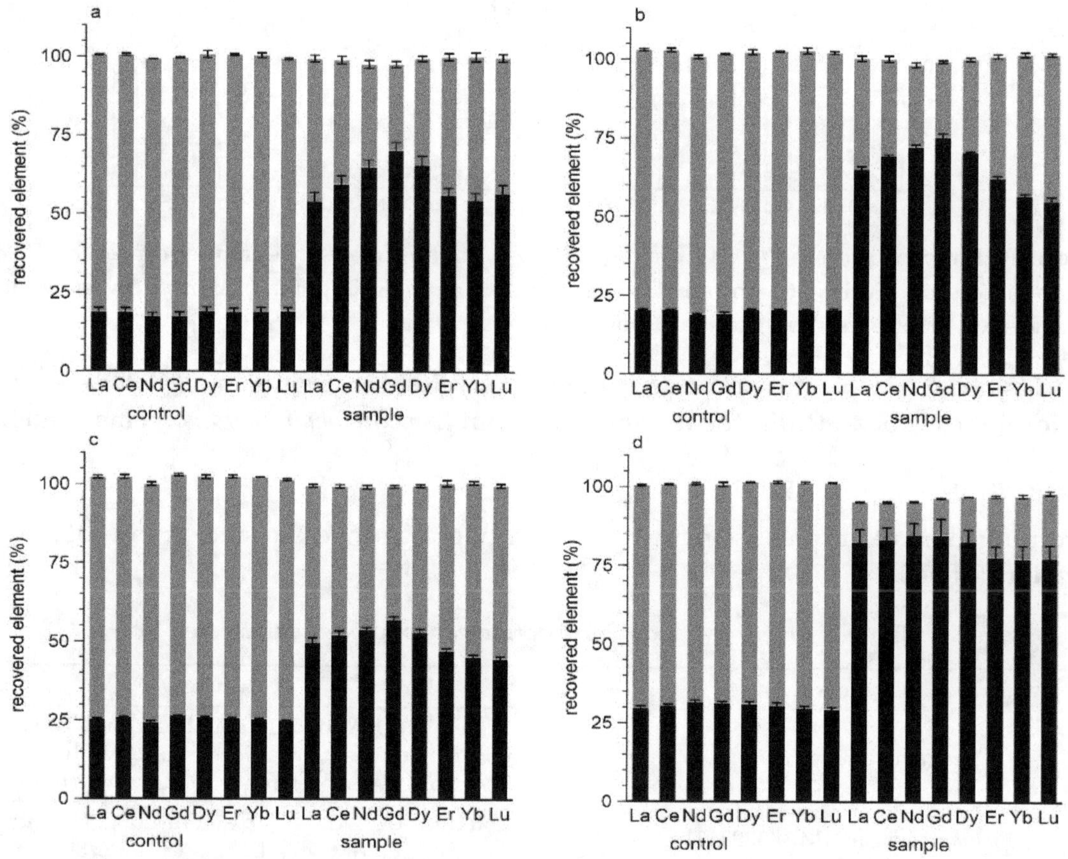

Figure 5. Recovery experiment: biosorption of *Pichia* sp. immobilized in agarose, biomass 2 g/L, pH 2.5, samples were taken after 24 h. Bars show mean of two parallel assays, error bars show standard deviation, (**a–d**) pH 1.5, 2.5, 3.5 and 4.5, respectively.

The reason for the reduction of biosorption at pH 3.5 remained open, due to the fact, that the biomass was diluted after centrifugation, experimental bias (e.g., clumping of biomass) cannot be completely excluded. Nevertheless, preference for the middle rare elements was confirmed.

To test the REE biosorption capability under application related conditions wastewater from a sewage plant was used. Concerning the middle rare elements, the annual emission of gadolinium (used as contrast agent) by medical hospitals and ambulances was estimated to be 1160 kg in Germany for 1996/1997, resulting in an average gadolinium concentration in surface water of 0.026 μg/L [30]. Gadolinium load resulting from magnetic resonance imaging is well investigated in Germany and Switzerland [31–33]. Recently, the first evidence that anthropogenic gadolinium enters human food chain was found [34]. Therefore, to elucidate whether biosorption is applicable here, an experiment with two samples (pH 8.1) from pre- and post-clarification wastewater from a sewage treatment plant in Germany was carried out. The amount of REEs was determined with ICP-MS. The gadolinium concentration was 3.8 and 3.9 nmol/L for the two water samples, respectively (0.6 μg/L). The amount of all other REEs together was below 0.6 nmol. For experiments, two approaches were selected: first, the samples were used without any modification, second, the samples were acidified to pH 2.5. Additionally, it was tested whether filtration (0.2 μm) of the samples had an influence.

The results did not show any biosorption using the not acidified samples. It is known that rare earth ions form complexes with hydroxide or carbonate at pH higher than approximately 5.5 [35]. Biosorption of the acidified samples was strongly depended on the REE/BM ratio, increasing the biomass resulted in biosorption of ~75% for both (pre- and post-clarification) samples. Post-clarification samples exhibited a better biosorption, this might be due to the lack of competing cations or complexing organic molecules. At the higher REE/BM ratio, a better biosorption within the unfiltered samples was found (Figure 6a,c). In conclusion, *Pichia* sp. seems suitable to recover gadolinium from wastewater at appropriate conditions.

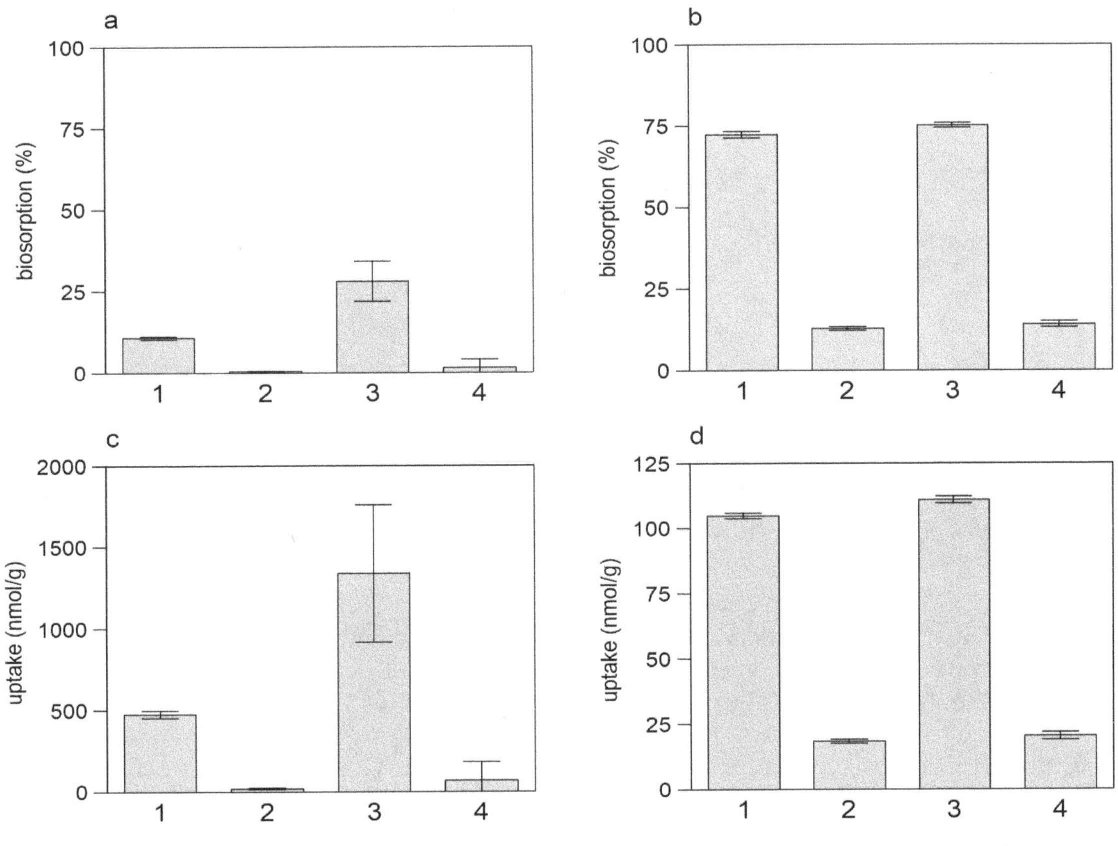

Figure 6. Biosorption and amount of uptake of gadolinium by *Pichia* sp. from acidified wastewater. (**a**,**c**) REE BM/ratio: 3.9 μmol/g; (**b**,**d**) REE/BM ratio: 0.13 μmol/g; samples 1, 3: post clarification wastewater; samples 2, 4: pre clarification wastewater; samples 1, 2: 0.2 μm filtered; samples 3, 4: not filtered. Error bars show standard deviation of two parallels.

Concerning *Fusarium* sp., *Catenulostroma chromoblastomyces* and *Pichia* sp., in various pH tests, decreased biosorption at a pH of 2.5 or 3.5, respectively, was seen (Figure 5, Figure S5, for *Fusarium* sp. not shown).

Because fungal cell walls are complex macromolecular structures predominantly consisting of chitins, glucans, mannans and proteins, but also containing other polysaccharides, lipids and pigments, e.g., melanin [3], it is conceivable, that at low pH the lanthanide binding tags of the eukaryotic cell wall integrity are destroyed. Complex interactions of protons with for example chitin [36] may mask lanthanide binding tags. The observed increasing biosorption at pH below 2.5–3.5 could also be explained by absorption on released intracellular matter (e.g., proteins or walls of cell organelles).

On the other hand, for the green algae *Galderia sulphuraria*, an active incorporation of REEs into the cell at pH 1.0–1.5 was suggested [15]. Further research to clarify this observed phenomenon is needed.

In addition to fungi, REE biosorption was tested for the red algae *G. sulphuraria* and the green algae *M. gracilis* for different cultivation conditions as described in the literature [12,15]. However, significant biosorption capabilities were not detected for both algae.

3.3. Tests with Dead vs. Living Biomass

The use of dead biomass instead of to living biomass has several advantages: storage of biomass is easier, versatility is better, the degree of uptake is higher, it might be possible to reuse the biomass, recovery with toxic substances is possible [29]. We tested this for *Bacillus subtilis* and one fungus (*Fusarium* sp.) with increasing REE concentrations and better results were achieved with living biomass vs. dead biomass (Figure 7).

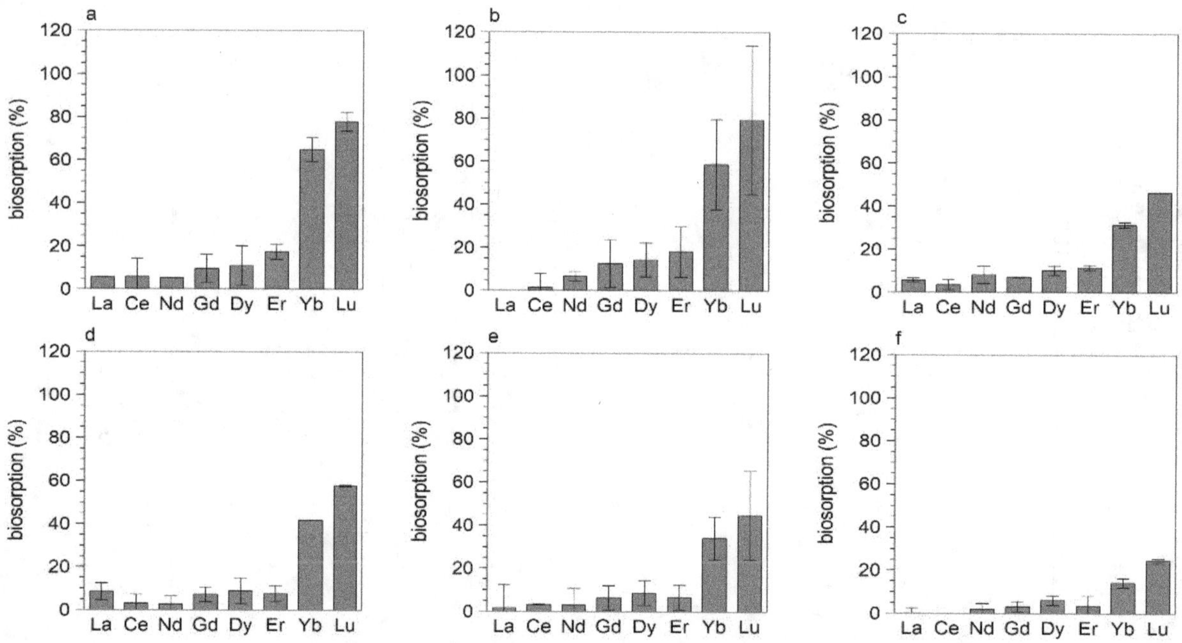

Figure 7. Comparison of REE biosorption by *B. subtilis* with living (**a–c**) and dead (**d–f**) biomass, (**a**,**d**): concentration of REEs each 15 µmol/L, (**b**,**e**): concentration of REEs each 30 µmol/L, (**c**,**f**): concentration of REEs each 60 µmol/L, BM 4g/L, pH 2.5.

Higher biosorption with living biomass was also achieved with *Fusarium* sp. (Figure 8).

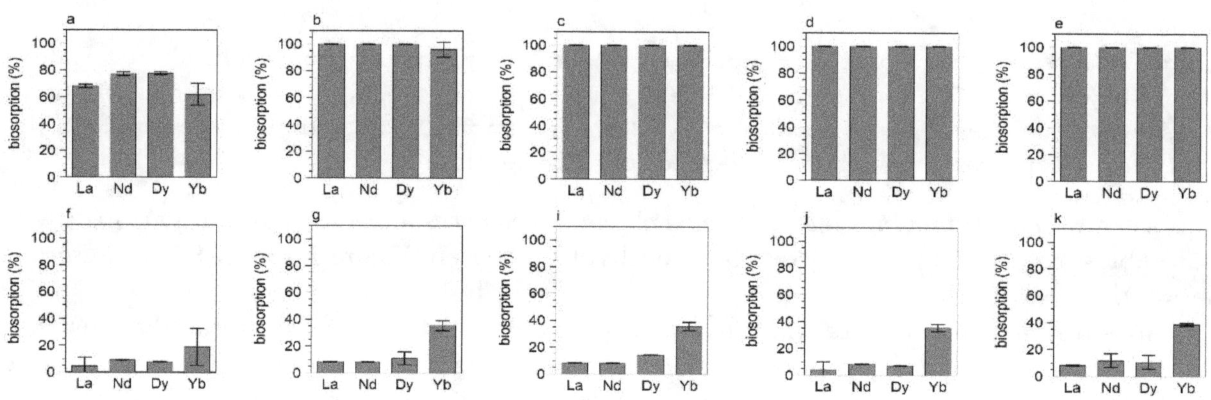

Figure 8. Comparison of REE biosorption by *Fusarium* sp. with living (**a–e**) and dead (**f–k**) biomass at different time intervals, (**a**,**f**): 0 h, (**b**,**g**): 2.5 h, (**c**,**i**): 5.5 h, (**d**,**j**): 8 h, (**e**,**k**): 25 h. Biomass was 5 g/L, pH was 2.5.

These results are in accordance with those of the red algae *Galderia sulphuraria* which also performed better biosorption with living vs. dead BM [15].

4. Conclusions

At low pH, a careful selection of microbial species for biosorption of REE is necessary. The importance of the REE/BM ratio for application strategies was shown. Biosorption preferences were shown for different REEs at comparable application conditions (pH, REE/BM ratio). Competing effects were found for *B. subtilis* for interaction of eight REEs but not for the addition of zirconium to the eight REEs at acidic pH. These results confirm and complement previous investigations with three elements [11]. For the chosen approach (to elucidate biosorption capabilities of microorganisms without further genetic or chemical modifications at low pH), the better biosorption results with live versus dead biomass confirmed the importance of cell integrity for biosorption. Due to discrimination between HREEs and LREEs, the gram-positive *B. subtilis* seems suitable for biosorption of HREEs. Due to the lower need of biomass to achieve a relevant biosorption of all REEs, eukaryotes—especially fungi and yeasts such as *Pichia* sp.—should be given priority. Nevertheless, gram-negative marine bacteria show a high potential for biosorption of all REEs (*L. methylohalidivorans*) and HREEs (*P. inhibens*). *Pichia* sp. could be used successfully to recover anthropogenic gadolinium from wastewater. For successful application strategies, attention should be paid on REE/BM ratios, competition experiments and enlightenment of biosorption mechanisms based on the biochemical and physical background.

Supplementary Materials:
Table S1: Comparison of the relative biosorption of four REEs by *B. subtilis* with the relative biosorption of eight REEs by *B. subtilis* at pH 2.5. Figure S1: Relative biosorption and amount of REE uptake by *B. subtilis*. Figure S2: Relative biosorption by *L. methylohalidivorans* and *P. inhibens* at different amounts of biomass. Figure S3: REE biosorption by *Pichia* sp. at two different amounts of biomass. Figure S4: Relative biosorption by *B. subtilis* immobilized in 1.5% and 2% agarose. Figure S5: Relative biosorption by *C. chromoblastomyces*.

Author Contributions: Conceptualization: A.B. and A.S.; Data curation: A.B.; Formal Analysis: A.B.; Investigation: A.B. and S.F.R; Supervision: A.S.; Validation: A.B.; Visualization: A.B.; Writing—original draft: A.B.; Writing—review & editing: A.B., S.F.R. and A.S. All authors have read and agreed to the published version of the manuscript.

Acknowledgments: We thank Martine Berger from ICBM Oldenburg (Institut für Chemie und Biologie des Meeres) for providing biomass of *L. methylohalidivorans* and *P. inhibens* and Ruiyong Zhang (BGR) for providing the strain *Pezicomycotina* sp. We also thank Claudia Rathmann for delivery of wastewater samples. Many thanks for straightforward help with lab work to Gudrun Mengel-Jung (BGR).

References

1. Tsezos, M. Biosorption: A mechanistic approach. *Adv. Biochem. Eng. Biotechnol.* **2014**, *141*, 173–209. [CrossRef]
2. Fomina, M.; Gadd, G.M. Biosorption: Current perspectives on concept, definition and application. *Bioresour. Technol.* **2014**, *160*, 3–14. [CrossRef] [PubMed]
3. Gadd, G.M. Biosorption: Critical review of scientific rationale, environmental importance and significance for pollution treatment. *J. Chem. Technol. Biotechnol.* **2008**, *84*, 13–28. [CrossRef]
4. Michalak, I.; Chojnacka, K.; Witek-Krowiak, A. State of the art for the biosorption process–a review. *Appl. Biochem. Biotechnol.* **2013**, *170*, 1389–1416. [CrossRef] [PubMed]
5. Ilyas, S.; Lee, J.-C. Biometallurgical recovery of metals from waste electrical and electronic equipment: A review. *ChemBioEng Rev.* **2014**, *1*, 148–169. [CrossRef]
6. Andrès, Y.; Gérente, C. Removal of rare earth elements and precious metal species by biosorption. In *Microbial Biosorption of Metals*; Kotrba, P., Mackova, M., Macek, T., Eds.; Springer: Dordrecht, The Netherlands, 2011; pp. 179–196. ISBN 978-94-007-0443-5.

7. Akcil, A. *Critical and Rare Earth Elements. Recovery from Secondary Resources*; CRC Press, Taylor et Francis Group: Boca Raton, FL, USA; London, UK; New York, NY, USA, 2020; ISBN 9780367086473.
8. Andrès, Y.; MacCordick, H.J.; Hubert, J.-C. Adsorption of several actinide (Th, U) and lanthanide (La, Eu, Yb) ions by *Mycobacterium smegmatis*. *Appl. Microbiol. Biotechnol.* **1993**, *39*. [CrossRef]
9. Karavaiko, G.I.; Kareva, A.S.; Avakian, Z.A.; Zakharova, V.I.; Korenevsky, A.A. Biosorption of scandium and yttrium from solutions. *Biotechnol. Lett.* **1996**, *18*, 1291–1296. [CrossRef]
10. Texier, A.C.; Andrès, Y.; Cloirec, P.L. Selective Biosorption of Lanthanide (La, Eu) Ions by *Mycobacterium smegmatis*. *Environ. Technol.* **1997**, *18*, 835–841. [CrossRef]
11. Texier, A.-C.; Andrès, Y.; Le Cloirec, P. Selective biosorption of lanthanide (La, Eu, Yb) ions by *Pseudomonas aeruginosa*. *Environ. Sci. Technol.* **1999**, *33*, 489–495. [CrossRef]
12. Palmieri, M.C.; Garcia, O.; Melnikov, P. Neodymium biosorption from acidic solutions in batch system. *Process Biochem.* **2000**, *36*, 441–444. [CrossRef]
13. Haferburg, G.; Merten, D.; Büchel, G.; Kothe, E. Biosorption of metal and salt tolerant microbial isolates from a former uranium mining area. Their impact on changes in rare earth element patterns in acid mine drainage. *J. Basic Microbiol.* **2007**, *47*, 474–484. [CrossRef] [PubMed]
14. Ngwenya, B.T.; Mosselmans, J.F.W.; Magennis, M.; Atkinson, K.D.; Tourney, J.; Olive, V.; Ellam, R.M. Macroscopic and spectroscopic analysis of lanthanide adsorption to bacterial cells. *Geochim. Cosmochim. Acta* **2009**, *73*, 3134–3147. [CrossRef]
15. Minoda, A.; Sawada, H.; Suzuki, S.; Miyashita, S.-I.; Inagaki, K.; Yamamoto, T.; Tsuzuki, M. Recovery of rare earth elements from the sulfothermophilic red alga *Galdieria sulphuraria* using aqueous acid. *Appl. Microbiol. Biotechnol.* **2015**, *99*, 1513–1519. [CrossRef] [PubMed]
16. Horiike, T.; Yamashita, M. A new fungal isolate, *Penidiella* sp. strain T9, accumulates the rare earth element dysprosium. *Appl. Environ. Microbiol.* **2015**, *81*, 3062–3068. [CrossRef]
17. Bonificio, W.D.; Clarke, D.R. Rare-earth separation using bacteria. *Environ. Sci. Technol. Lett.* **2016**, *3*, 180–184. [CrossRef]
18. Takahashi, Y.; Châtellier, X.; Hattori, K.H.; Kato, K.; Fortin, D. Adsorption of rare earth elements onto bacterial cell walls and its implication for REE sorption onto natural microbial mats. *Chem. Geol.* **2005**, *219*, 53–67. [CrossRef]
19. Takahashi, Y.; Yamamoto, M.; Yamamoto, Y.; Tanaka, K. EXAFS study on the cause of enrichment of heavy REEs on bacterial cell surfaces. *Geochim. Cosmochim. Acta* **2010**, *74*, 5443–5462. [CrossRef]
20. Markai, S.; Andrès, Y.; Montavon, G.; Grambow, B. Study of the interaction between europium (III) and *Bacillus subtilis*: Fixation sites, biosorption modeling and reversibility. *J. Colloid Interface Sci.* **2003**, *262*, 351–361. [CrossRef]
21. Saitoh, N.; Fujimori, R.; Nakatani, M.; Yoshihara, D.; Nomura, T.; Konishi, Y. Microbial recovery of gold from neutral and acidic solutions by the baker's yeast *Saccharomyces cerevisiae*. *Hydrometallurgy* **2018**, *181*, 29–34. [CrossRef]
22. Guillard, R.R.L.; Lorenzen, C.J. Yellow-green algae with chlorophyllide C1,2. *J. Phycol.* **1972**, *8*, 10–14. [CrossRef]
23. Webster, G.; Newberry, C.J.; Fry, J.C.; Weightman, A.J. Assessment of bacterial community structure in the deep sub-seafloor biosphere by 16S rDNA-based techniques: A cautionary tale. *J. Microbiol. Methods* **2003**, *55*, 155–164. [CrossRef]
24. Altschul, S.F.; Gish, W.; Miller, W.; Myers, E.W.; Lipman, D.J. Basic local alignment search tool. *J. Mol. Biol.* **1990**, *215*, 403–410. [CrossRef]
25. Hall, T.A. BioEdit: A user-friendly biological sequence alignment editor and analysis program for Windows 95/98/NT. *Nucleic Acids Symp. Ser.* **1999**, *41*, 95–98.
26. Mullen, M.D.; Wolf, D.C.; Ferris, F.G.; Beveridge, T.J.; Flemming, C.A.; Bailey, G.W. Bacterial sorption of heavy metals. *Appl. Environ. Microbiol.* **1989**, *55*, 3143–3149. [CrossRef] [PubMed]
27. Moriwaki, H.; Koide, R.; Yoshikawa, R.; Warabino, Y.; Yamamoto, H. Adsorption of rare earth ions onto the cell walls of wild-type and lipoteichoic acid-defective strains of *Bacillus subtilis*. *Appl. Microbiol. Biotechnol.* **2013**, *97*, 3721–3728. [CrossRef] [PubMed]
28. Park, D.M.; Reed, D.W.; Yung, M.C.; Eslamimanesh, A.; Lencka, M.M.; Anderko, A.; Fujita, Y.; Riman, R.E.; Navrotsky, A.; Jiao, Y. Bioadsorption of rare earth elements through cell surface display of lanthanide binding tags. *Environ. Sci. Technol.* **2016**, *50*, 2735–2742. [CrossRef]

29. Vijayaraghavan, K.; Yun, Y.-S. Bacterial biosorbents and biosorption. *Biotechnol. Adv.* **2008**, *26*, 266–291. [CrossRef]
30. Kümmerer, K.; Helmers, E. Hospital effluents as a source of gadolinium in the aquatic environment. *Environ. Sci. Technol.* **2000**, *34*, 573–577. [CrossRef]
31. Tepe, N.; Romero, M.; Bau, M. High-technology metals as emerging contaminants: Strong increase of anthropogenic gadolinium levels in tap water of Berlin, Germany, from 2009 to 2012. *Appl. Geochem.* **2014**, *45*, 191–197. [CrossRef]
32. Kulaksız, S.; Bau, M. Anthropogenic dissolved and colloid/nanoparticle-bound samarium, lanthanum and gadolinium in the Rhine River and the impending destruction of the natural rare earth element distribution in rivers. *Earth Planet. Sci. Lett.* **2013**, *362*, 43–50. [CrossRef]
33. Vriens, B.; Voegelin, A.; Hug, S.J.; Kaegi, R.; Winkel, L.H.E.; Buser, A.M.; Berg, M. Quantification of element fluxes in wastewaters: A nationwide survey in Switzerland. *Environ. Sci. Technol.* **2017**, *51*, 10943–10953. [CrossRef] [PubMed]
34. Schmidt, K.; Bau, M.; Merschel, G.; Tepe, N. Anthropogenic gadolinium in tap water and in tap water-based beverages from fast-food franchises in six major cities in Germany. *Sci. Total Environ.* **2019**, *687*, 1401–1408. [CrossRef] [PubMed]
35. Luo, Y.-R.; Byrne, R.H. Carbonate complexation of yttrium and the rare earth elements in natural waters. *Geochim. Cosmochim. Acta* **2004**, *68*, 691–699. [CrossRef]
36. Roy, J.C.; Salaün, F.; Giraud, S.; Ferri, A.; Chen, G.; Guan, J. Solubility of chitin: Solvents, solution behaviors and their related mechanisms. In *Solubility of Polysaccharides*; Xu, Z., Ed.; InTech: London, UK, 2017; ISBN 978-953-51-3649-1.

3

Geochemistry of Rare Earth Elements in Bedrock and Till, Applied in the Context of Mineral Potential in Sweden

Martiya Sadeghi *, Nikolaos Arvanitidis and Anna Ladenberger

Department of Mineral Resources, Geological Survey of Sweden, Box 670, 751 28 Uppsala, Sweden; Nikolaos.arvanitidis@sgu.se (N.A.); Anna.Ladenberger@sgu.se (A.L.)

* Correspondence: martiya.sadeghi@sgu.se

Abstract: The Rare Earth Element (REE) mineralizations are not so "rare" in Sweden. They normally occur associated and hosted within granitic crystalline bedrock, and in mineral deposits together with other base and trace metals. Major REE-bearing mineral deposit types are the apatite-iron oxide mineralizations in Norrbotten (e.g., Kiruna) and Bergslagen (e.g., Grängesberg) ore regions, the various skarn deposits in Bergslagen (e.g., Riddarhyttan-Norberg belt), hydrothermal deposits (e.g., Olserum, Bastnäs) and alkaline-carbonatite intrusions such as the Norra Kärr complex and Alnö. In this study, analytical data of samples collected from REE mineralizations during the EURARE project are compared with bedrock and till REE geochemistry, both sourced from databases available at the Geological Survey of Sweden. The positive correlation between REE composition in the three geochemical data groups allows better understanding of REE distribution in Sweden, their regional discrimination, and genetic classification. Data provides complementary information about correlation of LREE and HREE in till with REE content in bedrock and mineralization. Application of principal component analysis enables classification of REE mineralizations in relation to their host. These results are useful in the assessment of REE mineral potential in areas where REE mineralizations are poorly explored or even undiscovered.

Keywords: rare earth elements; Sweden; lithogeochemistry; till geochemistry

---

1. Introduction

The Rare Earth Element (REE) occurrences in Sweden are widely distributed all over the country but some regions are more endowed than others. Major REE-bearing mineral deposit types, found mainly in Norrbotten and Bergslagen ore regions, are the apatite-iron oxide mineralizations in Kiruna and Grängesberg, the skarn mineralizations of the Riddarhyttan-Norberg belt, the alkaline igneous rock-associated in Norra Kärr, and the hydrothermal mineralizations in Olserum and Alnö deposit areas [1] (see Figure 1). Distribution of REEs in Swedish soil and bedrock has previously been investigated by Sadeghi and Andersson [2] with the aim to identify the main changes in REE geochemistry related to geology and weathering, taking also into account the current baseline level for REE in soil and bedrock over Sweden. The Geochemical Atlas of Sweden provides a harmonized, countrywide database with modern baseline geochemical data from C horizon (element concentration in the weathered parent rock horizon of soil profile) in till [3] and REEs in different solid media (topsoil, subsoil, and stream sediments) have been investigated using the Forum of European Geological Surveys (FOREGS) database in order to identify the REEs regional background values [4]. In Sadeghi et al. [5], the Geochemical Mapping of Agricultural and Grazing Land Soil (GEMAS) project data were investigated with focus on REEs in two solid media (topsoil from agricultural (Ap) and grazing land (Gr) soil) to identify the background values of REEs both in Sweden and in Europe. The Ap samples were collected from regularly plowed fields at a depth interval of 0 to 20 cm and the Gr samples were collected from soil under permanent grass cover at a 0 to 10 cm depth range.

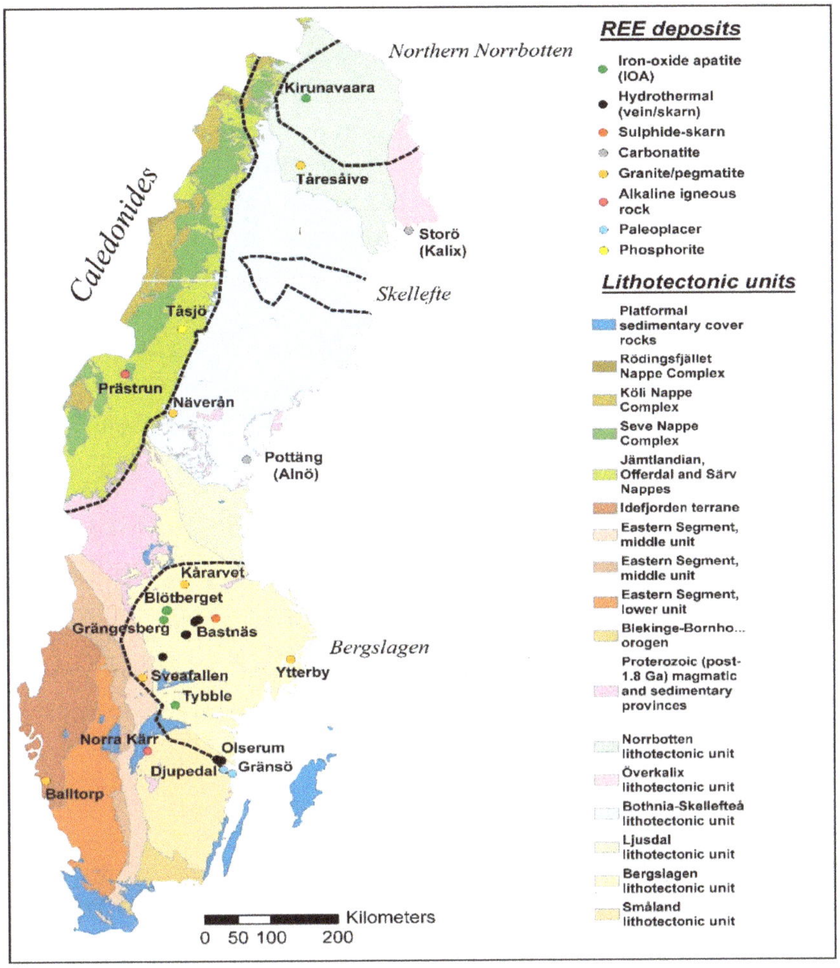

Figure 1. Simplified map of major lithotectonic units and ore districts in Sweden, with a selection of Rare Earth Elements (REEs) deposits, prospects, and occurrences (modified from [1] and bedrock database at Geological Survey of Sweden)

REE deposits can be referred to regolith, basinal, metamorphic and magmatic associations based on a mineral-systems approach [6]. Various deposit types form either directly from the crystallization of the melt and/or fluids predominantly derived from the melt. The magmatic deposits can be divided into orthomagmatic and hydrothermal types. Basinal deposit types are inferred to be formed through mechanical (e.g., placer) and chemical (e.g., phosphorite) sedimentary processes, and from diagenetic fluids generated in sedimentary basins. Deposit types of the regolith association require an REE-bearing rock source to form feasible secondary contents of REE. The REE deposits are formed either due to enrichment of REE in the residual material and/or from local remobilization of REE. Deposit types of the metamorphic association are generated during regional and/or contact metamorphism and involve related metamorphic fluids.

This study analyzes, explores and evaluates the composition of REE mineralized samples collected during the EURARE project (2013–2017) enabling their genetic classification into specific deposit types. This information is integrated with the Geological Survey of Sweden (SGU)'s lithogeochemistry and till geochemistry databases to better approach and interpret the distribution of REE anomalous provinces at the scale of the country.

2. Overview of REE Mineralizations in Sweden

2.1. REE Mineralizations in Granitic Pegmatites and Granitoids

Partial melting of crustal material produces felsic melts that are enriched in REEs. Elements in pegmatites that can be enriched up to minable (ore) grades include Li, Cs, Be, Sn, Nb, Ta, U, Y, Zr, and REEs. This long and diverse list of elements, to which a significantly high content of B, P, F, Rb, Bi, Hf, etc., could be added, make fractionated pegmatites among the most mineralogically complex deposit types on the Earth [7]. Pegmatitic facies associated with syenite-alkaline granite complexes constitute important deposits of REEs, U, Nb and Zr, and less commonly, Be and P [7].

In Sweden, granitic pegmatites, typically of a moderately to highly fractionated character, often host variable amounts of minerals rich in trace elements, including Be, Li, Nb, Ta, Sn, U, Th, as well as REEs. [8,9]. One notable granitic pegmatite, that shows locally relatively high REE content, (a porphyritic granite to gneiss with age 1300 Ga so-called "RA-granite") is located in Balltorp (Eastern Segment lithotectonic unit) (Figure 1) in southwestern Sweden [10]. The RA-granite is a Be-F-Nb-REE-Sn-Ta-Th-U-Zr-anomalous gneissic granite that does show some potential but has so far not been systematically explored (cf. [10] and references therein). Moreover, several other granitic to syenitic rocks in southern Sweden and elsewhere exhibit increased REE contents. The most important granitic pegmatites and granitoids with potential for REE mineralization are, Tåresåive (highest contents of REE_{tot} = 94,900 ppm) in northern Sweden (Norrbotten lithotectonic unit), Näverån in Central Sweden (close to the Caledonides), as well as granitoids in Bergslagen (e.g., Ytterby; highest contents of REE_{tot} = 30,624 ppm), and granite in the southwestern part of Sweden, (e.g., Balltorp; highest contents of REE_{tot} = 8800 ppm), shown in Figure 1. A contrasting case to these primary enrichments is the granitoid-hosted but epigenetic, shear zone-related mineralization at Näverån (highest contents of REE_{tot} = 6929 ppm) in central Sweden [1].

2.2. REE Mineralization in Alkaline Intrusive Rocks and Carbonatites

Several REE deposits associated with peralkaline complexes (syenitic rocks) are usually enriched in U and Th and have a relatively high Heavy Rare Earth Element (HREE) content when compared with carbonatites [11]. In Sweden, Norra Kärr deposit represents one of the most advanced and promising REE mining projects in northern Europe. This concentrically zoned alkaline intrusion, with nepheline-syenite, contains eudialyte-group minerals highly enriched in REEs with up to 8.5 wt% (La + Ca + Nd + Y) [1], showing also a strong resource potential in heavy REEs.

Carbonatites are common hosts of orthomagmatic-type REE mineralizations that tend to be variably enriched in REEs, Sr, Ba, U, Th, Nb, Ta, P, and F. Although the overall REE content of carbonatites may vary, their shape on the chondrite-normalized plot almost invariably displays high Light Rare Earth Element (LREE) content and no negative Eu anomalies [12]. Barium and Sr are generally abundant, whereas U, Th, Nb, and P, are variably abundant in the mineralized carbonatite. REE-bearing mineral phases have the highest REE content but major carbonate minerals, including calcite and dolomite, may also contain substantial contents of REEs [13]. Sövitic carbonatitic melts tend to show the highest levels of REE enrichment. They also show significant enrichment in LREE (LREE/HREE ratio of ~40, compared to ~7 of alkaline and felsic melts).

In Sweden, the Alnö complex consists of alkaline and carbonatite intrusive rocks located on the island of Alnö, in the Bothnian Bay and on the mainland north to northwest of Alnö, all along the coast of east-central Sweden [14–16]. The carbonatite, and specifically the calcite-dominated søvite, have significantly elevated total REE contents, between ca 500 and 1500 ppm [17]. Alnö intrusive complex and associated co-magmatic rocks on the mainland have also been the focus of exploration for REEs, including that carried out by the Boliden company in the 1970s. The main lithology ijolite contains clinopyroxene with LREEs up to 100 times chondritic values [18,19]. The Söråker intrusion located on the mainland north of Alnö comprises Ca- and Mg-carbonatite dikes and melilitolites. Another small carbonatite occurrence is known from the Middle Allochthon in the Caledonides where calc-silicate rocks at Prästrun in west-central Sweden host a REE-Nb-(U-Th) mineralization (highest contents of REE_{tot} = 2300 ppm). This intrusive-hosted REE mineralization was first described at Prästrun.

Alkaline lamprophyre is another lithology that tends to have elevated REE contents. Chemically, potassic lamprophyre dikes are often characterized by relatively low SiO_2 and TiO_2, and high MgO and K_2O contents. In addition, they show high content of large-ion lithophile trace elements (e.g., Rb, Sr, and Ba) and LREE, but low content of high-field-strength elements (e.g., Nb, Ta, Zr, Hf, and Ti) [20]. An example of the REE-enriched lamprophyre is the 1.1 Ga alkaline ultramafic dykes up to 1-metre-wide, outcropping along the northernmost part of the coast in the Gulf of Bothnia, (Kalix) known as Storön deposit (Figure 1; [20,21]). The dikes represent olivine- and mica-rich lamprophyres and silico-carbonatites [21].

2.3. REE Mineralization in Apatite-Iron Oxide Deposits

At present, over forty iron oxide apatite deposits are known in northern Sweden [22–24]. The main (at over 2000 Mt) deposit in the Norrbotten district is the famous Kiruna (Kiirunavaara) deposit and active mine (cf. [24]). A hydrothermal (including replacement) and (ortho-) magmatic origin of iron oxide apatite deposits in Norrbotten have been suggested in numerous studies (e.g., [25–31] and references therein). The main REE host minerals in this deposit are apatite, typically a fluorapatite, and monazite, occurring often as inclusions in apatite [32–35]. Apatite abundance and composition in the ore varies, e.g., at Kiruna, where the phosphorus content in the upper parts of the ore is ca. 2%, equivalent to almost 11% apatite. REE enrichment is represented mainly by LREEs, Ce and Y anomalies observed as early as the 1930s by Geijer [34]. REE geochemistry of the host rocks in the northern district shows two patterns, one with a negative Eu anomaly and strong enrichment in LREEs at Kiruna and area close to this mine and the second one without a well pronounced Eu anomaly e.g., at the Malmberget and the areas close to this mine (Figure 2) [36].

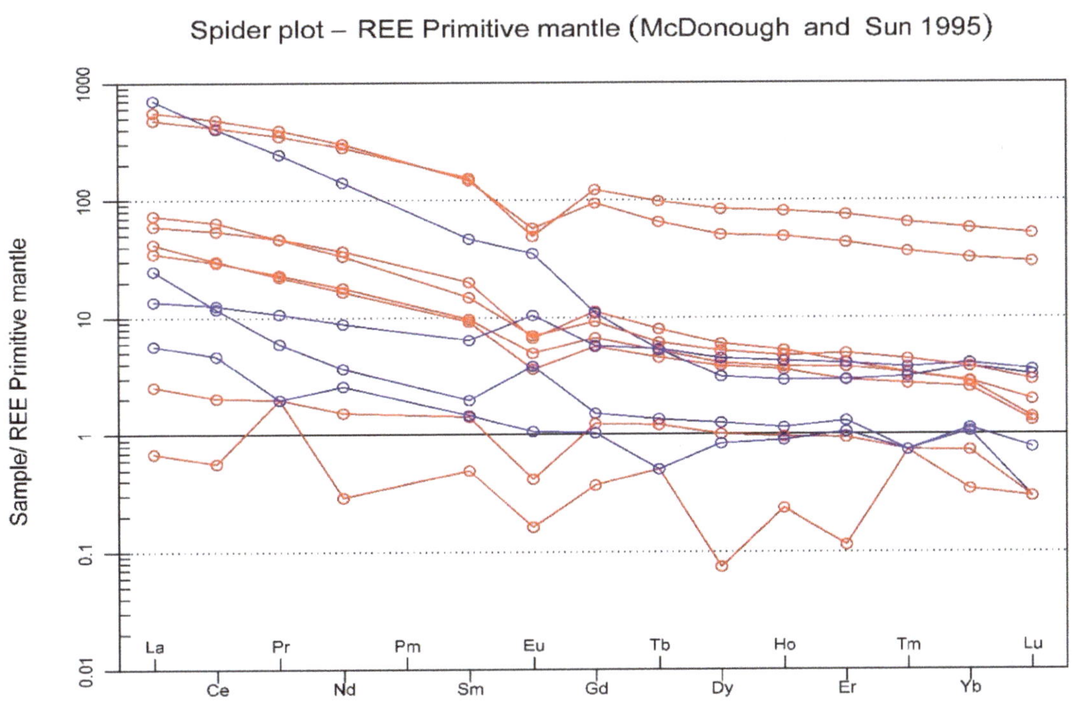

Figure 2. REE spider plot [37] of host rocks from the northern districts; Kiruna mine and occurrences close to this area (red) display a distinct Eu trough and a relative enrichment in LREEs compared to the HREEs, and while the samples from the Malmberget mine and occurrences nearby show extensive metamorphism and deformation overprint compared to the Kiruna area (Blue). Data sources: SGU and [36].

In central Sweden, (the Bergslagen district), a number of iron oxide apatite deposits are known with the biggest being Grängesberg and Blötberget, which were mined in historical times (Grängesberg mine closed in 1989) and both represent Kiruna-type deposits. The previous studies reveal that they still constitute a rather important potential reserve of REEs and phosphorus, both from the main ore and from mining waste [38]. The deposits are predominantly hosted by 1.9 Ga volcanosedimentary rocks variably metamorphosed and deformed. The ore is usually enriched in LREEs and to a minor degree in U, Th, Sm, Tb, Yb, and Y. The majority of REE enrichment is located in phosphate minerals such as apatite, xenotime, and monazite and to a lesser degree in silicates such as allanite and epidote [1,30,33,35].

2.4. Fe-REE Hydrothermal (Bastnäs-Type) Mineralizations

Evaluation of the REE geochemistry of hydrothermal mineral deposits, and the processes by which they are concentrated, is a complex task [39]. However, these mineralizations can be grouped on the basis of their geochemical and mineralogical characteristics. The degree of REE enrichment in a deposit is a function of the concentration of REEs in the fluid, the water-rock ratio, the efficiency of the precipitation process and the nature and amount of co-precipitated phases [39]. Oreskes and Einaudi [40] suggested that the high abundance of fluorocarbonates and the lack of Ca minerals in REE deposits (e.g., Olympic Dam) indicated that the fluids behind hydrothermal REE mineralizations may have been F- and CO_2-rich and that the REEs were transported as Cl and/or F complexes. Lottermoser [41] suggested that the association of REE with U minerals indicated that U and REE were complexed by the same ligands and inferred that those were CO_3, F, or SO_4.

The Bastnäs-type Fe-REE deposits in Sweden are early Proterozoic, skarn-hosted iron oxide (magnetite-dominated), locally polymetallic (±Cu, Au, Co, Bi, Mo) mineralizations that in part carry very REE-rich mineral assemblages (cf. [42]). They are located in the Bergslagen district and are characterized by the occurrence of locally abundant REE-rich silicate minerals such as cerite-(Ce) and allanite (*sensu lato*) but also include REE fluorocarbonates such as bastnäsite-(Ce).

3. Materials and Methods

In this study the lithogeochemical and till geochemistry databases have been used for re-interpretation of the REE baseline in different rock types and till in Sweden, and for the evaluation of REE mineral potential. More than 20,000 rock samples were collected and analyzed as part of the bedrock mapping program at the Geological Survey of Sweden (SGU; www.sgu.se). Within the EURARE project (www.eurare.org), field work activities were carried out in several REE-mineralized areas and were followed up by detailed investigations on selected mineralizations. The modal mineral composition, petrographical features and chemical compositions carried out for more than 200 rock samples [1]. The lithogeochemical analyses were carried out at the ACME Lab (Vancouver, BC, Canada) and ALS Scandinavia AB (Luleå, Sweden). At the ACME Lab the samples were analyzed for major, minor, and trace elements by ICP-emission spectrometry following a lithium metaborate/tetraborate fusion and dilute nitric digestion. At the ALS, trace elements including the full rare earth element suites were obtained from fused beads followed by acid digestion and measured by either ICP-AES or an inductively coupled plasma sector field mass spectrometer (ICP-SFMS).

The till database used in this study contains over 2500 samples analyzed by aqua regia digestion followed by ICP-MS. These national-wide results have been published as the Geochemical Atlas of Sweden [3] and are available for the public online (https://www.sgu.se/mineralnaring/geokemisk-kartlaggning/geokemisk-atlas/).

The open-source software Geochemical Data Toolkit in R (GCDKIT-version 4.1) [43] which is built using the freeware R language, has been applied for the data processing and biplot diagrams in this paper (http://www.gcdkit.org).

Principal component (PC) analysis is a conventional multivariate technique that is often used for studying geochemical data [4,44–47]. PC analysis reduces a large number of variables to a smaller number, allowing the user to determine the components (groups of variables) that account for variation in multivariate data [47]. PC analysis has often been used to process and interpret geochemical and other types of spatial data e.g., [4,5,47,48]. PC analysis builds on the correlation (covariance) matrix, which measures the interrelationships among multiple variables. The first PC (PC1) explains most of the variance within the original data, and each subsequent PC (PC2–n) explains progressively less of the variance. A multivariate dataset can usually be reduced to two or three PCs that account for the majority of the variance within the dataset. PC analysis was performed on average values of the different mineralization types as discussed above. Thirty-seven elements were chosen for PC analysis based on correlation coefficients. The Statistica software (version 13.2, Dell Software, Aliso Viejo, CA, USA) has been used for the principal component analysis (https://statistica.software.informer.com/).

The Geochemical Atlas of Sweden [3] is a country-wide harmonized database with modern baseline geochemical data from the C horizon in till. The database encompasses 2578 till samples from SGU archive as well as new sampled till collected mainly in the mountainous areas of western Sweden. Since the C horizon is considered as an anthropogenically undisturbed layer, the major geochemical signature in till should originate from underlying bedrock, its lithology, mineralogy, and potential mineralization. Secondary processes such as ice transport, leaching, and biological activity have minor impact on till chemical composition. During weathering, REEs are generally not very mobile, but this varies depending on the host mineral and local pH. REE mobility is further controlled by adsorption onto iron oxides, phosphates, and clay minerals [2,49]. In general, till can be used as proxy for the underlying bedrock in areas where outcrops are not available. Till can be treated then as composite rock material and its composition can be often interpolated to the average parent rock.

4. Results and Discussion

4.1. REE Content in Bedrock

Lithogeochemistry data from the SGU database have been used to examine links between REE contents in bedrock with known REE mineralizations. The main intention was to establish a tool for future exploration and predictive mapping surveys. Generally, primary REEs are associated with magmatic (granite and pegmatite) and alkaline rocks. Rocks enriched in monazite and zircon often contain higher contents of rare earth elements. In sedimentary rocks, higher contents of rare earth elements occur in shale and greywacke. There are no obvious differences in the geochemical distribution between the LREE group (La, Ce, Pr, Nd, and Sm) and the HREE group (Eu-Lu + Y) in the lithogeochemical data. However, their relative contents may differ for each element, probably due to the mineral composition and type of the host rock. For example, Eu is often found in rock-forming minerals such as plagioclase and in accessory minerals, mainly in allanite, bastnäsite, monazite, apatite, zircon, and fluorite. Typically, Eu can replace Sr and Ca in rock-forming minerals, therefore, its enrichment in plagioclase is common. This results in Eu enrichment in plagioclase-rich rocks in northernmost Sweden, such as gabbro and granodiorite.

The data with a higher content of the total REE and Y (above 500 ppm) were extracted from the database and compared with the extracted data with a content of Y more than 100 ppm (Figure 3). Our results show a linear correlation and match between the samples with a higher content of REE + Y in total with Y content, and more importantly, there is a direct link between Y content in samples with existing known mineralizations in Sweden (Figure 3).

4.2. Geochemistry of REE Mineralization in Sweden Inferred from the EURARE Project

During the EURARE field campaign, approximately 200 samples were taken for petrography and mineralogy studies and for geochemical analysis. The location of the samples and type of mineralization are shown in Figure 4, and the geochemical results are provided in Table 1.

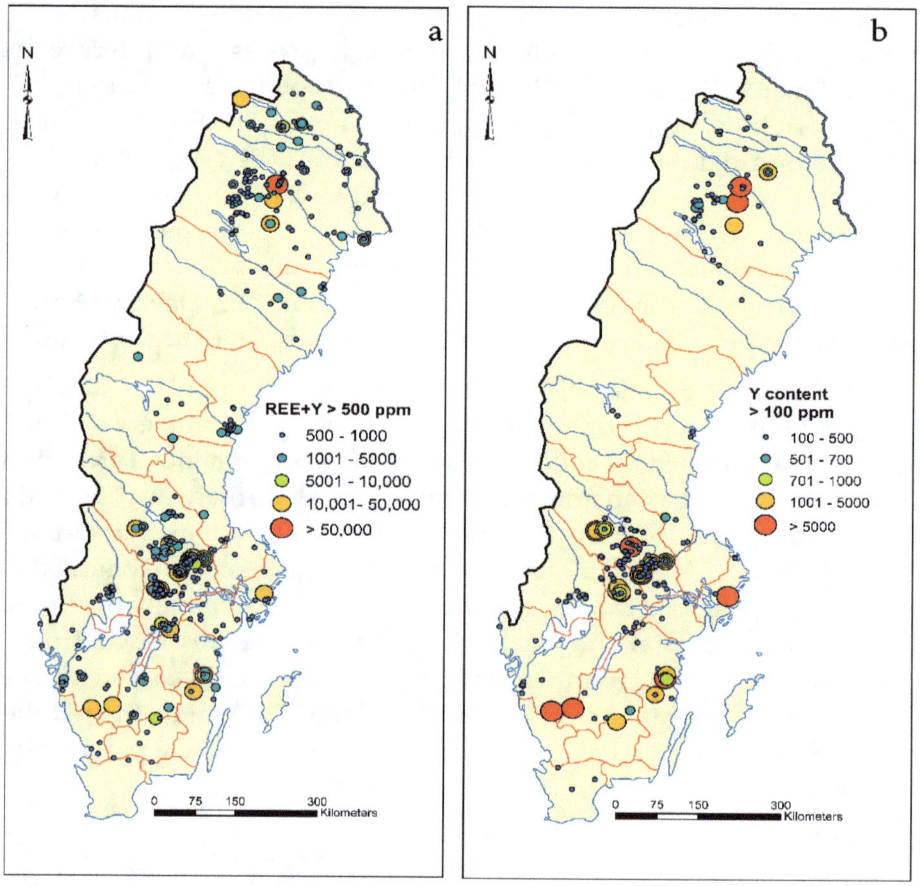

Figure 3. Simplified map of lithogeochemical data in Sweden: (**a**) total REE + Y > 500 ppm and (**b**) Y content >100 ppm (source of data: lithogeochemical database at the Geological survey of Sweden).

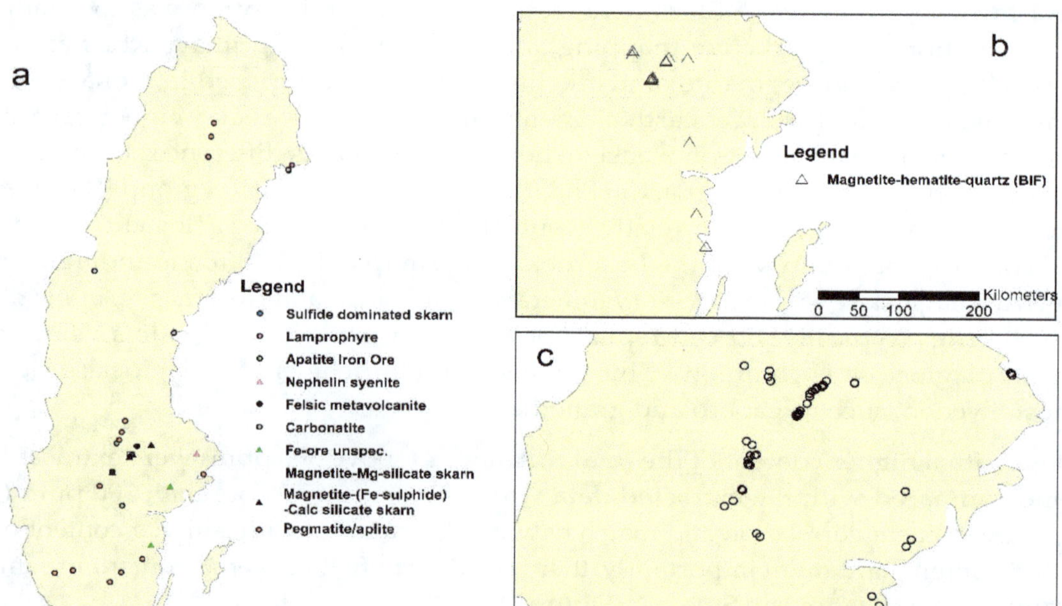

Figure 4. Simplified location map of samples in the EURARE project: (**a**) an overview of sample locations and types, excluding magnetite-calc silicate skarn and BIF ones; (**b**) magnetite-hematite-quartz samples; (**c**) magnetite-calc silicate skarn samples.

Table 1. Summary of samples analyzed with a total REE (TotREE) content above 1% (10,000 ppm). N and E are SwerefTM coordinates.

Occurrence/Deposit	Sample	N	E	Type	TotREE (ppm)	LREE (ppm)	HREE (ppm)	Th (ppm)	U (ppm)	P (ppm)
Ingridstorp	KES155002A	6,392,246	559,703	Pegmatite/aplite	47,172	43,007	4165	>1000	266	70
Stålklockan	PNY130005	6,635,349	532,852	Magnetite-Calc silicate skarn	45,512	34,001	11,511	2.01	58.5	90
Tåresåive	ING090076A	7,421,196	709,317	Pegmatite/aplite	42,651	33,056	9596	>1000	1505	140
Bastnäs	BAST140001	6,634,470	532,926	Magnetite-Calc silicate skarn	38,078	33,149	4929	20.7	66.4	70
Bastnäs	PNY130001	6,634,425	533,003	Magnetite-Calc silicate skarn	37,907	33,134	4773	9.64	92	60
Gyttorp	Gyttorp	6,597,427	497,660	Magnetite-Calc silicate skarn	37,069	33,233	3836	50.4	1120	<10
Mörkens	MÖRK2	6,662,160	577,465	Magnetite-Calc silicate skarn	36,921	33,225	3697	3.36	18.85	40
Djupedalsgruvan	JSM150007A	6,425,442	578,440	Fe-ore unspec.	34,643	22,746	11,898	158	9	2640
Ingridstorp	KES155003A	6,358,023	372,718	Pegmatite/aplite	33,895	19,577	14,318	>1000	1330	1260
Ytterby	YBY140002	6,592,332	690,226	Pegmatite/aplite	30,624	13,624	17,000	>1000	1185	6200
Rödbergsgruvan	Rödbergsgruvan A	6,597,194	494,237	Magnetite-Calc silicate skarn	30,536	29,379	1158	1.53	4.15	80
Östanmossa	PNY130015	6,660,543	551,764	Magnetite-Calc silicate skarn	30,319	28,850	1469	1.21	3.7	40
Djupedalsgruvan	JSM150008A	6,425,442	578,440	Fe-ore unspec.	29,990	16,626	13,365	214	8.39	3360
Holmtjärn	HOLM140001	6,691,407	514,383	Pegmatite/aplite	29,550	12,550	17,000	>1000	>10,000	450
Ytterby	YBY140003	6,592,332	690,226	Pegmatite/aplite	26,419	10,239	16,180	>1000	1270	1470
Djupedalsgruvan	JSM150006A	6,425,442	578,440	Fe-ore unspec.	25,904	11,133	14,771	134	21.5	690
Mörkens	MÖRK3	6,662,160	577,465	Magnetite-(Fe-sulphide)-Calc silicate skarn	24,971	21,636	3335	0.43	2.67	30
Gruvhagen	GRUVHAG1	6,662,377	577,026	Magnetite-Calc silicate skarn	24,574	23,618	956	3.34	0.77	10
Ingelsbo	KES155006A	6,364,317	411,300	Pegmatite/aplite	22,192	5192	17,000	>1000	>10,000	<10
Ytterby	YBY140001	6,592,332	690,226	Pegmatite/aplite	18,695	3801	14,894	>1000	2360	6280
Reunavaare	REU140001	7,342,033	697,981	Pegmatite/aplite	18,207	13,158	5049	>1000	438	30
Tybble	SGUR10007	6,517,096	516,016	Iron oxide apatite	17,428	17,274	154	6.41	457	1400
Ingridstorp	KES155001A	6,392,246	559,703	Pegmatite/aplite	16,320	16,026	294	257	53.9	90
Flakaberget	FLA140001	7,389,427	703,658	Pegmatite/aplite	15,556	1627	13,929	>1000	1600	520
Johanna	PNY130013	6,658,660	550,883	Magnetite-Calc silicate skarn	14,008	12,966	1042	0.72	8.68	30
Sveafallen	SVEAFALLEN	6,563,279	467,518	Pegmatite/aplite	13,220	12,579	641	>1000	16	1460
Mörkens	MÖRK1	6,662,160	577,465	Magnetite-(Fe-sulphide)-Calc silicate skarn	11,077	9203	1874	0.09	2.47	40

The geochemical analyses of the samples collected during the EURARE project show that 68 samples have a total REE content above 1000 ppm and of those 27 have more than 1% total REE.

The samples with more than 1% REEs represent mineralizations classified as: iron oxide apatite (1 sample), Fe-ore unspecified (3 samples), magnetite-calc-silicate skarn (9 samples), magnetite-(Fe-sulphide)-Calc silicate skarn (2 samples), and granitic pegmatite (12 samples). These results are summarised in Table 1.

A bivariate plot of total REE content vs. LREE (Figure 5a) and vs. HREE (Figure 5b) show that there is a general bimodal linear correlation between total REE and LREE (Figure 5a), one trend is characterized by LREE enrichment (for magnetite-skarn mineralization) and a second one by LREE depletion, especially in some late magmatic pegmatite REE mineralizations and some related to iron mineralizations. The comparison of total REEs with HREEs also shows clear decoupling between two groups, the first composed of low HREE mineralization types (e.g., various iron ores, skarn mineralizations, carbonatite, and sulphide-dominated skarn ores) and these with elevated HREE content are mainly represented by late magmatic pegmatites and unspecified iron ore types. It is unclear how the magmatic rocks differ in primary composition, so they host both low- and high-level HREE mineralizations. BIF deposits (magnetite-hematite-quartz) do not show a very high content of REE_{tot} and generally are most likely to have elevated LREE rather than HREE. The number of two samples of sulfide dominated skarn is limited and, therefore, it is excluded from interpretation of the results to avoid any over interpretation.

A bivariate plot of La vs. La/Yb (Figure 6) shows a weak but overall positive correlation between the two variables, except for a few samples, which show higher HREE contents. This suggests that those samples with a high LREE content also have relatively high LREE/HREE ratios. They mostly represent REE mineralizations associated with granitic pegmatites. One sample from iron oxide-apatite deposits contains REE-bearing minerals such as xenotime and monazite and shows higher contents of HREEs (Figure 6).

Thorium and uranium content and their ratio are useful for recognizing geochemical facies and also the content of radioactive elements in different types of mineralization which may have an impact on future exploitation. Figure 7a shows Th content versus total REEs. In general, skarn and iron oxide-apatite mineralizations with higher contents of total REEs ($REE_{tot} > 1\%$) show a positive correlation with low contents of Th ($Th < 50ppm$). A few samples from iron oxide mineralizations show higher content of Th up to 200 ppm and total REE contents of up to 3.5%. Two samples of granitic pegmatites show higher contents of Th, whereas those of total REEs are low, which means that the content of Th in those samples is not directly related to REE-bearing minerals and may also correspond to the primary origin of the host rocks and its later deformation. For example, rocks with sedimentary origin have usually higher Th/U ratios than magmatic rocks, and these ratios can be further modified during regional metamorphism [11,12,49].

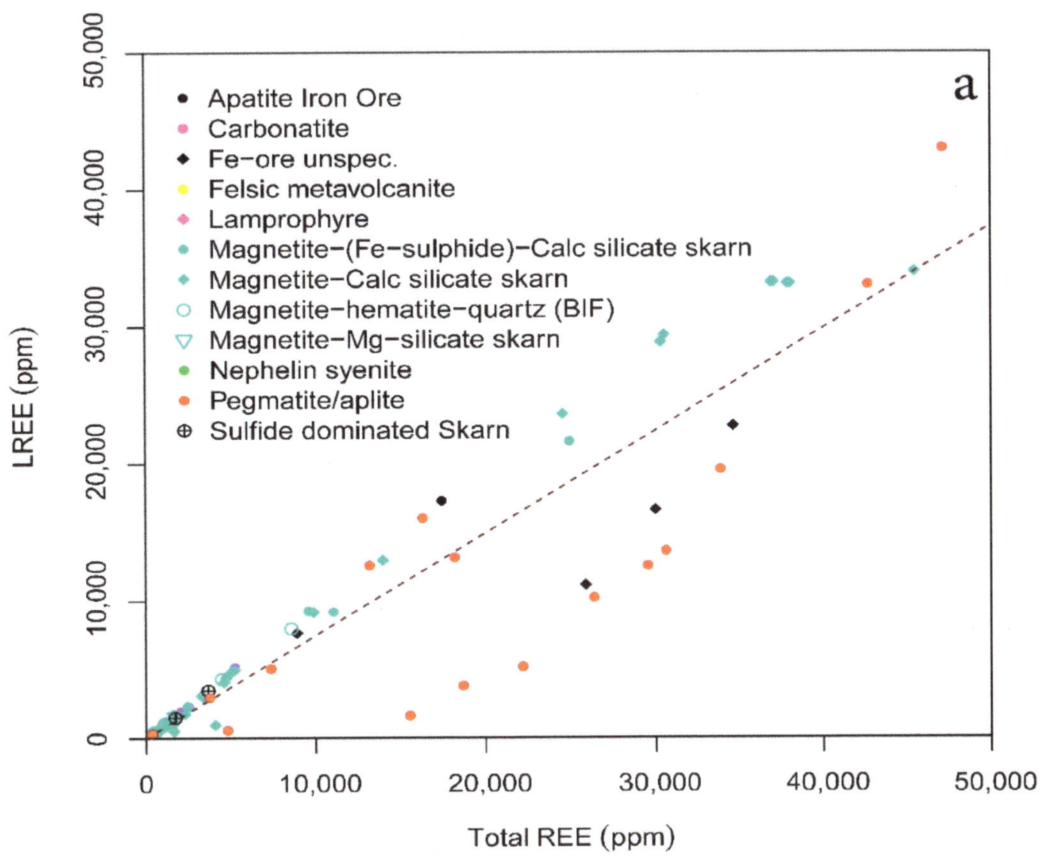

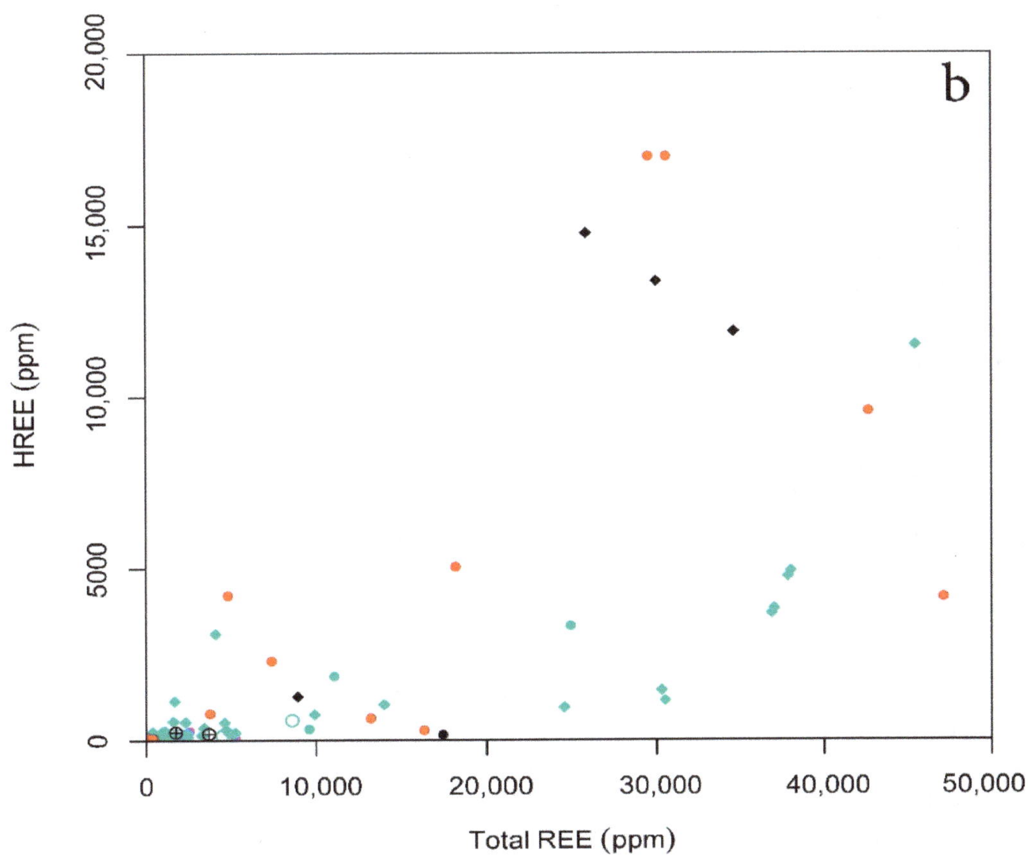

Figure 5. Bivariate diagrams of total REEs versus LREEs (**a**) and HREEs (**b**) for EURARE project samples.

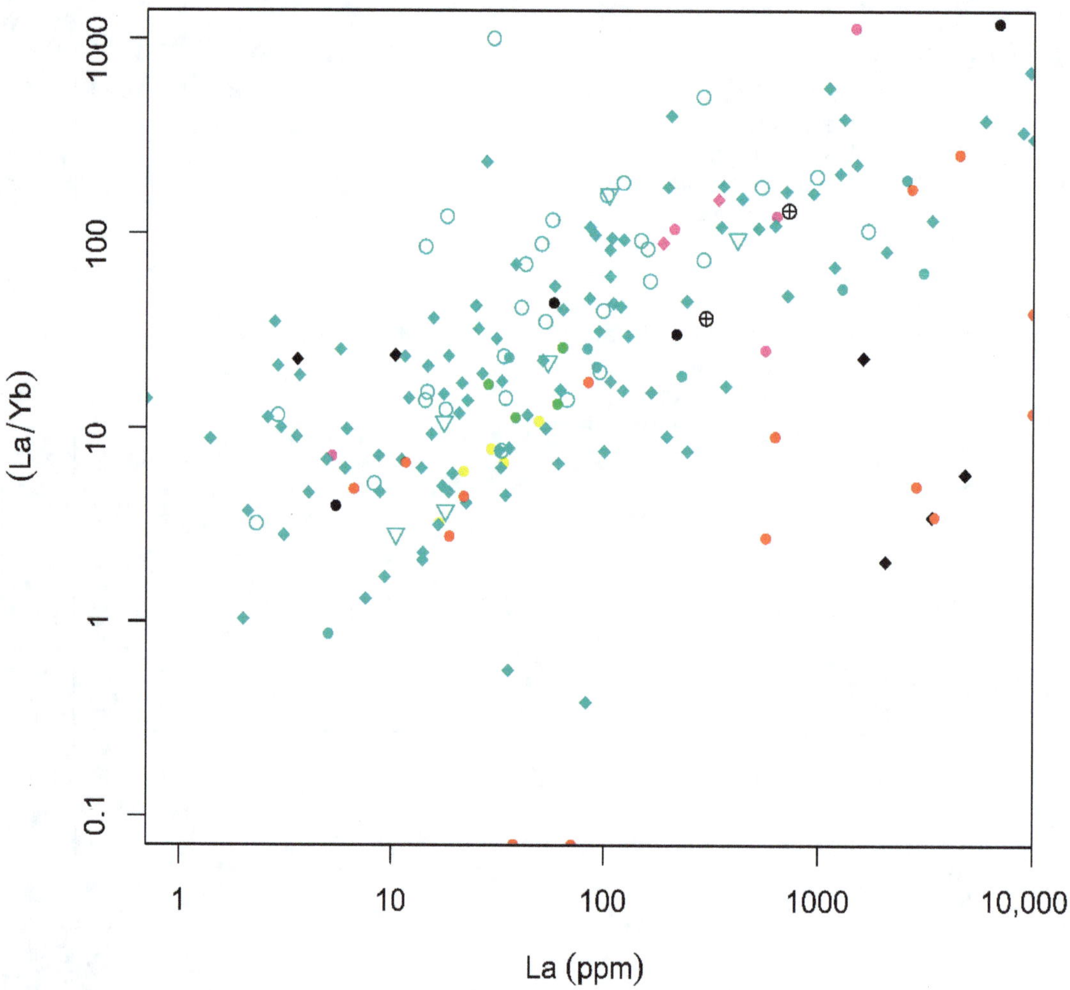

Figure 6. Bivariate plot of La (ppm) versus La/Yb for the EURARE samples (see the legend in Figure 5).

The U/total REE (Figure 7b) of the EURARE samples shows that high U contents (U > 400 ppm) are typical of REE mineralization hosted in granitic pegmatites. An additional few samples collected from magnetite-calc silicate skarn type deposits show higher contents of U (>1000 ppm). In the sample from magnetite-calc silicate skarn, one sample shows a high content of total REE correlates with high U content which may indicate the U enrichment in REE-bearing minerals.

Comparison of total REE with Th/U ratios based on two analyzed samples from lamprophyres (Figure 7c) shows that the samples with low contents of total REE ($REE_{tot} < 400$) typically have higher Th than U contents (Th/U > 20). The Th/U ratio is generally low in the samples from skarn and iron oxide-apatite REE mineralizations (Th/U < 10) with the exception of four samples from iron oxide-apatite REE mineralizations (at the Djupedalsgruvan) that show high contents of both Th and REE (and contain xenotime and allanite minerals) (Figure 7c).

Plots of quantitative values of selected REEs (Nd, Pr, and Y) and phosphorus compared to binary plots of LREE versus HREE show at least two contradicting trends. For example, in Figure 8a samples with high Nd contents show two opposite trends; one positive with well correlated LREE and Nd contents accompanied by low HREE and the second negative trend where samples with decreasing Nd contents plot towards a lower LREE content and a higher HREE. Few exceptions of samples with high content of HREE and Nd content (>4000 ppm) can also be observed (Figure 8a).

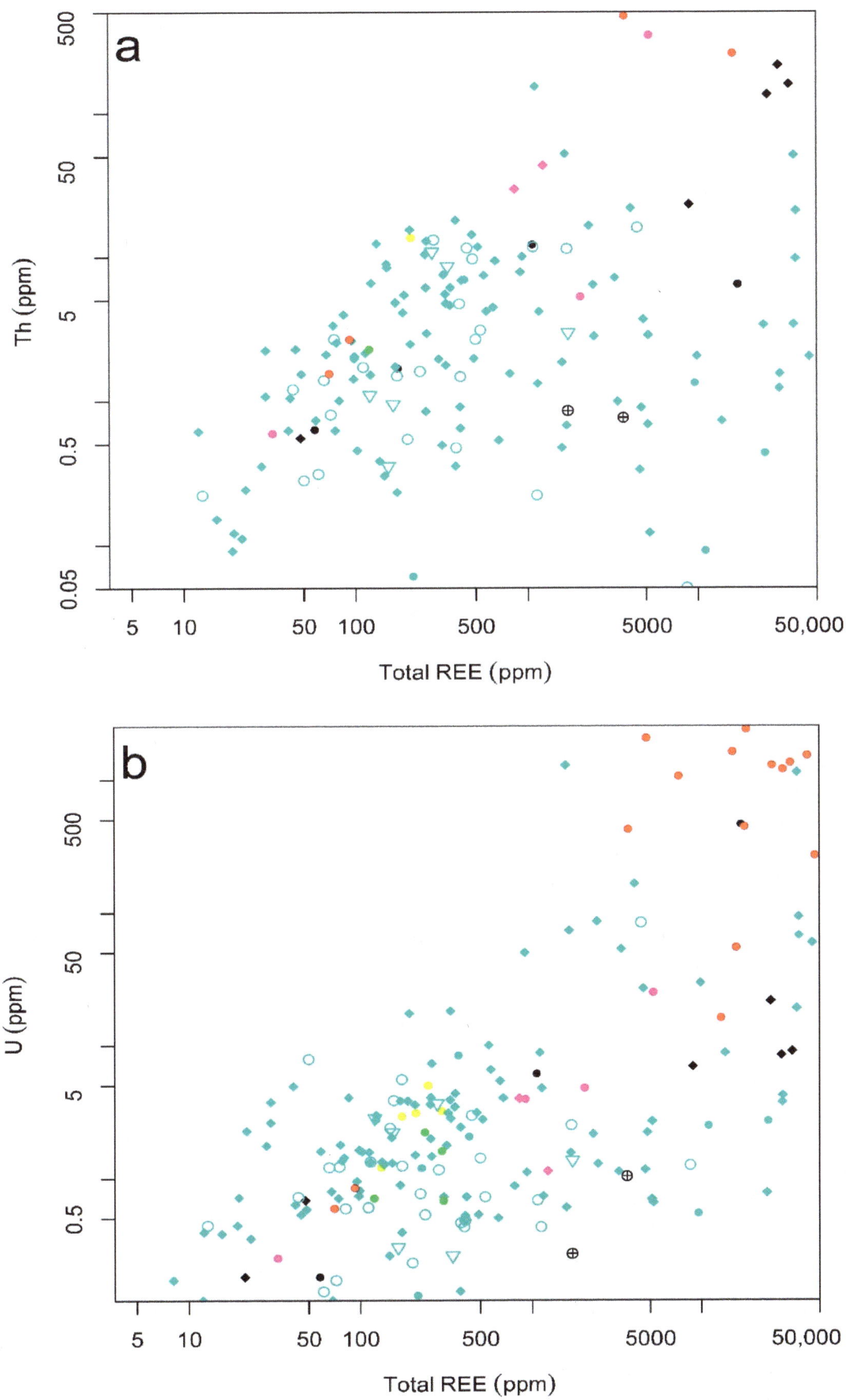

Figure 7. *Cont.*

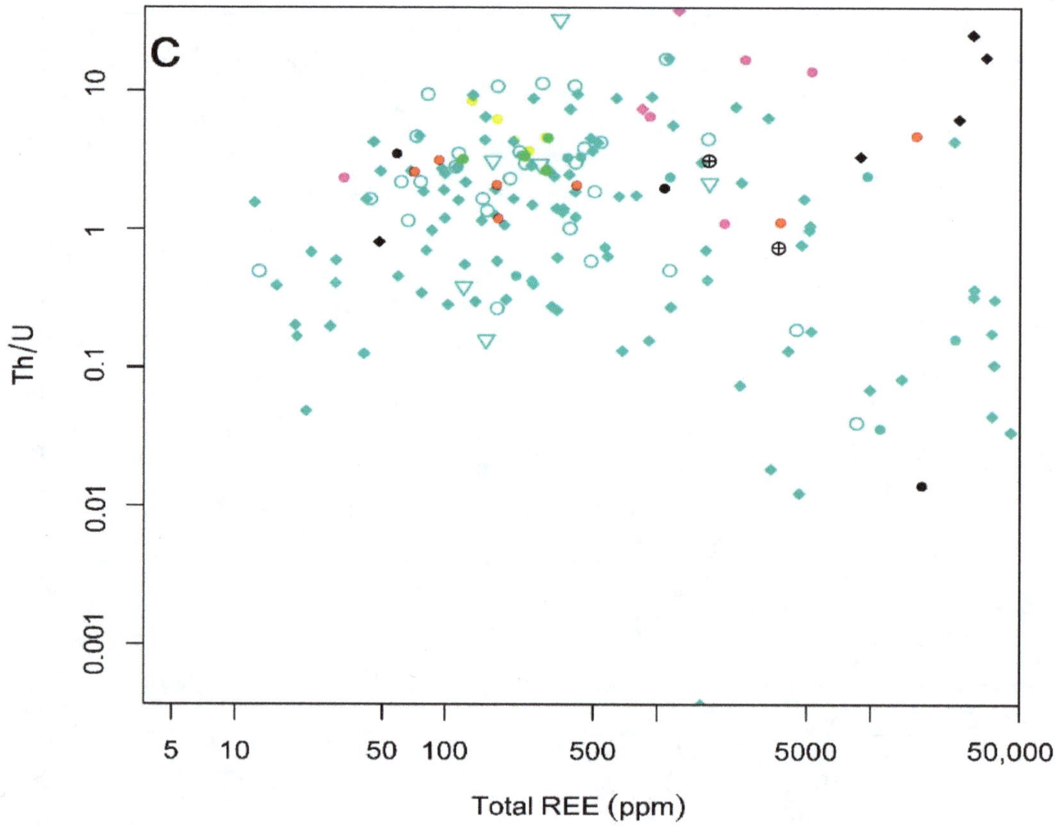

Figure 7. Binary plots of (**a**) total REE vs. Th (ppm); (**b**) total REE vs. U (ppm); (**c**) total REE vs. Th/U; for the EURARE samples (see the legend in Figure 5).

In a similar type of plot, quantitative content of Pr in the samples shows a semi-concordant trend with Nd, and this may reflect that the minerals bearing Pr and Nd are the ones containing HREE, such as xenotime-(HREE) and fergusonite-(HREE) as well as gadolinite-(REE).

Figure 8c shows linear positive correlation of high Y samples with a higher content of HREE. Since Y has a similar geochemical behavior as HREE the low HREE contents follow low Y contents in EURARE samples. High LREE samples are less affected by Y content with weak negative correlation. Samples with low and high Y content when compared to LREE might reflect various degrees of host magma differentiation or secondary remobilization during regional metamorphism (competitive re-crystallization and crystallization of metamorphic minerals) [11,12,49].

Comparison of phosphorus content, indicative of the presence of phosphate minerals shows two different clusters (Figure 8d). The first cluster shows that some high P samples have low contents of both LREEs and HREEs, which can be interpreted as apatite mineralization in iron ore with no REE enrichment. The second cluster shows that samples with high P content correlate with high HREE content, which may indicate the presence of minerals other than apatite as a host of HREEs, such as xenotime.

Within the sample set collected during the EURARE project, there are several iron ore samples (Fe-ore unspec.) that cannot be easily classified (more mineralogical and petrological investigations are needed). In an attempt to classify them, the classical PM-normalized REE diagram has been used [37] (McDonough and Sun 1995) (Figure 9). At a first glance, their REE patterns suggest the presence of two genetic groups. The first group with REE enrichment (>100 ppm) shows a similar trend to the iron oxide-apatite- and skarn-type REE mineralizations, with a negative Eu anomaly and relative HREE enrichment. The second group shows a weak positive Eu anomaly and steep decreasing trends from LREEs to HREEs accompanied by a rather low total of REEs (<100 ppm). These two samples may have an alkaline or peralkaline origin but a more detailed study is needed (Figure 9).

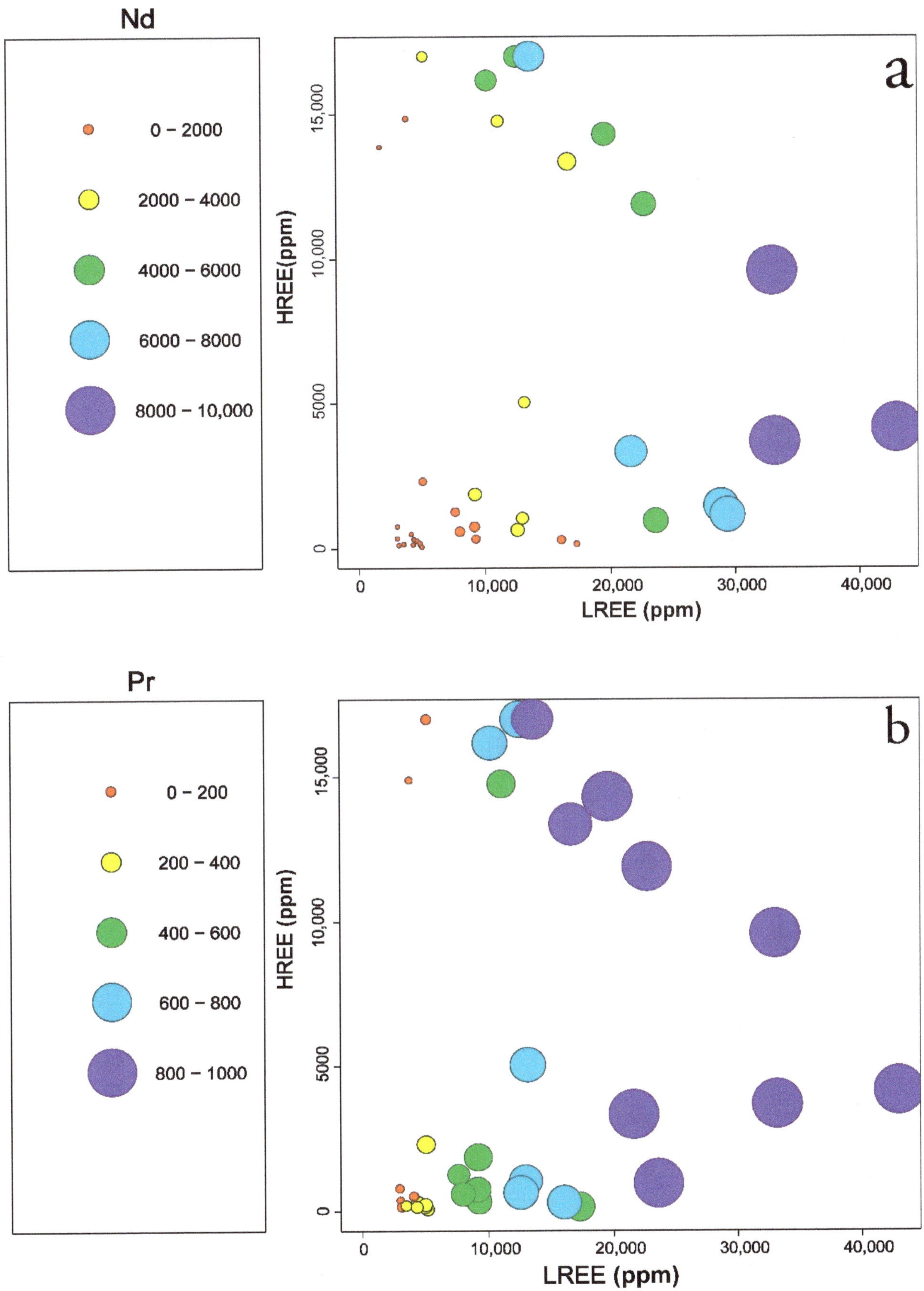

Figure 8. *Cont.*

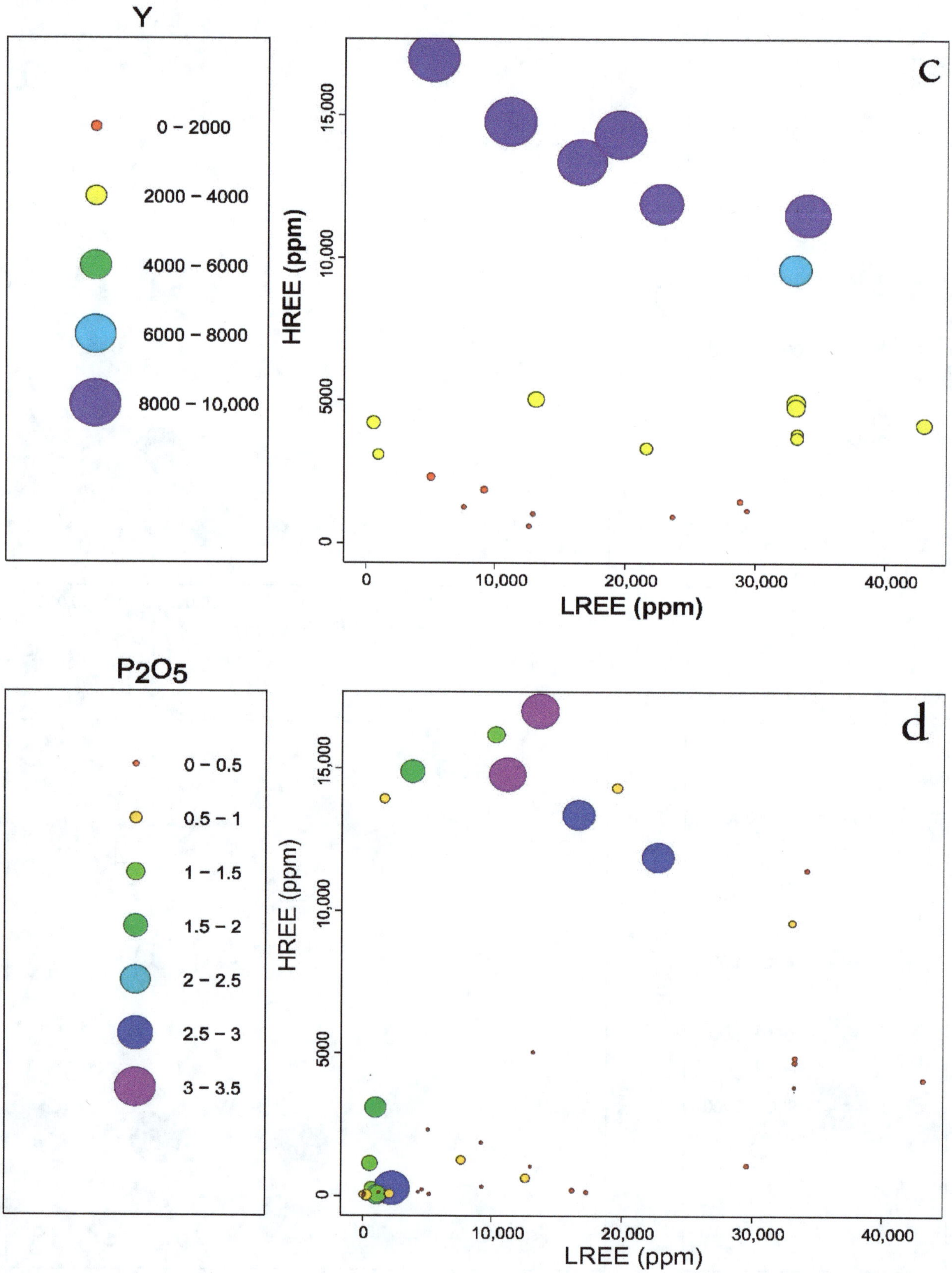

Figure 8. Plots of quantitative values of selected elements compared to LREEs and HREEs in the EURARE samples: (**a**) Nd (ppm); (**b**) Pr (ppm); (**c**) Y (ppm); (**d**) P_2O_5 (wt%). Samples with contents less than background values are excluded.

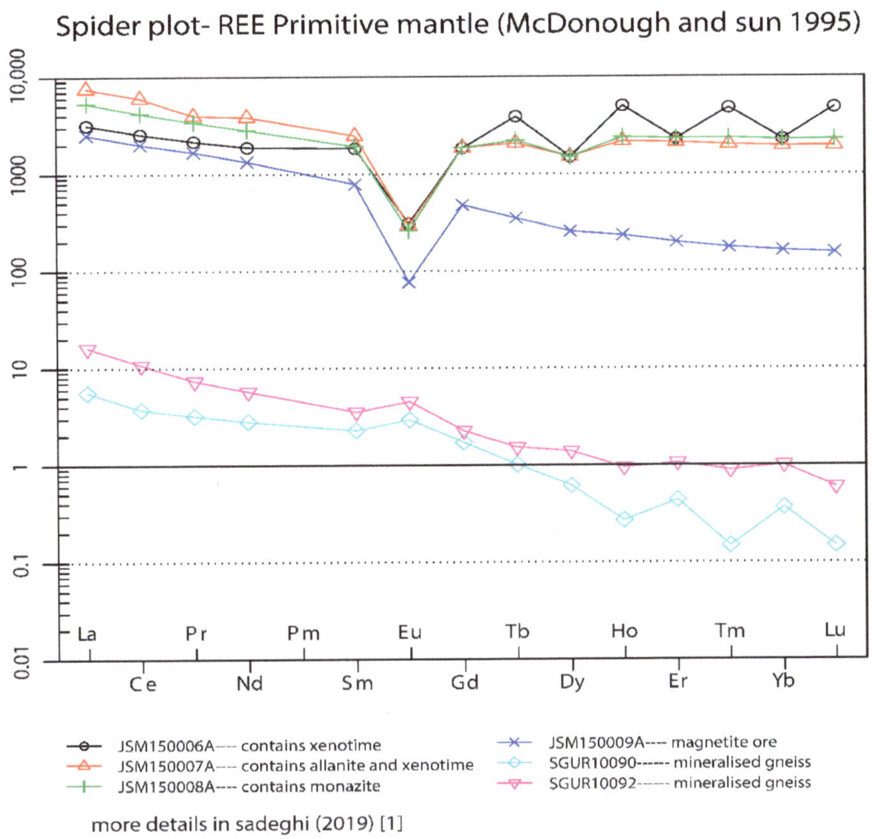

Figure 9. Primitive mantle-normalized [37] REE diagram of iron oxides of unknown origin.

4.3. Principal Component (PC) Analysis

The results of PC analysis are reported in Table 2 for the PCs with eigenvalue >1 and show that PC1 and PC2 explain 30% and 17%, respectively, of the total variance among the REE-mineralized samples. It implies that in PC1 with an association of REEs, and Co, Fe, and Mn most likely represent magnetite calc-silicate skarn samples, while an association of Cs-Hf-Nb-Sr-Ta-Th-Zr-Al-Ba in PC2 likely indicates REE mineralizations related to granitic pegmatite bearing Nb-Ta mineralization.

Table 2. Principal component loading, eigenvalues, % variance explained, and cumulative % variance for PC1 to PC8.

Value Number	Eigenvalues of Correlation Matrix and Related Statistics			
	Eigenvalue	**% Total (Variance)**	**Cumulative % (Eigenvalue)**	**Cumulative % Variance**
PC1	17.56	29.77	17.56	29.77
PC2	9.99	16.93	27.55	46.69
PC3	4.65	7.88	32.20	54.57
PC4	3.41	5.78	35.61	60.35
PC5	2.89	4.90	38.50	65.25
PC6	2.20	3.74	40.70	68.98
PC7	1.94	3.30	42.64	72.27
PC8	1.86	3.16	44.51	75.43

The PC3 defines an association of Cs-Rb-K-Li, which probably represents alkaline and/or felsic rocks including granitic pegmatite and related polymetallic mineralization (Table 3).

Table 3. Explanation of the five principal components for elements analyzed, accounting for 65% of the total variance.

Component	Centered Log-Ratio (clr)-Transformed Data		
	% of Variance Explained	Association	Interpretation
PC1	29.77	(i) REEs (ii) Fe-Mn	(i) REEs associations (ii) Magnetite calc-silicate skarn
PC2	16.93	(i) REEs (ii) Cs-Hf-Nb-Sr-Ta-Th-Zr-Al-Ba	(i) REEs associations (ii) Granitic pegmatit-bearing Nb-Ta mineralization
PC3	7.88	(i) Cs-Rb-K-Li (ii) Ag-Cd-Bi-Co-Cu-Zn-C	(i) alkaline/granitic rocks (ii) Base metals component
PC4	5.78	(i) C-Hf-Sr-Ba-Ca (ii) Fe- -Mg-Tl	(i) Carbonatite/lamprophyr? (ii) Magnetite calc-silicate skarn?
PC5	4.9	(i) Hf-Zr-In (ii) V-P-Ni-Cr	(i) Felsic component (ii) Mafic component

Plots of pairs of PCs show that it is likely possible to distinguish between some different types of REE mineralization or to distinguish between samples with a higher content of REEs and those with low one (Figure 10). For example, in the plot of PC1 versus PC2 (Figure 10) there are several clusters representing an association of REEs and granitic pegmatite bearing Nb-Ta mineralization. A plot of PC3 versus PC4 distinguishes between REE mineralization associated with magnetite calc-silicate skarn, alkaline, and/or granitic rocks and polymetallic mineralization (Figure 10).

4.4. *Comparison of REE Till Geochemistry with Lithogeochemistry of Underlying Bedrock*

Table 4 shows simple statistics of REE distribution in till. The median values are similar to upper crust values by Rudnick and Gao [49]. Due to the fact that Rudnick and Gao's values are calculated for total contents and till results originate from aqua regia digestion, one of the conclusions is that the Swedish till is enriched in REEs in comparison to the global values.

Table 4. Percentiles, minimum, and maximum values for the REE + Y + Sc from the till dataset. P50 is a median value. * abundance in the upper continental crust [49].

Element	n = 2578	UCC *	Min	P25	P50	P75	P90	P99	Max
La	ppm	31	8.1	25.6	31.6	39.6	49.9	85.5	198.9
Ce	ppm	63	16.3	60.1	74.8	96.0	123.7	211.8	388.1
Pr	ppm	7.1	1.9	6.3	7.7	9.4	11.8	20.6	38.1
Nd	ppm	27	7.0	23.3	28.5	34.9	43.4	73.1	129.3
Sm	ppm	4.7	1.4	4.4	5.4	6.7	8.3	13.3	24.2
Eu	ppm	1.0	0.1	0.6	0.8	1.0	1.2	1.9	5.5
Gd	ppm	4.0	1.1	3.6	4.4	5.4	6.6	10.4	22.3
Tb	ppm	0.7	0.2	0.5	0.6	0.8	1.0	1.5	3.6
Dy	ppm	3.9	0.9	2.9	3.5	4.3	5.3	8.5	22.6
Ho	ppm	0.8	0.1	0.6	0.7	0.9	1.1	1.7	4.8
Er	ppm	2.3	0.4	1.5	1.9	2.3	3.0	4.6	14.3
Tm	ppm	0.3	0.1	0.2	0.3	0.3	0.4	0.6	2.0
Yb	ppm	2.0	0.3	1.3	1.7	2.1	2.7	4.2	14.6
Lu	ppm	0.3	0.0	0.2	0.2	0.3	0.4	0.6	2.6
Y	ppm	21	3.7	14.9	18.3	22.6	27.9	47.1	163.2
Sc	ppm	14	0.6	3.5	4.5	5.6	6.8	9.6	21.2

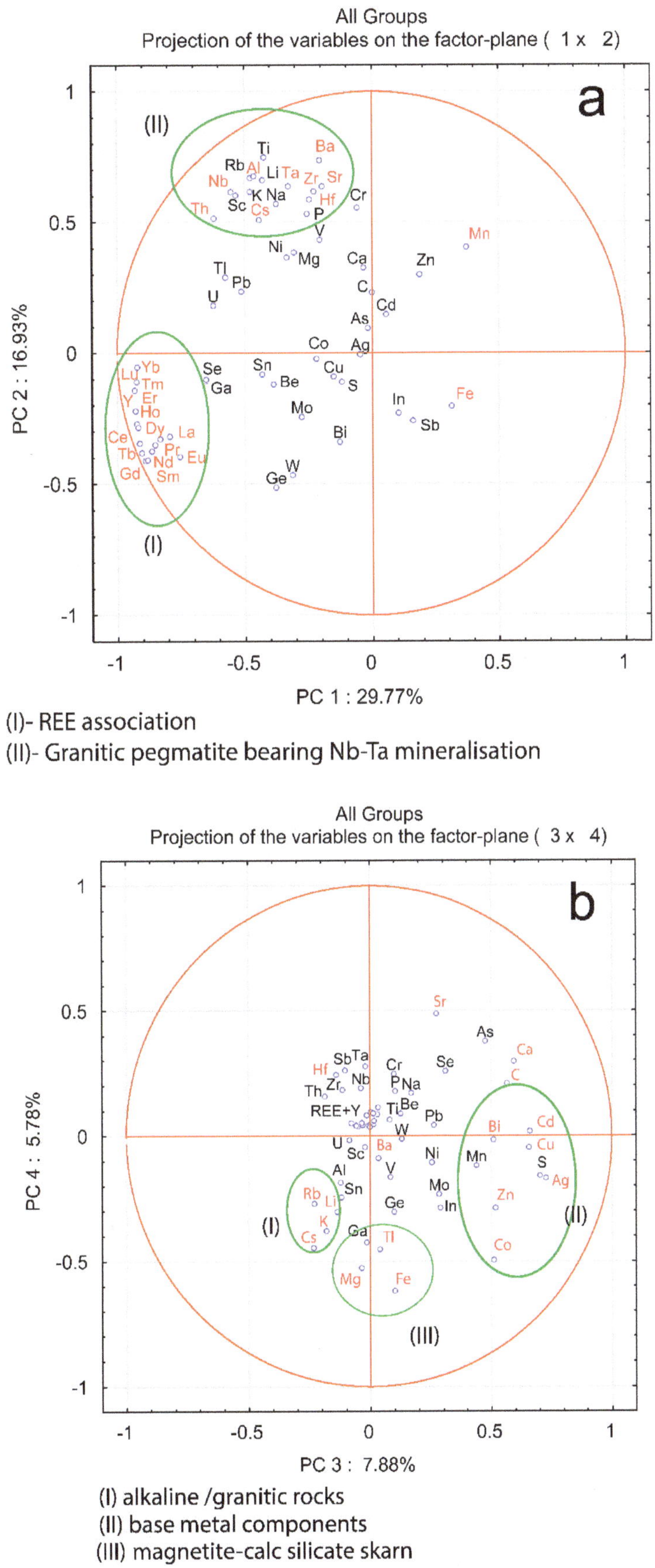

Figure 10. Biplots of (**a**) PC1 vs. PC2; (**b**) PC3 vs. PC4.

Maps of LREE and HREE contents in till (see [3]) show both similar and distinct distribution patterns. Generally, LREE (La-Sm) and HREE groups (Gd-Lu, Y) have similar distribution within the group, therefore, we use here the La till map as a representative for all LREEs and the Yttrium map as a representative for HREEs (Figure 11). Raster La and Y maps with regional anomalies show large-scale REE trends, while plotted lithogeochemistry results (circle symbols) underlie local spots and extreme values.

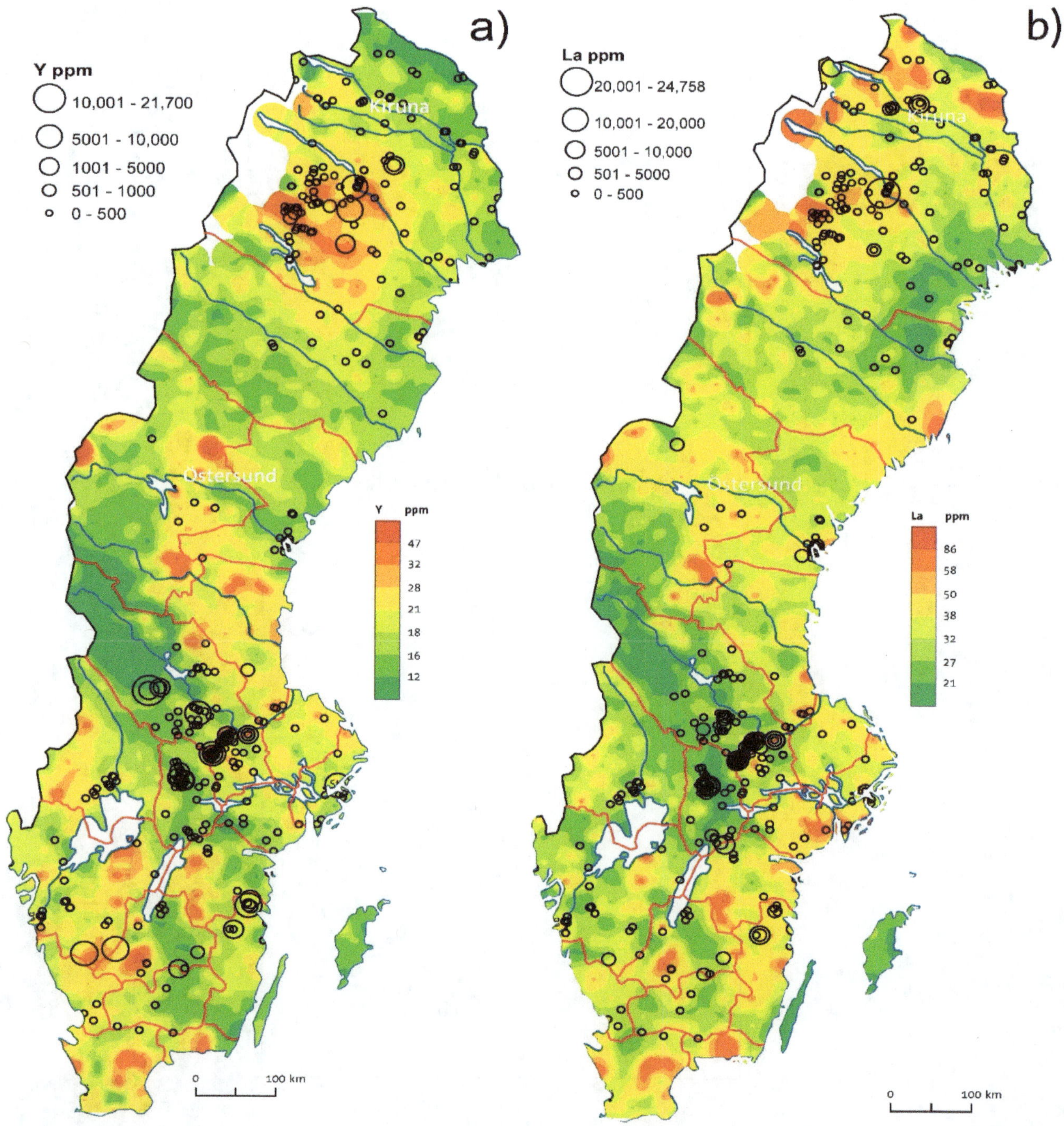

Figure 11. Geochemical maps comparing till geochemistry (raster, after Andersson et al., 2014 [3]) with lithogeochemistry (circle symbols) in Sweden: (**a**) Y in till and bedrock; (**b**) La in till and bedrock.

LREEs and Eu have major anomalies in northernmost Sweden (outlying Archean lithologies of magmatic origin) while these anomalies are absent in HREEs. Minor enrichment in LREEs visible in lithogeochemistry data reflects possible LREE enrichment in granite-pegmatite bedrock and apatite-iron deposits near Kiruna.

Further south, in central Norrbotten (Figures 1 and 11), the bedrock (granite-pegmatite and felsic metavolcanics rocks) and overlying till are highly enriched both in LREE and HREE and correlation with many REE (+W, Mn, and U) mineralizations, e.g., Tåresåive (REEs, Mo, U, Th, Nb, and Ta).

In central Sweden, LREE and HREE enrichment in till occurs in the Caledonides within the tectonic windows where many sulfide mineralizations and U deposits have been mapped. LREE enrichment at the Caledonian front and within the Lower Allochthon can be attributed to the REE enrichment in the fine-grained sedimentary rocks (black shale and phosphorite), e.g., Tåsjö deposit (Figures 1 and 11). On the Fennoscandian Shield side, few REE mineralizations are known and they possibly may be reflected in till with its prominent anomaly south of Östersund, possibly in relation to the so-called Rätan granite. NE from Östersund, in central Sweden, -there is a prominent HREE anomaly and it is probably related to the outcrops of younger granite (so-called Revsund granite). The lithogeochemistry database has poor coverage in central Sweden, however, Alnö carbonatitic intrusion by the coast is both visible in till and few collected rock samples with elevated LREE contents. In the Ljusdal (Ljusdal lithotectonic unit, Figure 1), high HREE (and elevated LREE) contents can be noticed in till and several small REE (W, Mn, U, and Fe-Mn) mineralizations are known in direct relation to the anomaly extension. Lithogeochemistry data are missing in this region.

In the Bergslagen ore district, there is a general decoupling between REE lithogeochemistry and REE content in till. The known REE mineralization line in western Bergslagen is not visible in till geochemistry. Point REE till anomalies can be observed in the NW part of Bergslagen (often related to iron mineralizations) and LREE high contents in till occur in southern Bergslagen by the coast.

In southern Sweden, LREE anomalies in the bedrock are often related to mineralizations and granitic lithologies. Younger granites of the Blekinge lithotectonic unit (Figure 1) and fluorite mineralization in southern Sweden also seem to be well reflected in till geochemistry. High contents of Y in bedrock and till in southern Sweden, south and south-west of the lake Vättern (Figure 11) seem to correlate with occurrences of highly metamorphosed rocks including occurrences of eclogites with Y-enriched garnet and pyroxene. Minor REE (W, Mn, and U) occurrences are known north of the Y till anomaly.

The comparison of till geochemical maps with lithogeochemistry and mineralizations shows clear correlation between REE anomalies in surficial deposits and the underlying bedrock and the presence of mineralization. Therefore, till can be effectively used for REE mineral exploration and mineral potential studies in areas where there is no sufficient information about the bedrock and mineralizations.

5. Conclusions

The analysis of REE geochemistry in mineralized samples compared to bedrock lithogeochemistry and till geochemistry reveals several regional trends in REE distribution in Sweden, which may have practical implications for mineral exploration:

1. Lithogeochemistry of the bedrock correlates with geochemical REE provinces and even single REE ore deposits;
2. REE geochemistry of mineralized EURARE samples allows geochemical classification of various REE occurrences in Sweden;
3. Till geochemistry presents the complex picture of REE distribution at the country scale. Many known REE locations are underlain by till anomalies. On the other hand, there are many high REE spots in till, which cannot be easily explained by known REE deposits. They may present interesting exploration targets;
4. Taking into account favorable geological and metallogenetic conditions in Sweden, there is a good potential for discovering new REE deposits;

5. Applied geochemical methods should be integrated in a multidisciplinary way with complementary data and methods, such as mineralogy, geophysics, and numerical modeling and statistics, to improve understanding of REE ore genesis and related geo-modeling. This holistic approach, also including ore geochemistry and mineralogy, will provide a well-documented REE mineral knowledge base, intelligence, and exploration methodology.

Author Contributions: Conceptualization, M.S. and N.A.; methodology, M.S. and A.L.; software, M.S. and A.L.; validation, M.S., A.L. and N.A. and.; formal analysis, M.S. and N.A.; investigation, M.S.; resources, M.S.; data curation, M.S.; writing—original draft preparation, M.S and A.L.; writing—review and editing, N.A. and A.L.; visualization, M.S and A.L. All authors have read and agreed to the published version of the manuscript.

Acknowledgments: We thank Erik Jonsson, Torbjörn Bergman, Magnus Ripa, Per Nysten, Johan Söderhielm, and Dick Claeson for their contributions to the EURARE project. The EURARE project was funded by the European Community Seventh Framework Programme (FP7/2007–2013) under grant agreement no. 309373. The project, number 35242, was co-funded by the Geological Survey of Sweden (SGU). The authors thank two anonymous reviewers for their constructive and valuable comments and suggestions.

References

1. Sadeghi, M. (Ed.) Rare earth elements distribution, mineralisation and exploration potential in Sweden. In *Sveriges Geologiska Undersökning, Rapporter Och Meddelanden*; Geological Survey of Sweden: Uppsala, Sweden, 2019; Volume 146, p. 184.
2. Sadeghi, M.; Andersson, M. *Sällsynta Jordartsmetaller i Sverige, Förekomst och Utbredning i berg och jord "the Rare Earth Element Distribution over Sweden" (in Swedish and English)*; Report No 2015-21; Geological Survey of Sweden: Uppsala, Sweden, 2015; p. 98.
3. Andersson, M.; Carlsson, M.; Ladenberger, A.; Morris, G.; Sadeghi, M.; Uhlbäck, J. *Geochemical Atlas of Sweden (Geokemisk Atlas over Sverige)*; Sveriges Geologiska Undersökning: Uppsala, Sweden, 2014; p. 210. ISBN 978-91-7403-258-1.
4. Sadeghi, M.; Morris, G.A.; Carranza, E.J.M.; Ladenberger, A.; Andersson, M. Rare earth element distribution and mineralisation in Sweden: An application of principal component analysis to FOREGS soil geochemistry. *J. Geochem. Exp.* **2013**, *133*, 160–175. [CrossRef]
5. Sadeghi, M.; Petrosino, P.; Ladenberger, A.; Albanese, S.; Andersson, M.; Morris, G.; Lima, A.; De Vivo, B. Ce, La and Y concentration in agricultural and grazing land soils of Europe—An exercise for mapping and exploration. *J. Geochem. Exp.* **2013**, *133*, 202–213. [CrossRef]
6. Hudson, M. *2007: Drilling Expands Near-Surface Uranium at Tåsjö, Sweden*; Mawson Resources Ltd.: Vancouver, BC, Canada, 2007.
7. London, D. Rare-element granitic pegmatites. In *Reviews in Economic Geology*; Society of Economic Geologists, Inc.: Littleton, CO, USA, 2016; Volume 18, pp. 165–193.
8. Sundius, N. Kvarts, fältspat och glimmer samt förekomster där av i Sverige. In *Sveriges Geologiska Undersökning* C; Geological Survey of Sweden: Uppsala, Sweden, 1952; Volume 520, p. 231.
9. Lundegårdh, P.H. *Nyttosten i Sverige*; Almqvist & Wiksell: Stockholm, Sweden, 1971; p. 271.
10. Holmqvist, A. Be-F-Nb-REE-Sn-Ta-Th-U-Zr-mineraliserad granit i Västsverige. In *Sveriges Geologiska AB, Division Prospektering, PRAP-Rapport 89026*; Geological Survey of Sweden: Uppsala, Sweden, 1989; p. 14 + appendices.
11. Dostal, J. Rare metal deposits associated with alkaline/peralkaline igneous rocks. In *Reviews in Economic Geology*; Society of Economic Geologists, Inc.: Littleton, CO, USA, 2016; Volume 18, pp. 33–54.
12. Verplanck, P.L.; Mariano, A.N.; Mariano, A., Jr. Rare earth element ore geology of carbonatites. In *Reviews in Economic Geology*; Society of Economic Geologists, Inc.: Littleton, CO, USA, 2016; Volume 18, pp. 5–32.
13. Mariano, A.N. Nature of economic mineralization in carbonatites and related rocks. In *Carbonatites: Genesis and Evolution*; Bell, K., Ed.; Unwin Hyman: London, UK, 1989; pp. 149–176.
14. Kresten, P. A magnetometric survey of the Alnö complex. *Geol. Fören. Stockh. Förh.* **1976**, *98*, 364–365. [CrossRef]

15. Kresten, P. The Alnö complex: Discussion of the main features, bibliography and excursion guide. In Proceedings of the Nordic Carbonatite Symposium, Sundsvall, Sweden, 20–27 May 1979; p. 67.
16. Kresten, P. The Alnö area (Alnöområdet). In *Beskrivning till Berggrundskartan över Västernorrlands län*; Lundqvist, T., Gee, D., Kumpulainen, R., Karis, L., Kresten, P., Eds.; Sveriges Geologiska Undersökning ser Ba nr 31: Uppsala, Sweden, 1990; pp. 238–278.
17. Hornig-Kjarsgaard, I. Rare earth elements in sövitic carbonatites and their mineral phases. *J. Petrol.* **1998**, *39*, 2105–2121. [CrossRef]
18. Svensson, U. *Sällsynta Jordartsmetaller på Alnön. Boliden Mineral AB, Årsrapport 1971, Bilaga 36*; Boliden Mineral AB: Boliden, Sweden, 1972; p. 5 + appendix (unpublished report).
19. Chai, F.; Zhang, Z.; Mao, J.; Parat, A.; Wang, L.; Dong, L.; Ye, H.; Chen, l.; Zheng, R. Lamprophyre or Lamproite Dyke in the SW Tarim Block?—Discussion on the Petrogenesis of These Rocks and Their Source Region. *J. China Univ. Geosci.* **2006**, *17*, 13–24. [CrossRef]
20. Kresten, P.; Åhman, E.; Brunfelt, A.O. Alkaline ultramafic lamprophyres and associated carbonatite dykes from the Kalix area, northern Sweden. *Geol. Rundsch.* **1981**, *70*, 1215–1231. [CrossRef]
21. Kresten, P.; Rex, D.C.; Guise, P.G. 40Ar-39Ar ages of ultramafic lamprophyres from the Kalix area, northern Sweden. In *Radiometric Dating Results 3. SGU C830*; Lundqvist, T., Ed.; Geological Survey of Sweden: Uppsala, Sweden, 1997.
22. Jonsson, E. Epigenetic REE-U-Th-anomalous Fe oxide mineralisation in the Narken area, NE Sweden. In Proceedings of the 33rd International Geological Congress, Oslo, Norway, 6–14 August 2008. Abstracts, MRD-11.
23. Geijer, P. *The Iron Ores of the Kiruna Type: Geographical Distribution, Geological Characters, and Origin*; Sveriges Geologiska Undersökning C: Uppsala, Sweden, 1931; Volume 367, p. 39.
24. Geijer, P. *The Rektorn Ore Body at Kiruna*; Sveriges Geologiska Undersökning C: Uppsala, Sweden, 1950; Volume 514, p. 18.
25. Parák, T. Kiruna iron ores are not "intrusive-magmatic ores of the Kiruna type". *Econ. Geol.* **1975**, *70*, 1242–1258. [CrossRef]
26. Frietsch, R. On the magmatic origin of iron ores of the Kiruna type. *Econ. Geol.* **1978**, *73*, 478–485. [CrossRef]
27. Hitzman, M.W.; Oreskes, N.; Einaudi, M.T. Geologic characteristics and tectonic setting of Proterozoic iron oxide (Cu-U-Au-REE) deposits. *Precambrian Res.* **1992**, *58*, 241–287. [CrossRef]
28. Nyström, J.O.; Henriquez, F. Magmatic features of iron ores of the Kiruna type in Chile and Sweden: Ore textures and magnetite geochemistry. *Econ. Geol.* **1994**, *89*, 820–839. [CrossRef]
29. Williams, P.J.; Barton, M.D.; Johnson, D.A.; Fontboté, L.; de Haller, A.; Mark, G.; Oliver, N.H.S.; Marschik, R. Iron oxide copper-gold deposits: Geology, space-time distribution, and possible modes of origin. In *Economic Geology, 100th Anniversary Volume*; Society of Economic Geologists: Littleton, CO, USA, 2005; pp. 371–405.
30. Jonsson, E.; Troll, V.R.; Högdahl, K.; Harris, C.; Weis, F.; Nilsson, K.P.; Skelton, A. Magmatic origin of giant central Swedish "Kiruna-type" apatite-iron oxide ores. *Sci. Rep.* **2013**, *3*, 1–8. [CrossRef] [PubMed]
31. Bergman, S.; Kübler, L.; Martinsson, O. *Description of Regional Geological and Geophysical Maps of Northern Norrbotten County (East of the Caledonian Orogen)*; Sveriges Geologiska Undersökning Ba: Uppsala, Sweden, 2001; Volume 56, p. 110.
32. Harlov, D.E.; Andersson, U.B.; Förster, H.-J.; Nyström, J.O.; Dulski, P.; Broman, C. Apatite-monazite relations in the Kiirunavaara magnetite-apatite ore, northern Sweden. *Chem. Geol.* **2002**, *191*, 47–72. [CrossRef]
33. Jonsson, E.; Harlov, D.; Majka, J.; Högdahl, K.; Persson-Nilsson, K. Fluorapatite-monazite-allanite relations in the Grängesberg apatite-iron oxide ore district, Bergslagen, Sweden. *Am. Mineral.* **2016**, *101*, 1769–1782. [CrossRef]
34. Pan, Y.; Fleet, M.E. Composition of the fluorapatite-group minerals: Substitution mechanisms and controlling factors. *Rev. Mineral. Geochem.* **2002**, *48*, 13–49. [CrossRef]
35. Jonsson, E.; Persson Nilsson, K.; Hallberg, A.; Högdahl, K. The Palaeoproterozoic apatite-iron oxide deposits of the Grängesberg area: Kiruna-type deposits in central Sweden. In *NGF Abstracts and Proceedings, 29th Nordic Geological Winter Meeting, Oslo, Norway*; Nakrem, H.A., Harstad, A.O., Haukdal, G., Eds.; Norsk Geologisk Forening: Trondheim, Norway, 2010; Volume 1, pp. 88–89.
36. Lundh, J. A Lithogeochemical Study of Northern Sweden and the Kiruna and Malmberget Iron-Apatite Ore Deposits (Dissertation). 2014. Available online: http://urn.kb.se/resolve?urn=urn:nbn:se:uu:diva-227039 (accessed on 24 June 2014).

37. McDonough, W.F.; Sun, S.S. The composition of the Earth. *Chem. Geol.* **1995**, *120*, 223–253. [CrossRef]
38. Hallberg, A.; Albrecht, L.; Jonsson, E.; Olsson, A. The Grängesberg apatite-iron deposit—biggest in Bergslagen, Sweden. In Proceedings of the 27th Nordic Geological Winter Meeting, Bulletin of the Geological Society of Finland, Oulu, Finland, 9–12 January 2006; Special issue 1. p. 45.
39. Samson, I.M.; Wood, S.A. The rare earth elements: Behavior in hydrothermal fluids and concentration in hydrothermal mineral deposits, exclusive of alkaline setting. In *Rare elements Geochemistry and Mineral Deposits, Geological Association of Canada short course Notes*; Linnen, R.I., Samson, I.M., Eds.; Society of Economic Geologists: Littleton, CO, USA, 2005; Volume 17, pp. 269–297.
40. Oreskes, N.; Einaudi, M.T. Origin of rare earth element-enriched hematite breccias at the Olympic Dam Cu-U-Au-Ag deposit, Roxby Downs, South Australia. *Econ. Geol.* **1990**, *85*, 1–28. [CrossRef]
41. Lottermoser, B.G. Rare earth elements in Australian uranium deposits. In *Uranium–Past and Future Challenges, Proceeding of the 7th International Conference on Uranium Mining and Hydrogeology*; Merkel, B.J., Arab, A., Eds.; Springer: Berlin, Germany, 1995; pp. 25–30.
42. Holtstam, D.; Andersson, U.B. The REE minerals of the Bastnäs-type deposits, south-central Sweden. *Can. Mineral.* **2007**, *45*, 1073–1114. [CrossRef]
43. Janoušek, V.; Farrow, C.M.; Erban, V. Interpretation of whole-rock geochemical data in igneous geochemistry: Introducing Geochemical Data Toolkit (GCDkit). *J. Petrol.* **2006**, *47*, 1255–1259. [CrossRef]
44. Carranza, E.J.M. Geochemical Anomaly and Mineral Prospectivity Mapping in GIS. In *Handbook of Exploration and Environmental Geochemistry*; Elsevier: Amsterdam, The Netherlands, 2008; Volume 11.
45. Grunsky, E.C. The interpretation of geochemical survey data. *Geochem. Exp. Environ. Anal.* **2010**, *10*, 27–74. [CrossRef]
46. Güller, C.; Thyne, G.; McGray, J.E.; Turner, A.K. Evaluation of graphical and multivariate statistical methods for classification of water chemistry data. *Hydrogeol. J.* **2002**, *10*, 455–474. [CrossRef]
47. Harris, J.R.; Grunsky, E.C.; Wilkinson, L. Developments in the effective use of lithogeochemistry in regional exploration programs: Application of GIS technology. In *Proceedings of the Exploration 97, Fourth Decennial International Conference on Mineral Exploration*; Gubins, A.G., Ed.; Prospectors and Developers Association of Canada: Toronto, ON, Canada, 1997; pp. 285–292.
48. Hode Vuorinen, J.; Hålenius, U.; Whitehouse, M.J.; Mansfeld, J.; Skelton, A.D.L. Compositional variations (major and trace elements) of clinopyroxene and Ti-andradite from pyroxenite, ijolite and nepheline syenite, Alnö Island, Sweden. *Lithos* **2005**, *81*, 55–77. [CrossRef]
49. Rudnick, R.L.; Gao, S. Composition of the Continental Crust. In *The Crust*; Rudnick, R.L., Ed.; Elsevier-Pergamon: Oxford, UK, 2003; pp. 1–64. [CrossRef]

4

Concentration of Rare Earth Elements (Sc, Y, La, Ce, Nd, Sm) in Bauxite Residue (Red Mud) Obtained by Water and Alkali Leaching of Bauxite Sintering Dust

Leonid Chaikin [1], Andrei Shoppert [1,*], Dmitry Valeev [2,3], Irina Loginova [1] and Julia Napol'skikh [1]

[1] Department of Non-Ferrous Metals Metallurgy, Ural Federal University Named after the First President of Russia B.N. Yeltsin, 620002 Yekaterinburg, Russia; l.i.chaikin@urfu.ru (L.C.); i.v.loginova@urfu.ru (I.L.); anapolskikh512@gmail.com (J.N.)

[2] Laboratory of New Metallurgical Processes (#24), A.A. Baikov Institute of Metallurgy and Materials Science, Russian Academy of Sciences, 49, Leninsky Prospect, 119334 Moscow, Russia; dvaleev@imet.ac.ru

[3] Scientific Research Centre "Thermochemistry of Materials", National University of Science & Technology "MISIS", 4, Leninsky Prospect, 119049 Moscow, Russia

* Correspondence: a.a.shoppert@urfu.ru

Abstract: One of the potential sources of rare-earth elements (REE) is the industrial waste known as red mud (bauxite residue), in which the majority of REE from the initial bauxite are concentrated via the Bayer process. Therefore, the studies of the subject, both in Russia and outside, focus almost exclusively on red mud processing. This article looks into the possibility of REE concentration into red mud by leaching an intermediate product of the bauxite sintering process at Russian alumina refineries, namely electrostatic precipitator (ESP) dust. The experimental works were performed by X-ray diffraction (XRD)and electron probe microanalysis (EPMA) of the sinter and sinter dust. The determination of major and rare-earth elements in the sinter from the rotary kilns and in the ESP dust before and after leaching was carried out by X-ray fluorescence (XRF) and plasma mass spectrometry (ICP-MS). The study showed that it is possible to obtain red mud that contains three times more REE than traditional waste red mud after two-stage leaching ESP dust in the water at 95 °C followed by leaching in an alkaline-aluminate liquor at 240 °C. The shrinking core model was used to study the kinetics of leaching of the original ESP dust and water-treated dust in alkaline-aluminate liquor. The study showed the change in the limiting stage of the alkaline leaching process after water treatment, with the activation energy growing from 24.98 to 33.19 kJ/mol.

Keywords: rare-earth elements; scandium; alumina production; sinter processes; red mud; electrostatic precipitation dust; alkali leaching; kinetics

---

1. Introduction

The volumes of production/consumption of rare and rare-earth metals, which are critical for modern industries, are steadily increasing [1–4]. In this group of metals—scandium, yttrium, lanthanum, and lanthanides—scandium is the most scarce and commercially attractive one. The increased interest in scandium is associated with its application in various industries [5–7]. Small amounts of scandium are found in ores of aluminum, cobalt, iron, molybdenum, nickel, phosphate, titanium, tungsten, uranium, and others [8–16].

At the moment, phosphogypsum—waste from the processing of apatite concentrates [17,18]—and red mud (bauxite residue) are considered to be the most promising sources of scandium [19–24]. Phosphogypsum is also considered to be the most promising source of other rare earth metals [6], as it may contain up to 1% of rare earth elements in total [25]. More than 250 million tons of phosphogypsum has been accumulated in Russia [26] with an annual increase by over 14 million tons. However, despite the large number of works devoted to the processing of phosphogypsum, there is still no commercially viable technology for its processing [27], which is explained by isomorphic co-crystallization of REE with gypsum, and therefore the need for its complete release [28].

Bauxite residue generated in the recovery and processing of bauxites is a source of environmental pollution [29,30] on the one hand and a promising object for obtaining valuable elements [31] on the other hand. Bauxite residue contains a high amount of scandium (70–260 ppm) that is close to its primary resources [32]. Large amounts of this waste form what may be called man-made deposits, which can be recycled into building materials, pigments, coagulants for wastewater treatment, and which can be used to re-extract alumina, extract iron concentrates, titanium, rare and rare earth metals [33–35]. In the industrial production of alumina, almost all Sc is concentrated in red mud. According to available estimates, the global reserves of scandium in the industrial waste range from 1 to 3 million tons, with 70–80% of it being contained in bauxite processing waste [36]. The full processing of the bauxite residue currently generated makes it possible to extract 6600–20,400 t of scandium per year. Therefore, a large number of studies have been dedicated to the recovery of rare-earth metals from various types of bauxite residue [37] obtained by processing various bauxites in the Bayer method and its variations. However, the low content of scandium in the red mud, the high content of alkali and alkaline earth metals, and the simultaneous recovery of iron together with REE [38] or low extraction efficiency [39,40] render the existing methods for extracting scandium from this type of industrial waste non-viable.

It should be noted that most studies of the separation of REE from alumina products have focused on the processing of red mud. In this article, we made an attempt to study the possibility of obtaining the REE concentrate from electrostatic precipitator (ESP) dust, which is an intermediate product of kiln sintering in alumina production. The dust is generated in a rotary kiln (Figure 1), where low-quality bauxite (the mass ratio of Al_2O_3 to SiO_2 (silicon modulus) is less than 7) is sintered with soda ash in order to convert the alumina-containing mineral into a water-soluble sodium aluminate. Iron and silicon-containing minerals react similarly with soda to form ferrite and sodium silicate, respectively. The technological process, especially at the drying and decarbonization stages, generates a large number of fine particles less than 100 μm (25–40% of the charge), which are carried away with the exhaust gases. To clean the exhaust gases of kilns from dust particles, a three-stage capture system is used. The first stage is the dust chamber where most coarse dust particles precipitate; the second stage is cyclones, where medium dust particles are removed from the exhaust gases through vortex separation; and the final stage is electrostatic precipitator. The last stage captures the finest particle fraction (less than 20 μm) using the force of an induced electrostatic charge.

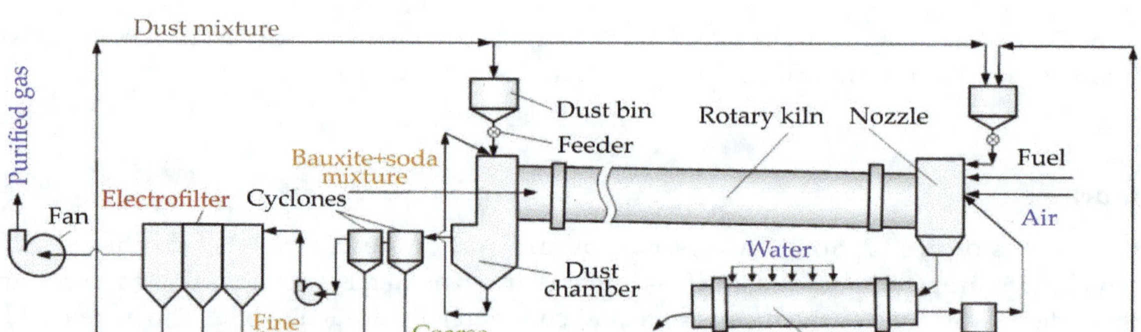

Figure 1. Flow chart of the sintering process on the RUSAL-Kamensk-Uralsky alumina refinery.

It is impossible to remove all dust from the technological process since that would upset the heat balance of the furnace and would greatly complicate the movement of the charge in the drying zone.

On the other hand, the ESP dust acts mostly as a dead weight, since a number of dust circulations through the kiln are larger than through all other units of the gas treatment system. Electrostatic precipitators capture only the smallest particles, which often pass very quickly through the hot zones of the furnace, and the sintering process fails to complete. Ultimately, they acquire the charge of the electrostatic precipitator and are released into the atmosphere. The physicochemical properties of sintering dust and the possibility of their leaching together with bauxite had already been investigated in our previous work [41]. It was shown that ESP dust can be highly reactive and can be removed from the process with further leaching to extract useful components. Also, the high recovery rate of ESP dust means significant losses of secondary heat. The proportion of ESP dust in the total mass of dust is up to 15%, which means that alumina refineries in Russia produce more than 30 thousand tons of ESP dust per year. At the same time, ESP dust contains about 50 ppm of scandium [42]. Therefore, more than 1.5 t of scandium and even more of other rare earth metals can be obtained annually from this intermediate product.

In this work we studied the possibility of leaching ESP dust with water and mother liquor of the Bayer process in order to extract rare-earth elements into valuable components such as alumina, caustic alkali, and soda, and, at the same time, concentrate them in the red mud resulting from leaching.

2. Materials and Methods

2.1. Solid Phase Characterization

The phase and quantitative composition of the ESP dust from sintering kilns of the Kamensk-Uralsky alumina refinery and leaching products were determined by X-ray diffraction (XRD) on a Rigaku D/MAX-2200 diffractometer (Rikagu Corp., Tokyo, Japan) using the PDF-2 database (International Center for Diffraction Data) and by X-ray fluorescence (XRF) using an Axios MAX X-ray fluorescence spectrometer (Malvern Panalytical Ltd., Almelo, The Netherlands). The content of REE micro impurities in the feedstock and leachate was determined using inductively coupled plasma mass spectrometry (ICP-MS) on a PerkinElmer NexION 300S instrument (PerkinElmer Inc., Waltham, MA, USA).

Electron probe micro-analysis (EPMA) was performed using a Cameca SX 100 microanalyzer (CAMECA Instruments, Inc., Madison, WI, USA) equipped with an energy-dispersive X-ray spectroscopy analysis (EDS) module Bruker XFlash 6 (Bruker Nano GmbH, Berlin, Germany).

2.2. Reagents

All reagents used in this study were of analytical purity. Distilled water was obtained using a GFL distiller (GFL mbH, Burgwedel, Germany). A model alkali liquor obtained was obtained by dissolving NaOH in distilled water. Then, the necessary amount of $Al(OH)_3$ was dissolved in the resulting hot caustic alkali liquor to achieve the desired ratio of NaOH and Al_2O_3 in the model liquor. After complete dissolution of aluminum hydroxide, the liquor was brought to the required concentration in caustic alkali. The aluminate liquor had the following chemical composition: Na_2O_k (caustic alkali) = 240 g/dm^3, Al_2O_3 = 120 g/dm^3. The sinter and ESP dust were obtained from the RUSAL-Kamensk-Uralsky alumina refinery (56.304530, 61.980334; Kamensk-Uralsky, Russia) where low-grade bauxite is processed via combined Bayer-Sintering process. Bauxite with the high silica content is sent in rotary kiln together with sodium carbonate to obtain water-soluble sinter at 1150 °C. The chemical and phase composition of the ESP dust and the sinter are different, since dust passes through the hot zone of the sintering kiln faster and many processes fail to complete; also, the difference may be caused by the difference in the distribution of minerals into different fractional classes. Table 1 shows the chemical composition of the ESP dust of the Kamensk-Uralsky alumina refinery, against the chemical composition of the bauxite sinter.

Table 1. The chemical composition of electrostatic precipitator (ESP) dust and bauxite sinter, wt. %.

Products	Al_2O_3	SiO_2	Na_2O	MgO	CaO	Fe_2O_3	LOI [1]	μ_{Si} [2]	ΣREE
ESP dust	25.5	2.5	28.3	0.4	2.5	12.2	24.8	10.2	755
Bauxite sinter	35.0	5.0	33.2	0.8	4.8	16.7	0.5	7.0	567

[1] LOI—loss on ignition at 1000 °C. [2] μ_{Si} (silicon modulus)—mass ratio of Al_2O_3 to SiO_2.

The results of X-ray diffraction analysis of the electrostatic precipitator dust of the RUSAL-Kamensk-Uralsky alumina refinery is shown in Figure 2.

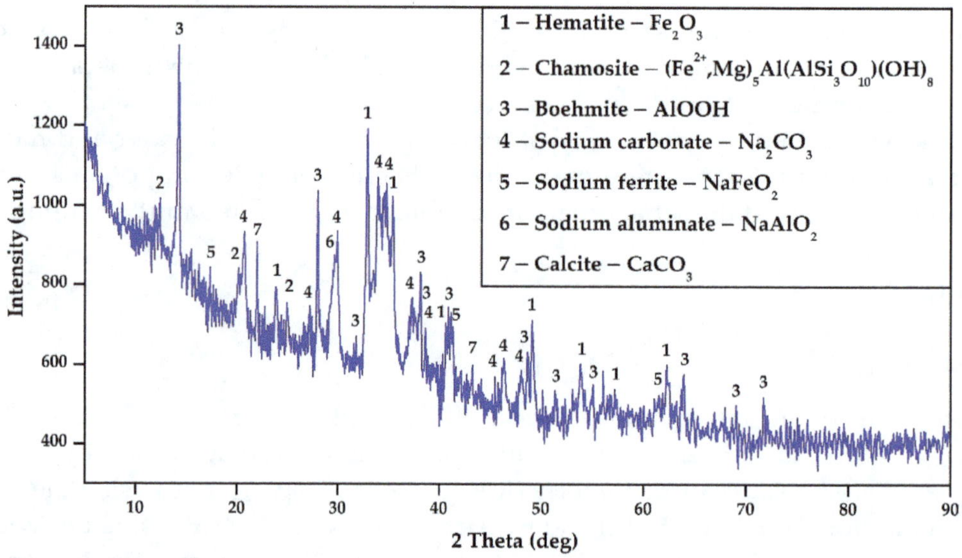

Figure 2. X-ray diffraction analysis of the electrostatic precipitator dust of the RUSAL-Kamensk-Uralsky alumina refinery in Russia.

2.3. Experimentation

The leaching of the required weight of ESP dust in distilled water was carried out at L:S = 10:1 at 95 °C for 60 min in a 0.5 dm^3 Lenz Minni thermostatic reactor (Lenz Laborglas GmbH & Co. KG, Wertheim, Germany) fitted with an overhead stirrer and ports for a refrigerator and samplers.

Autoclave leaching of the ESP dust sample at L:S = 3:1 was carried out at 160, 200, and 240 °C in an aluminate mother liquor for 15–90 min in a 1 dm^3 Parr autoclave (Parr Instrument, Moline, IL, USA). Leaching in an aluminate liquor at 90 °C also used a thermostatic Lenz Minni reactor.

After water and alkaline leaching, the resulting red mud was separated from the aluminate liquor by filtration on a Buchner funnel. After washing and drying of the bauxite residue for 8 h at 110 °C, the content of rare-earth elements and other components was measured. The solid phase obtained after water leaching of ESP dust is designated as ESPDW, and that obtained by autoclave leaching with an alkaline aluminate liquor at 240 °C for 90 min is designated as ESPDA.

3. Results and Discussion

It can be seen from Table 1 that the chemical composition of ESP dust and bauxite sinter are different. The high LOI in the dust is associated with the incomplete decomposition of soda and calcite; however, there is also a significant decrease in the content of silica, magnesium, and calcium in dust relative to alumina and an increase in sodium content. A decrease in silica content in dust leads to a significant increase in the silicon modulus (μ_{Si}). The results of X-ray diffraction analysis (Figure 2) confirm the presence of sodium aluminate in the ESP dust; however, most of the soda remains unreacted.

According to the previous research [19,21,43,44], the REE and Sc are possibly adsorbed on goethite, hematite and in the channels of aluminosilicates in the typical Bayer bauxite residue. The primary container of these elements could be cancrinite [45]. However, during the sintering process iron minerals are transformed into sodium ferrite and aluminosilicate—into sodium silicate and sodium aluminate (Figure 2). Therefore, REE and Sc could be liberated through the sintering process, which could lead to an increase in their concentration in the by-products and increased leachability than in typical Bayer bauxite residue.

Figure 3 shows the content of rare-earth elements in ESP dust and sinter. The obtained data show that the ESP dust from the sintering process contains more REE than the sinter itself. Moreover, if we detract LOI (consisting of water and CO_2 carbonates, which are removed from sinter at 1000 °C and which remain in ESP dust, according to Table 1), the REE content in the calcined ESP dust will be approximately 70–80% higher than in the sinter. The obtained data is in good agreement with the results presented previously in the patent [42], where it was shown that ESP dust contains about 50 ppm of scandium. This may be due to the repeated circulation of fine dust in the process and mineralogical transformation. Consequently, ESP dust can be a source of REE, and we can consider the option of removing it from the cycle in order to extract valuable components and increase the efficiency of the sintering kiln.

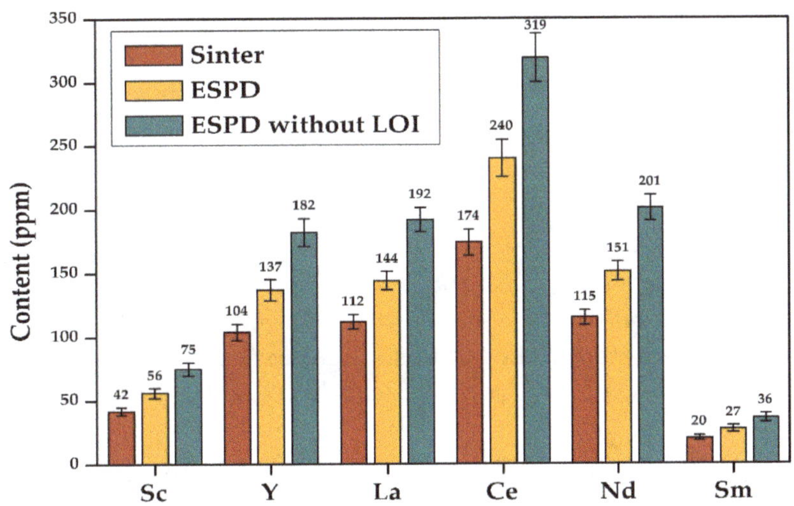

Figure 3. Content of rare earth elements in electrostatic precipitator dust and bauxite sinter, ppm (LOI—loss on ignition at 1000 °C).

3.1. *Water Leaching of ESP Dust*

The XRD analysis of ESP dust (Figure 2) shows a high content of soluble minerals (such as sodium silicate and sodium aluminate). Tests were carried out to leach ESP dust with water for 1 h at 95 °C, which is necessary for the complete conversion of sodium salts into liquor. The output of the solid phase (ESPDW) after leaching was 40% of the initial weight of ESP dust. Table 2 shows the chemical composition of the resulting red mud, and the degree of recovery of the main components into the liquor is shown in Figure 4. Part of the aluminum in electrostatic precipitator dust was converted to sodium aluminate; it is therefore easily leached with water. Notably, aqueous leachate has a higher silicon modulus than the original dust, as silicon has a degree of recovery higher than aluminum. This indicates that the degree of transformation of silica to sodium silicate is higher than alumina to sodium aluminate. Na_2O is almost completely leached, which indicates that with water leaching for less than 1 h the components in the resulting liquor do not have time to form a disilication product according to the following Equation (1):

$$6Na_2SiO_3 + 6NaAl(OH)_4 + Na_2X \rightarrow Na_6[Al_6Si_6O_{24}]\cdot Na_2X + 12NaOH + 6H_2O, \quad (1)$$

where X represents various inorganic anions, most often sulfate, carbonate, chloride, aluminate, etc. [46].

Table 2. The chemical composition of electrostatic precipitator dust after water leaching, wt. %.

Product	Al_2O_3	SiO_2	Na_2O	MgO	CaO	Fe_2O_3	LOI	μ_{Si}
ESPDW	39.8	3.1	1.7	1.0	6.2	30.5	18.3	12.8

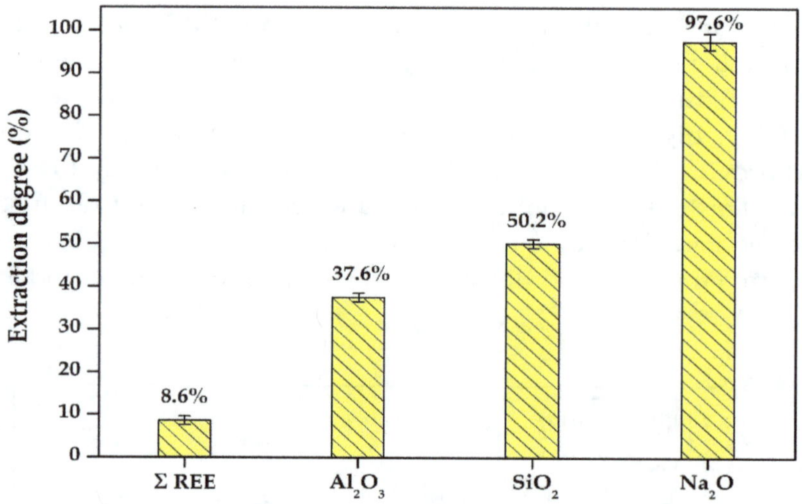

Figure 4. The extraction degree of the main components into the liquor by water leaching of electrostatic precipitator dust.

It can also be seen in Figure 4 that REE (Sc, Y, La, Ce, Nd) are slightly extracted during water leaching, because of their interaction with alkali and soda [47]. As shown in some studies [39,40], the degree of scandium recovery from red mud to soda liquor usually does not exceed 20%.

As it was stated in the introduction section, scandium is the most valuable REE in red mud. It was previously shown that scandium in red mud is mainly associated with iron minerals [48]. However, some researches show [45] that cancrinite could accumulate Sc as well. Therefore, to study the association of rare-earth metals with different phases, we performed ESPD and ESPDW surface mapping using EPMA (Figure 5).

It can be seen from Figure 5 that scandium in ESP dust is mainly associated with iron minerals, and to a lesser extent—with silicon compounds. It could be associated with the fact that iron minerals (hematite) are not fully transformed into sodium ferrite during sintering because small particles of dust pass very quickly through the hot zones of the furnace and a disilication product is not formed yet. Almost the same picture is observed in ESPDW; however, the distribution of scandium is more uniform than in ESP dust. This may be due to the formation of a disilication product (sodalite or cancrinite), which can adsorb the REE. This suggests that destruction of the hematite matrix is required at first for the complete recovery of REE from ESPD, as is the case with the typical Bayer process red mud. This also explains the low degree of REE recovery at the water leaching stage, since hematite cannot be dissolved by sodium carbonate liquor.

3.2. *Kinetics of ESP Dust Leaching by Alkaline Aluminate Liquor*

To study the mechanism and effect of temperature on the leaching of initial ESPD and ESPDW with the alkaline aluminate mother liquor, experiments were carried out to measure the aluminum extraction degree from the liquor with variable duration and temperature of the process. Aluminum was chosen as an indicator of leaching efficiency as it was the only element extracted from the liquor during alkali leaching of ESPDW. The temperature range was selected, taking into account that, after the extraction of readily soluble sodium salts, the alumina in the material is mainly represented by boehmite, the leaching of which requires a temperature of more than 160 °C [49]. The experimental results are shown in Figure 6.

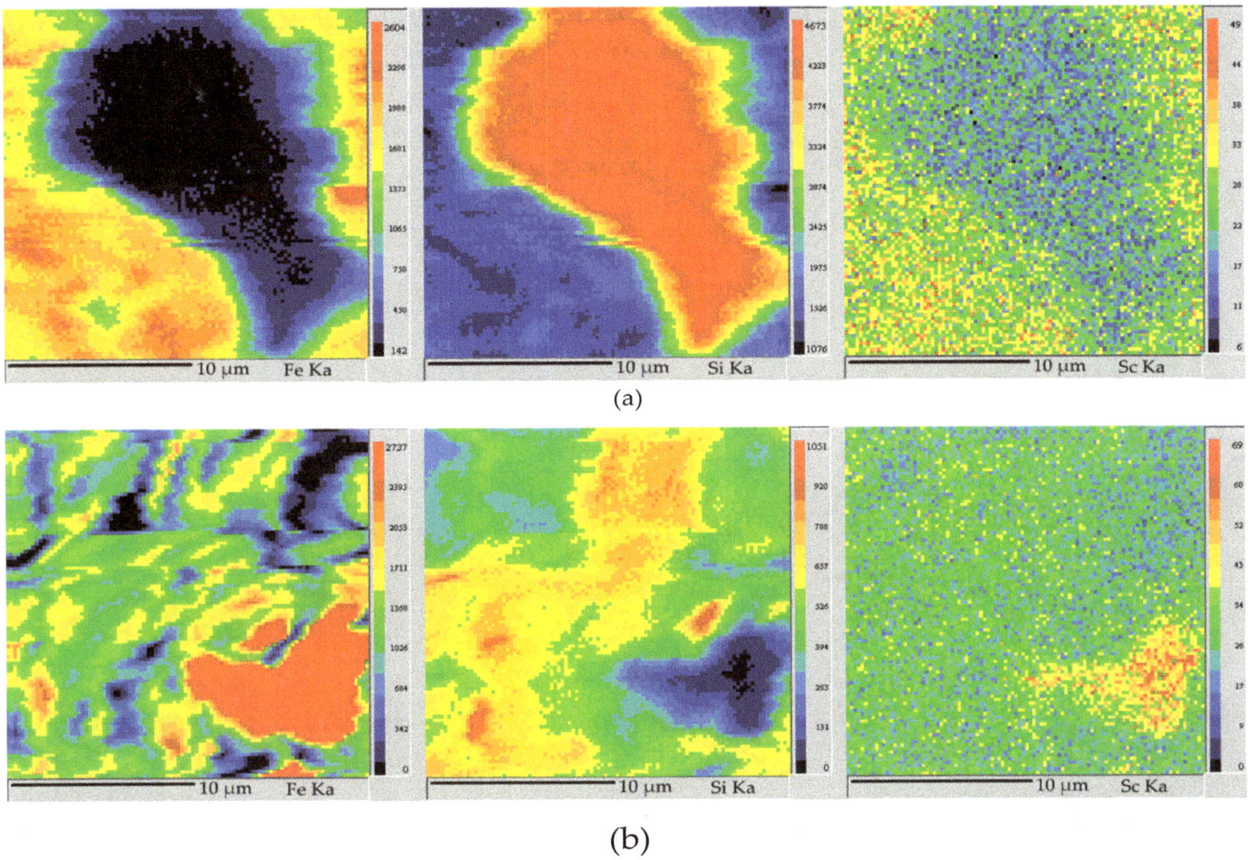

Figure 5. Mapping of electrostatic precipitator dust surface (**a**) and water-leached electrostatic precipitator dust surface (**b**) using the electron probe micro-analyzer.

It can be seen from Figure 6 that the kinetics of aluminum extraction into the liquor is quite high at all temperatures for the first 30 min for ESP dust. Even at 90 °C, more than 40% of aluminum is extracted from ESP dust after 30 min, apparently because of a certain degree of conversion of boehmite to sodium aluminate in the sintering kiln. However, after pre-leaching in water (ESPDW), sodium aluminate had already been extracted into the first-stage liquor; therefore, the rate of leaching of aluminum from ESPDW at 90 °C is significantly lower. Although at higher temperatures, due to a lower silica content, the efficiency of alumina recovery from ESPDW increases, reaching 90% after 90 min of leaching at 240 °C.

To study the leaching mechanism, the obtained kinetic curves were processed using the shrinking core model [50]. We studied six kinetic equations [51] describing the process in various modes, from kinetic to intra-diffusion; however, the models shown below (Equations (2) and (3)) proved to be most promising for the process description:

$$1 - 3(1 - X)^{2/3} + 2(1 - X) = k_1 t, \quad (2)$$

$$1/3\ln(1 - X) + ((1 - X)^{-1/3} - 1) = k_2 t, \quad (3)$$

where X is the degree of aluminum recovery into the liquor at a time t, k_i is the apparent rate constant. Equation (2) describes the process in the intra-diffusion area, while Equation (3) describes the process limited by interfacial transfer and diffusion through the product layer. The plot of $1/3\ln(1 - X) +$

$((1 - X)^{-1/3} - 1)$ versus t for ESP dust leaching gives a straight line (Figure 7a) with the determination coefficient R^2 the highest among all models used (more than 0.98 for all temperatures except 90 °C), which indicates that leaching, in this case, is most likely limited by interfacial transfer and diffusion through the product layer. It can be concluded that during leaching, a disilication product is formed around the core of the boehmite (Figure 8), which slows down the leaching process. Also, a film of sodium titanate can form on the surface of boehmite, which is known [52] to reduce the rate of dissolution of aluminum hydroxides significantly. For water-leached dust, the highest determination coefficient is observed, when using Equation (2) (Figure 7c), which implies that the process is limited by diffusion through the layer of the product or unreacted matter.

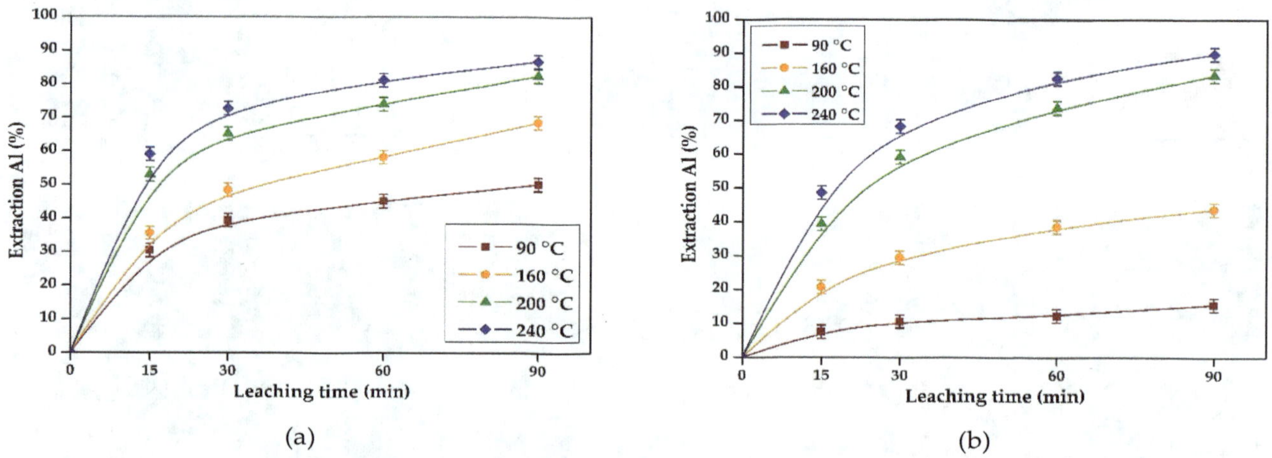

Figure 6. Effect of leaching time and temperature for Al extraction from raw ESP dust (**a**), from water-leached ESP dust (**b**).

Using the obtained values of the apparent rate constant (k_i) in Figure 7b,d and the Arrhenius equation (Equation (4)), we determined the values of the apparent activation energy for leaching ESPD and ESPDW (Figure 6) to be 24.98 kJ/mol and 33.19 kJ/mol, respectively.

$$k_i = A\exp(-E_a/RT), \tag{4}$$

where A is the Arrhenius constant, R is the universal rate constant (8.314 J/mol·K), T is the temperature (K), E_a is the apparent activation energy (J/mol).

The obtained values of the activation energy also confirm diffusion limitation. However, for an intradiffusion stage, the activation energy should be in the range of 8–22 kJ/mol. The higher values, in this case, maybe due to the fact that higher activation energy is required for the dissolution of the boehmite since a temperature of more than 160 °C is required for its extraction into alkali liquor. In the second case, the activation energy is higher, since there is no easily soluble phase of aluminum left after water leaching and a low degree of leaching efficiency of boehmite is observed at 90 °C.

Figure 8 shows XRD patterns of ESPDW and ESPDA, from which it can be seen that alkaline leaching results in disappearing of boehmite peaks and appearing instead of the peaks corresponding to cancrinite; iron, in contrast to the Bayer red mud, is represented by both hematite and hydroxide phases.

The yield of solid residue (red mud) from the leaching of ESPD in alkaline aluminate liquor at 240 °C for 90 min was 29.8% of the initial weight of dust. At the same time, the yield of red mud (ESPDA), after ESPDW leaching with alkali liquor at 240 °C for 90 min was 21.0%. As a result, the degree of REE enrichment of red mud in the second case was higher. Figure 9 shows the content of rare earth metals in these products.

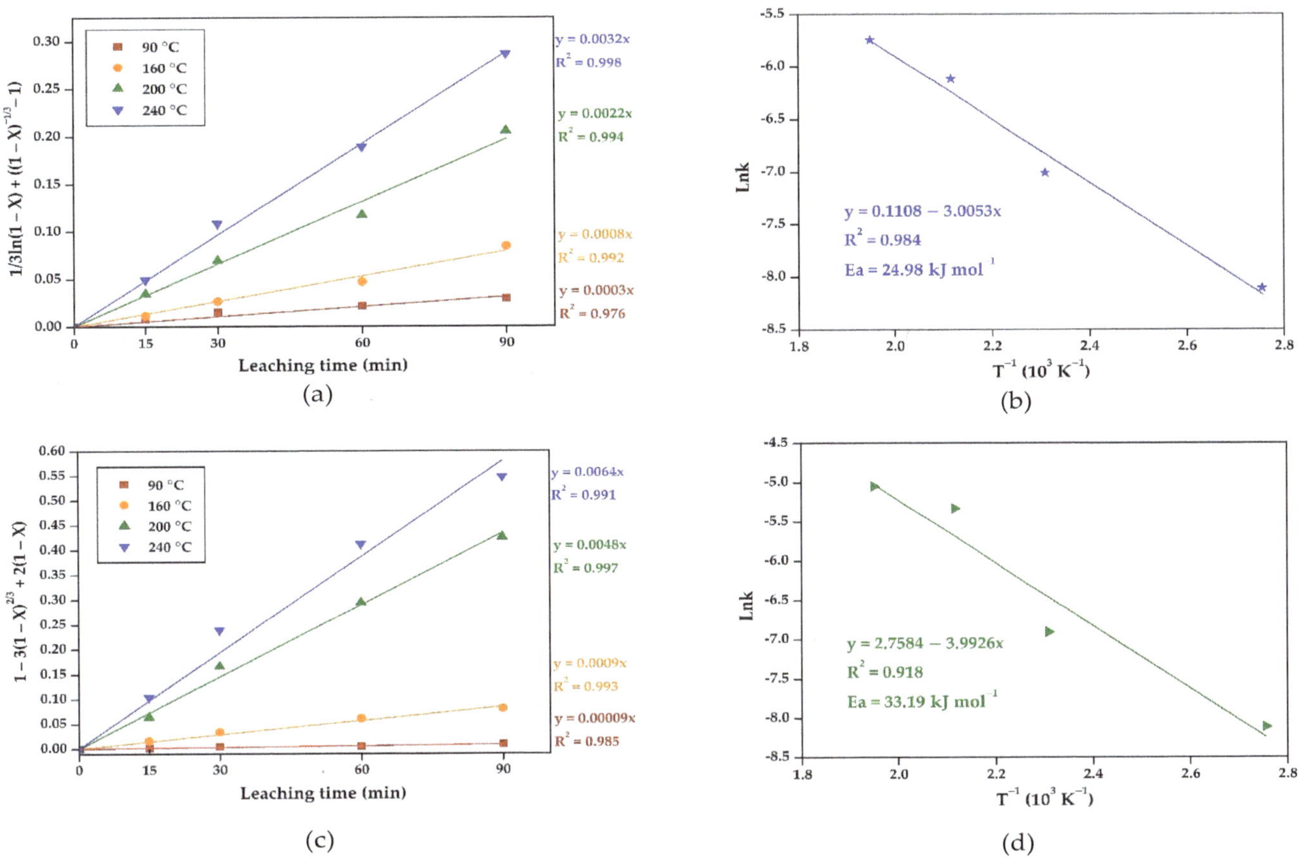

Figure 7. Results of substituting the data for ESPD leaching to Equation (3) (**a**); substituting the data for ESPDW leaching to Equation (2) (**c**); dependence ln$k$ – T^{-1} for leaching the original ESP dust (**b**) and water-treated ESP dust (**d**).

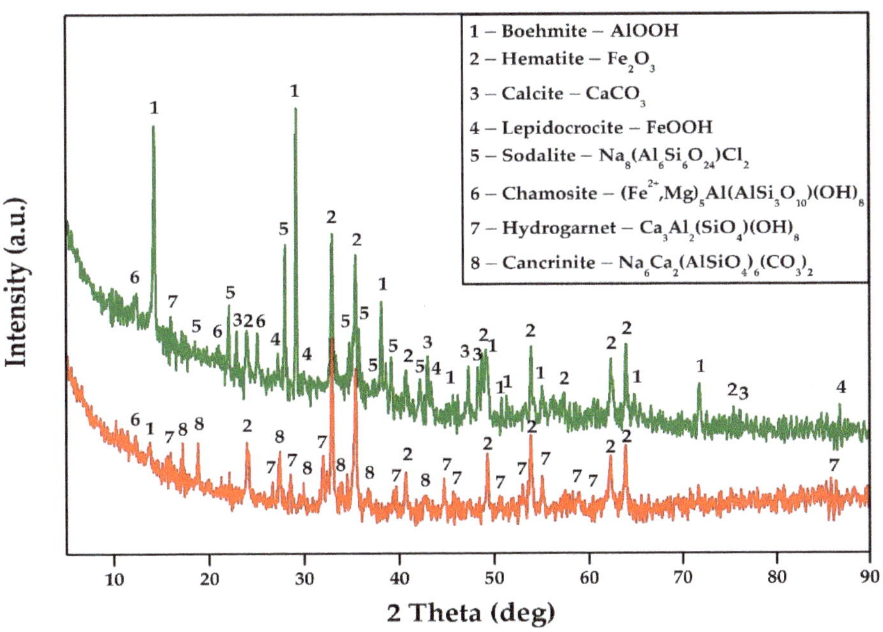

Figure 8. XRD pattern of water-leached electrostatic precipitator dust (**green**) and alkali-leached electrostatic precipitator dust (**red**).

The data in Figure 9 show that the total amount of REE in the red mud after ESP dust leaching was more than 1700 ppm in the first process and over 3200 ppm in the second process. Thus, we demonstrated that, in principle, it is possible to concentrate rare-earth elements in red mud by leaching ESP dust with water and alkaline aluminate liquor. The REE content in the red mud obtained thereby is three times higher than in conventional red mud of the alumina refinery, which can significantly reduce the cost of obtaining the REE concentrate in the future. Moreover, the recovery of additional components from natural raw materials can improve the efficiency of processing of bauxite raw materials in general. Based on these findings, we propose the following ESPD processing scheme (Figure 10), which enables efficient recovery of both alumina and rare earth elements. The stage of leaching REE from ESPDA will be discussed in the next paper.

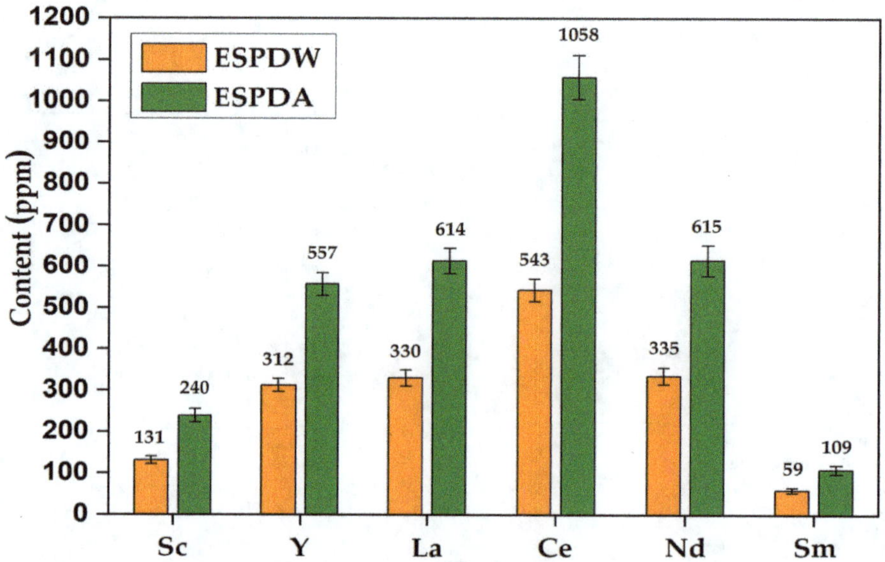

Figure 9. Content of rare earth elements in water-leached electrostatic precipitator dust and alkali-leached electrostatic precipitator dust.

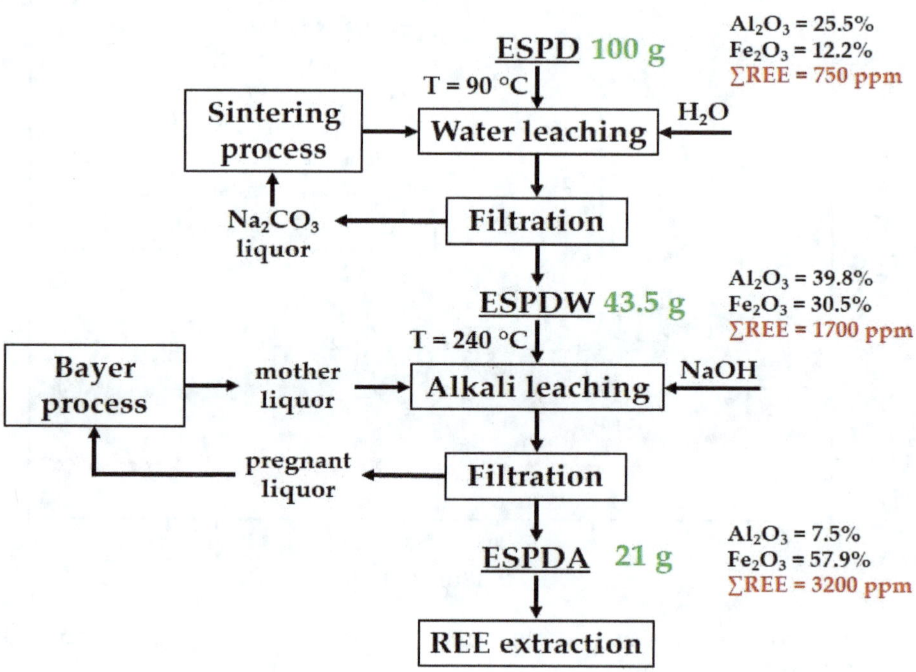

Figure 10. A flowsheet of rare-earth elements (REE) concentration from electrostatic precipitator dust by water/alkali leaching.

4. Conclusions

In this article, we investigated the possibility of concentrating rare-earth elements into red mud obtained by leaching dust from electrostatic precipitators used in bauxite charge sintering. We compared the chemical composition of the sinters and ESP dust formed during bauxite sintering in tubular rotary kilns. The sinter and the dust were analyzed for the content of rare earth elements. We also studied their association with various minerals of the materials used. The study of leaching of electric precipitation dust allows us to draw the following conclusions:

1. The chemical and phase composition of the electrostatic precipitation dust and sinter obtained in the same technological process are different. The content of rare-earth elements in the dust of electrostatic precipitators is 70–80% higher than in the sinter, which may be explained by a large number of cycles of ESP dust return to the kiln due to the very small size of particles.
2. Similar to the red mud of the Bayer process, scandium in ESP dust and in the solid residue after ESP dust leaching is associated to a greater extent with hematite, because of an insufficient transformation of latter into sodium ferrite in the sintering process and to a lower extent with disilication product (sodalite and cancrinite); the complete recovery of REE requires the destruction of the hematite at first.
3. The following conditions allow to obtain red mud from ESP dust that contains three times more REE than traditional waste red mud of the Ural Alumina Refinery: pre-leaching with water at 90 °C followed by autoclave leaching with an alkaline-aluminate liquor at 240 °C for 90 min.
4. The kinetic patterns of leaching alumina with the alkali liquor have been studied for the original electrostatic precipitation dust and water-leached dust. The change in the limiting stage of the process after water-treatment has been shown by an increase of the activation energy from 24.98 to 33.19 kJ/mol; changes in the form of the kinetic curves have also been demonstrated.

A technological scheme has been proposed for the removal of ESP dust from the cycle and its further comprehensive processing. Recovery of additional components from ESP dust will improve the efficiency of processing of bauxite raw materials in general and reduce the environmental footprint.

Author Contributions: Conceptualization, I.L. and L.C.; methodology, I.L.; validation, L.C., I.L.; formal analysis, A.S., D.V.; investigation, L.C., J.N.; resources, L.C., I.L.; data curation, I.L.; writing—original draft preparation, A.S. I.L.; writing—review and editing, A.S., D.V.; visualization, J.N., D.V.; supervision, L.C.; project administration, I.L.; funding acquisition, A.S. All authors have read and agreed to the published version of the manuscript.

Acknowledgments: RUSAL-Kamensk-Uralsky are acknowledged for providing materials. Technicians at Ural Branch of Russian Academy of Sciences are acknowledged for their assistance with XRD, XRF, EPMA, and ICP-MS analysis.

References

1. Qi, D. *Hydrometallurgy of Rare Earths: Extraction and Separation*; Elsevier: Amsterdam, The Netherlands, 2018; ISBN 978-0-12-813921-9.
2. Savchenkov, S.A.; Bazhin, V.Y.; Brichkin, V.N.; Kosov, Y.I.; Ugolkov, V.L. Production Features of Magnesium-Neodymium Master Alloy Synthesis. *Metallurgist* **2019**, *63*, 394–402. [CrossRef]
3. Wübbeke, J. Rare earth elements in China: Policies and narratives of reinventing an industry. *Resour. Policy* **2013**, *38*, 384–394. [CrossRef]
4. Savchenkov, S.; Bazhin, V.; Brichkin, V.; Povarov, V.; Ugolkov, V.; Kasymova, D. Synthesis of magnesium-zinc-yttrium master alloy. *Lett. Mater.* **2019**, *9*, 339–343. [CrossRef]
5. Røyset, J.; Ryum, N. Scandium in aluminium alloys. *Int. Mater. Rev.* **2005**, *50*, 19–44. [CrossRef]
6. Binnemans, K.; Jones, P.T.; Blanpain, B.; Van Gerven, T.; Yang, Y.; Walton, A.; Buchert, M. Recycling of rare earths: A critical review. *J. Clean. Prod.* **2013**, *51*, 1–22. [CrossRef]

7. Wang, W.; Cheng, C.Y. Separation and purification of scandium by solvent extraction and related technologies: A review. *J. Chem. Technol. Biotechnol.* **2011**, *86*, 1237–1246. [CrossRef]
8. Li, Z.; Din, J.; Xu, J.; Liao, C.; Yin, F.; Lǖ, T.; Cheng, L.; Li, J. Discovery of the REE minerals in the Wulong–Nanchuan bauxite deposits, Chongqing, China: Insights on conditions of formation and processes. *J. Geochem. Explor.* **2013**, *133*, 88–102. [CrossRef]
9. Ou, Z.; Li, J.; Wang, Z. Application of mechanochemistry to metal recovery from second-hand resources: A technical overview. *Environ. Sci. Process. Impacts* **2015**, *17*, 1522–1530. [CrossRef]
10. Rychkov, V.; Koukkari, P.; Kirillov, S.; Kirillov, E. Best Practices of Russia and Finland in Extracting REE from Fertilizer Waste. *KnE Mater. Sci.* **2017**, *2*, 168–173. [CrossRef]
11. Diallo, M.S.; Baier, G.; Moyer, B.A.; Hamelers, B. Critical Materials Recovery from Solutions and Wastes: Retrospective and Outlook. *Environ. Sci. Technol.* **2015**, *49*, 9387–9389. [CrossRef]
12. Ding, Y.; Wang, J.; Wang, G.; Xue, Q. Innovative Methodology for Separating of Rare Earth and Iron from Bayan Obo Complex Iron Ore. *ISIJ Int.* **2012**, *52*, 1772–1777. [CrossRef]
13. Wang, W.; Pranolo, Y.; Cheng, C.Y. Metallurgical processes for scandium recovery from various resources: A review. *Hydrometallurgy* **2011**, *108*, 100–108. [CrossRef]
14. Ayora, C.; Macías, F.; Torres, E.; Lozano, A.; Carrero, S.; Nieto, J.-M.; Pérez-López, R.; Fernández-Martínez, A.; Castillo-Michel, H. Recovery of Rare Earth Elements and Yttrium from Passive-Remediation Systems of Acid Mine Drainage. *Environ. Sci. Technol.* **2016**, *50*, 8255–8262. [CrossRef] [PubMed]
15. Ramasamy, D.L.; Puhakka, V.; Repo, E.; Sillanpää, M. Selective separation of scandium from iron, aluminium and gold rich wastewater using various amino and non-amino functionalized silica gels—A comparative study. *J. Clean. Prod.* **2018**, *170*, 890–901. [CrossRef]
16. Smirnov, A.L.; Titova, S.M.; Rychkov, V.N.; Bunkov, G.M.; Semenishchev, V.S.; Kirillov, E.V.; Poponin, N.N.; Svirsky, I.A. Study of scandium and thorium sorption from uranium leach liquors. *J. Radioanal. Nucl. Chem.* **2017**, *312*, 277–283. [CrossRef]
17. Rychkov, V.N.; Kirillov, E.V.; Kirillov, S.V.; Semenishchev, V.S.; Bunkov, G.M.; Botalov, M.S.; Smyshlyaev, D.V.; Malyshev, A.S. Recovery of rare earth elements from phosphogypsum. *J. Clean. Prod.* **2018**, *196*, 674–681. [CrossRef]
18. Lokshin, E.P.; Tareeva, O.A.; Elizarova, I.P. A study of the sulfuric acid leaching of rare-earth elements, phosphorus, and alkali metals from phosphodihydrate. *Russ. J. Appl. Chem.* **2010**, *83*, 958–964. [CrossRef]
19. Rivera, R.M.; Ounoughene, G.; Malfliet, A.; Vind, J.; Panias, D.; Vassiliadou, V.; Binnemans, K.; Van Gerven, T. A Study of the Occurrence of Selected Rare-Earth Elements in Neutralized–Leached Bauxite Residue and Comparison with Untreated Bauxite Residue. *J. Sustain. Metall.* **2019**, *5*, 57–68. [CrossRef]
20. Loginova, I.V.; Shoppert, A.A.; Chaikin, L.I. Extraction of Rare-Earth Metals During the Systematic Processing of Diaspore-Boehmite Bauxites. *Metallurgist* **2016**, *60*, 198–203. [CrossRef]
21. Borra, C.R.; Pontikes, Y.; Binnemans, K.; Van Gerven, T. Leaching of rare earths from bauxite residue (red mud). *Miner. Eng.* **2015**, *76*, 20–27. [CrossRef]
22. Mordberg, L.E. Patterns of distribution and behaviour of trace elements in bauxites. *Chem. Geol.* **1993**, *107*, 241–244. [CrossRef]
23. Anawati, J.; Azimi, G. Recovery of Strategic Materials from Canadian Bauxite Residue by Smelting Followed by Acid Baking–Water Leaching. In *Rare Metal Technology 2020*; Azimi, G., Forsberg, K., Ouchi, T., Kim, H., Alam, S., Baba, A.A., Eds.; The Minerals, Metals & Materials Series; Springer International Publishing: Cham, Switzerland, 2020; pp. 139–150. ISBN 978-3-030-36757-2.
24. Sinha, S.; Sinha, M.K.; Pandey, B.D. Extraction of lanthanum and cerium from Indian red mud. *Int. J. Miner. Process.* **2014**, *127*, 70–73. [CrossRef]
25. Anhaeusser, C.R.; Geological Society of South Africa (Eds.) *Mineral Deposits of Southern Africa: In 2 Vol. 1*; Geological Society of South Africa: Johannesburg, Southern Africa, 1986; ISBN 978-0-620-09438-2.
26. Lokshin, E.P.; Tareeva, O.A. Production of high-quality gypsum raw materials from phosphogypsum. *Russ. J. Appl. Chem.* **2015**, *88*, 567–573. [CrossRef]
27. Binnemans, K.; Jones, P.T.; Blanpain, B.; Van Gerven, T.; Pontikes, Y. Towards zero-waste valorisation of rare-earth-containing industrial process residues: A critical review. *J. Clean. Prod.* **2015**, *99*, 17–38. [CrossRef]
28. Lokshin, E.P.; Tareeva, O.A. Activation of leaching of rare earth elements from phosphohemihydrate. *Russ. J. Appl. Chem.* **2013**, *86*, 1638–1642. [CrossRef]
29. Kovács, T.; Horváth, M.; Csordás, A.; Bátor, G.; Tóth-Bodrogi, E. Tobacco plant as possible biomonitoring tool of red mud dust fallout and increased natural radioactivity. *Heliyon* **2020**, *6*, e03455. [CrossRef]

30. Alam, S.; Das, B.K.; Das, S.K. Dispersion and Sedimentation Characteristics of Red Mud. *J. Hazard. Toxic Radioact. Waste* **2018**, *22*, 04018025. [CrossRef]
31. Liu, Z.; Li, H. Metallurgical process for valuable elements recovery from red mud—A review. *Hydrometallurgy* **2015**, *155*, 29–43. [CrossRef]
32. Rivera, R.M.; Xakalashe, B.; Ounoughene, G.; Binnemans, K.; Friedrich, B.; Van Gerven, T. Selective rare earth element extraction using high-pressure acid leaching of slags arising from the smelting of bauxite residue. *Hydrometallurgy* **2019**, *184*, 162–174. [CrossRef]
33. Klauber, C.; Gräfe, M.; Power, G. Bauxite residue issues: II. options for residue utilization. *Hydrometallurgy* **2011**, *108*, 11–32. [CrossRef]
34. Qu, Y.; Li, H.; Tian, W.; Wang, X.; Wang, X.; Jia, X.; Shi, B.; Song, G.; Tang, Y. Leaching of valuable metals from red mud via batch and continuous processes by using fungi. *Miner. Eng.* **2015**, *81*, 1–4. [CrossRef]
35. Shoppert, A.; Loginova, I.; Rogozhnikov, D.; Karimov, K.; Chaikin, L. Increased as Adsorption on Maghemite-Containing Red Mud Prepared by the Alkali Fusion-Leaching Method. *Minerals* **2019**, *9*, 60. [CrossRef]
36. Sabirzyanov, N.A.; Yatsenko, S.P. *Gidrokhimicheskie Sposoby Kompleksnoĭ Pererabotki Boksita*; IKhTT UrO RAN: Ekaterinburg, Russia, 2006; ISBN 978-5-7691-1629-2. (In Russia)
37. Akcil, A.; Akhmadiyeva, N.; Abdulvaliyev, R.; Abhilash; Meshram, P. Overview on Extraction and Separation of Rare Earth Elements from Red Mud: Focus on Scandium. *Miner. Process. Extr. Metall. Rev.* **2018**, *39*, 145–151. [CrossRef]
38. Borra, C.R.; Blanpain, B.; Pontikes, Y.; Binnemans, K.; Van Gerven, T. Smelting of Bauxite Residue (Red Mud) in View of Iron and Selective Rare Earths Recovery. *J. Sustain. Metall.* **2016**, *2*, 28–37. [CrossRef]
39. Petrakova, O.V.; Panov, A.V.; Gorbachev, S.N.; Klimentenok, G.N.; Perestoronin, A.V.; Vishnyakov, S.E.; Anashkin, V.S. Improved Efficiency of Red Mud Processing through Scandium Oxide Recovery. In *Light Metals 2015*; Hyland, M., Ed.; John Wiley & Sons, Inc.: Hoboken, NJ, USA, 2015; pp. 91–96. ISBN 978-1-119-09343-5.
40. Yatsenko, S.P.; Pyagai, I.N. Red mud pulp carbonization with scandium extraction during alumina production. *Theor. Found. Chem. Eng.* **2010**, *44*, 563–568. [CrossRef]
41. Loginova, I.V.; Shoppert, A.A.; Chaikin, L.I. Effect of Adding Sintering Furnace Electrostatic Precipitator Dust on Combined Leaching of Bauxites and Cakes. *Metallurgist* **2015**, *59*, 698–704. [CrossRef]
42. Diev, V.N.; Sabirzyanov, N.A.; Skryabneva, L.M.; Yatsenko, S.P.; Anashkin, V.S.; Aminov, S.N.; Zavadskij, K.F.; Sysoev, A.V.; Ustich, E.P. Method for Scandium Extraction from Bauxite Treatment for Alumina Production. Russia Patent 2,201,988, 10 April 2003.
43. Vind, J.; Malfliet, A.; Blanpain, B.; Tsakiridis, P.; Tkaczyk, A.; Vassiliadou, V.; Panias, D. Rare Earth Element Phases in Bauxite Residue. *Minerals* **2018**, *8*, 77. [CrossRef]
44. Borra, C.R.; Blanpain, B.; Pontikes, Y.; Binnemans, K.; Van Gerven, T. Recovery of Rare Earths and Other Valuable Metals from Bauxite Residue (Red Mud): A Review. *J. Sustain. Metall.* **2016**, *2*, 365–386. [CrossRef]
45. M Tóth, T.; Schubert, F.; Raucsik, B.; Fintor, K. Mineralogical and Geochemical Constraints of the REE Accumulation in the Almásfüzitő Red Mud Depository in Northwest Hungary. *Appl. Sci.* **2019**, *9*, 3654. [CrossRef]
46. Smith, P. The processing of high silica bauxites—Review of existing and potential processes. *Hydrometallurgy* **2009**, *98*, 162–176. [CrossRef]
47. Pasechnik, L.A.; Shirokova, A.G.; Koryakova, O.V.; Sabirzyanov, N.A.; Yatsenko, S.P. Complexing Properties of Scandium(III) in Alkaline Medium. *Russ. J. Appl. Chem.* **2004**, *77*, 1070–1073. [CrossRef]
48. Anawati, J.; Azimi, G. Recovery of scandium from Canadian bauxite residue utilizing acid baking followed by water leaching. *Waste Manag.* **2019**, *95*, 549–559. [CrossRef] [PubMed]
49. Alex, T.C.; Kumar, R.; Roy, S.K.; Mehrotra, S.P. Towards ambient pressure leaching of boehmite through mechanical activation. *Hydrometallurgy* **2014**, *144–145*, 99–106. [CrossRef]
50. Levenspiel, O. *Chemical Reaction Engineering*, 3rd ed.; Wiley: New York, NY, USA, 1999; ISBN 978-0-471-25424-9.
51. Rogozhnikov, D.A.; Shoppert, A.A.; Dizer, O.A.; Karimov, K.A.; Rusalev, R.E. Leaching Kinetics of Sulfides from Refractory Gold Concentrates by Nitric Acid. *Metals* **2019**, *9*, 465. [CrossRef]
52. Wang, Y.; Li, X.; Zhou, Q.; Qi, T.; Liu, G.; Peng, Z.; Zhou, K. Effects of Si-bearing minerals on the conversion of hematite into magnetite during reductive Bayer digestion. *Hydrometallurgy* **2019**, *189*, 105126. [CrossRef]

5

Trace Element Distribution in Magnetite Separates of Varying Origin: Genetic and Exploration Significance

Demetrios G. Eliopoulos [1] and Maria Economou-Eliopoulos [2,*]

1 Institute of Geology and Mineral Exploration (IGME), Sp. Loui 1, Olympic Village, GR-13677 Acharnai, Greece; eliopoulos@igme.gr
2 Department of Geology and Geoenvironment, University of Athens, 15784 Athens, Greece
* Correspondence: econom@geol.uoa.gr

Abstract: Magnetite is a widespread mineral, as disseminated or massive ore. Representative magnetite samples separated from various geotectonic settings and rock-types, such as calc-alkaline and ophiolitic rocks, porphyry-Cu deposit, skarn-type, ultramafic lavas, black coastal sands, and metamorphosed Fe–Ni-laterites deposits, were investigated using SEM/EDS and ICP-MS analysis. The aim of this study was to establish potential relationships between composition, physico/chemical conditions, magnetite origin, and exploration for ore deposits. Trace elements, hosted either in the magnetite structure or as inclusions and co-existing mineral, revealed differences between magnetite separates of magmatic and hydrothermal origin, and hydrothermal magnetite separates associated with calc-alkaline rocks and ophiolites. First data on magnetite separates from coastal sands of Kos Island indicate elevated rare earth elements (REEs), Ti, and V contents, linked probably back to an andesitic volcanic source, while magnetite separated from metamorphosed small Fe–Ni-laterites occurrences is REE-depleted compared to large laterite deposits. Although porphyry-Cu deposits have a common origin in a supra-subduction environment, platinum-group elements (PGEs) have not been found in many porphyry-Cu deposits. The trace element content and the presence of abundant magnetite separates provide valuable evidence for discrimination between porphyry-Cu–Au–Pd–Pt and those lacking precious metals. Thus, despite the potential re-distribution of trace elements, including REE and PGE in magnetite-bearing deposits, they may provide valuable evidence for their origin and exploration.

Keywords: magnetite separates; trace elements; exploration; ophiolites; skarn; porphyry-Cu; Fe–Ni-laterites

---

1. Introduction

Magnetite is a widespread mineral, as disseminated or massive ore, in various geotectonic settings and rock-types, such as magmatic, hydrothermal, ophiolites, calk-alkaline rocks (porphyry-Cu deposits, skarn type), laterites, and metamorphosed rocks/ores, all formed under a variety of conditions [1–13]. It has an inverse cubic (space group *Fd3m*) spinel-type structure with the general stoichiometry AB_2O_4, where A and B are tetrahedral (Fe^{3+}) and octahedral (Fe^{3+} and Fe^{2+}) coordination sites, respectively [14]. In addition, the geochemistry of magnetite separates is also affected by the co-precipitation of other minerals [5–9]. Trace elements hosted either in the structure of magnetite, such as cations which can substitute on the A (Mg, Fe^{2+}, Ni, Mn, Co, or Zn) and B (Al, Cr, Fe^{3+}, V, Mn or Ga) sites [14,15], and/or in mineral inclusions. They may result in distinctive trace element signatures in magnetite separates and can be used as a prospective tool for many types of ore deposits, and/or a variety of trace-element discrimination diagrams can classify magnetite separates from different types of ore deposits [4–13]. However, it has been proposed that the discrimination diagram (Ti + V) versus (Ca + Al + Mn) can be used with caution as a petrogenetic indicator for trace element data in the case of magnetite separates which have unusual composition and/or have been re-equilibrated. [16].

The present study is focused on massive and disseminated magnetite separates from various geotectonic settings covering a wide range of formation conditions: Hydrothermal magnetite associated with calc-alkaline rocks (skarn type, porphyry-Cu–Au deposit), ophiolite complexes (associated with sulphides or apatite), magmatic magnetite from the magmatic sequence of ophiolites and ultramafic lavas/wherlites, magnetite from metamorphosed laterites, and magnetite separates from black coastal sand. Major and minor/trace element contents, including rare earth elements (REEs), in these magnetite separate types are presented and are compared to literature data. The purpose of this application of the magnetite chemistry and texture features is to provide a potential relationship between compositional variation and physico/chemical conditions, the magnetite origin, and exploration of ore deposits.

2. Materials and Methods

2.1. Mineral Chemistry

Polished sections (two-tree samples from each type) from magnetite ores (Tables 1–4) were carbon-or gold-coated and examined by a scanning electron microscope (SEM) using energy-dispersive spectroscopy (EDS). The SEM images and EDS analyses were carried out at the University of Athens (NKUA), using a JEOL JSM 5600 scanning electron microscope (Tokyo, Japan), equipped with the ISIS 300 OXFORD automated energy dispersive analysis system (Oxfordshire, UK) with the following operating conditions: Accelerating voltage 20 kV, beam current 0.5 nA, time of measurement (dead time) 50 s, and beam diameter 1–2 μm. The following X-ray lines were used: FeKα, NiKα, CoKα, CuKα, CrKα, AlKα,TiKα, CaKα, SiKα, MnKα, and MgKα. Standards used were pure metals for the elements Cr, Mn, Ni, Co, Zn, V, and Ti, and Si, MgO for Mg, and Al_2O_3 for Al. Contents of Fe_2O_3 and FeO for spinels were calculated on the basis of the spinel stoichiometry. The H_2O for (Co, Mn, Ni)-hydroxides was calculated to obtain 2 (OH) *pfu*.

Whole Rock Analysis

The studied magnetite samples were massive and disseminated mineralizations, derived from large (weighing approximately 2 kg) samples, which were crushed and pulverized in an agate mortar. The samples of magnetite separates from the Skouries deposit are representative of increasing depth from 43 to 363 m (the depth for the sample skou8b was 280 m) [4]. The multi-stage water flotation was applied and magnetite separates were concentrated magnetically from the water suspension of the rock powder. This portion was furthermore pulverized in the agate mortar at a size fraction −75 μm, and magnetite separates were separated furthermore from silicates. Thus the magnetite separate concentrates with weak silicates and potential very fine inclusions of various minerals (SEM/EDS images; Table 1) were used for bulk rock analysis.

Major and minor/trace elements in magnetite separate samples and two samples from the extinct fumaroles co-existing with native sulphur from the volcanic environment in the Aegean Sea (Santorini and Nisyros islands) were determined at the SGS Global—Minerals Division Geochemistry Services Analytical Laboratories Ltd., Vancouver, BC, Canada. The samples were dissolved using sodium peroxide fusion, combined ICP-AES and ICP-MS (Package GE_ICP91A50). Detection limits of the method are provided (Table S1). On the basis of the quality control report provided by Analytical Labs, the results of the reference material analysis in comparison to expected values, and the results from multistage analysis of certain samples, showed accuracy and precision of the method in good agreement with international standard (<10%).

3. Geological Outline and Mineralogical Features

3.1. Hydrothermal Magnetite Separates

3.1.1. Massive Magnetite Separates of Skarn-Type (Lavrion, Plaka Mine)

The ancient (earlier than 1000 B.C.), famous mine of Lavrion in Attica (Greece) is associated with a granodiorite intrusion, of Upper Miocene age, within the metamorphic -Cycladic Crystalline belt, which belongs to the carbonate replacement deposits (CRDs) [17]. It is well known for the exploitation of mixed massive sulphide mineralization composed by sphalerite, pyrite and galena (B.P.G), especially the production of silver [17–20]. Early hornfels formation was followed by the multistage development of a skarn deposit at the area of Plaka, around the granodiorite body, located into two stratigraphic horizons, near the contacts between the Kaessariani schist with lower and upper marble (Figure 1) [2].

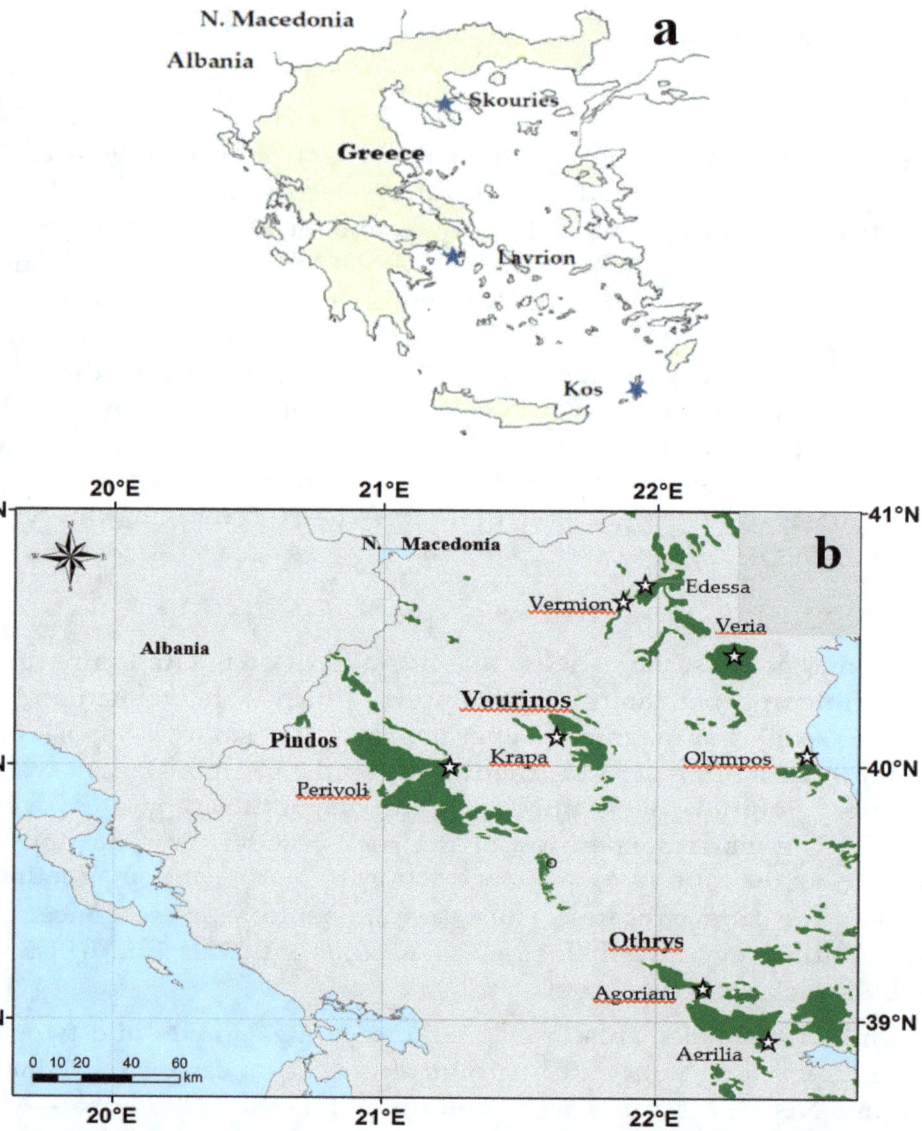

Figure 1. Sketch map showing the location of the studied magnetite separates hosted in calc-alkaline rocks (**a**). Ophiolitic rocks and metamorphosed Ni-laterite deposits (**b**).

Texture characteristics of the magnetite mineralization show multiple stages of the mineralization and that early magnetite and magnetite–hematite representing endoskarns (stage I) was followed by pyrite, pyrrhotite, and Fe-rich sphalerite, Ag-rich galena, arsenopyrite, forming exoskarns (stage II) [2,16] high-Fe sphalerite, chalcopyrite, and galena forming (Figure 2) [17–19]. Magnetite occurs as euhedral or anhedral crystals, homogeneous or having abundant porosity and solid inclusions, mostly silicate minerals. It is almost pure, having very low Si, Al, and Mn contents, while Ti and V are lower than the detection limit (Table 1).

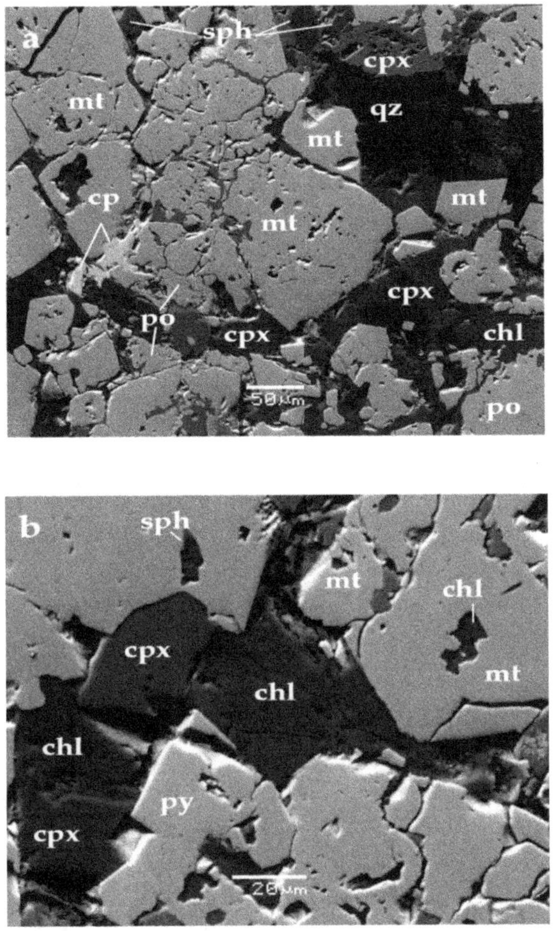

Figure 2. Back-scattered electron (BSE) images showing homogeneous magnetite separates (mt) in sharp contact with magnetite, having abundant porosity and solid inclusions, mostly silicate minerals, associated with clinopyroxene, quartz, chlorite, pyrrhotite, chalcopyrite, and sphlerite (**a**); magnetite is associated with silicates, sphalerite, and pyrite (**b**). Abbreviations: mt = magnetite; po = pyrrhotite; cp = chalcopyrite; py = pyrite; sph = sphalerite; cpx = clinopyroxene; qtz = quartz; chl = chlorite.

3.1.2. Disseminated Magnetite in Porphyry-Cu Deposit (Skouries)

The Skouries porphyry-Cu deposit is related to tertiary subvolcanic–porphyritic stocks, which belong to the Vertiskos Formation of the Serbo-Macedonian massif (SMM), Greece, and are controlled by deep fracture systems [21]. The typical alteration types of the porphyry-Cu intrusions are more or less presented in the Skouries intrusion, due to the repeated overprinting and intense silicification, with potassic being the predominant alteration type. Two mineral assemblages occurring as veinlets and disseminations can be distinguished: (1) Magnetite, reaching up to 10 vol.% (average 6 vol.%)—bornite–chalcopyrite, linked to pervasive potassic and propylitic alteration type, in the central parts of the deposit; and (2) chalcopyrite–pyrite, dominant at the peripheral parts of the deposit [4]. Chalcopyrite, and to a lesser extent bornite, contain occasional exsolutions of clausthalite (PbSe) (Figure 3b). There is a frequent association of magnetite separates and Cu-minerals (bornite and chalcopyrite) with inclusions of thorite, U-bearing thorite, rare earth element (REE)-enriched silicates of the epidote-group (allanite), Ce- and Th-rich monazites, and zircons identified as accessory minerals (Figure 3b–d; Table 1).

Table 1. Representative electron SEM/EDS analyses of magnetite in various rock-types (Figures 2–5, 7 and 8). Symbol n.d. = no detected.

	CALC-ALKALINE					OPHIOLITES							PLACER
Deposit type	Skarn-type			Hydrothermal		Hydrothermal				Magmatic			
Location	Lavrion			Skouries		Othrys	Edessa	Pindos	Vourinos	Vourinos	Othrys	Othrys	KOS
sample	Plaka P.L.mt1	Plaka P.L.mt2	SK43	SG-6	Skou8	Agoriani O.Ag.t	Samari EdK2	Perivoli Pi.Mt	Krapa VKMt	Krapa VKMt1	Agrilia 15Agr2	Agrilia O.Agr34	KOS1
wt.%													
SiO_2	1.7	2.6	0.9	0.7	0.4	0.5	2.3	0.3	0.5	0.9	0.3	0.7	0.5
Al_2O_3	0.5	0.9	n.d	n.d.	n.d.	0.6	0.2	n.d.	1.7	2.1	11.1	7.5	1.5
Cr_2O_3	n.d.	n.d.	2.1	0.9	n.d.	n.d.	0.1	n.d.	n.d.	n.d.	3.9	1.5	0.2
V_2O_5	n.d.	n.d.	0.7	0.5	n.d.	n.d.	n.d.	n.d.	1.4	1.8	0.7	0.6	0.5
Fe_2O_3	67.5	67.7	66.7	67.6	67.9	68.1	68.1	69.7	55.1	55.4	45.2	51.4	57.1
FeO	30.4	30.4	30.6	30.8	30.7	30.1	29.9	30.2	33.5	33.3	32.3	31.1	34.5
MgO	n.d.	0.3	n.d.	n.d.	0.5	0.9	n.d.	0.2	n.d.	0.9	3.1	3.5	n.d.
MnO	0.5	0.3	n.d.	n.d.	n.d.	n.d.	0.2	0.2	0.5	0.6	0.2	0.4	1.2
CoO	n.d.	n.d.	n.d.	n.d.	n.d.	n.d.	n.d.	n.d.	n.d.	n.d.	0.2	n.d.	n.d.
TiO_2	n.d.	n.d.	n.d.	n.d.	0.6	n.d.	n.d.	n.d.	6.9	4.4	3.6	4.6	5.1
Total	100.6	101.5	101	99.6	99.8	99.3	100.8	100.6	99.6	99.4	100.9	100.7	100.6

Table 2. Representative electron SEM/EDS analyses of fine-grained minerals associated with magnetite (Figure 3) from the Skouries porphyry-Cu deposit. Symbol n.d. = no detected.

wt.%	Monazite	Thorite	U-Thorite	Na, Ce-epidote	Allanite	Allanite
SiO_2	1.1	18.8	15.7	38.3	32.1	35.4
Al_2O_3	n.d.	n.d.	n.d.	27.2	16.2	14.1
FeO	n.d.	n.d.	0.6	8.3	14.9	17.4
CaO	0.8	n.d.	n.d.	22.2	12.4	13.6
P_2O_5	30.3	n.d.	n.d.	n.d.	n.d.	n.d.
ZrO_2	n.d.	n.d.	n.d.	n.d.	n.d.	n.d.
La_2O_3	17.2	n.d.	n.d.	0.9	5.1	3.9
Ce_2O_3	31.6	n.d.	n.d.	2.2	14.5	12.5
Nd_2O_3	10.5	n.d.	n.d.	n.d.	4.7	2.2
Pr_2O_3	2.8	n.d.	n.d.	n.d.	n.d.	n.d.
ThO_2	5.1	79.4	70.8	n.d.	n.d.	n.d.
UO_3	n.d.	n.d.	11.7	n.d.	n.d.	n.d.
Total	99.4	98.2	98.1	99.1	99.9	99.1

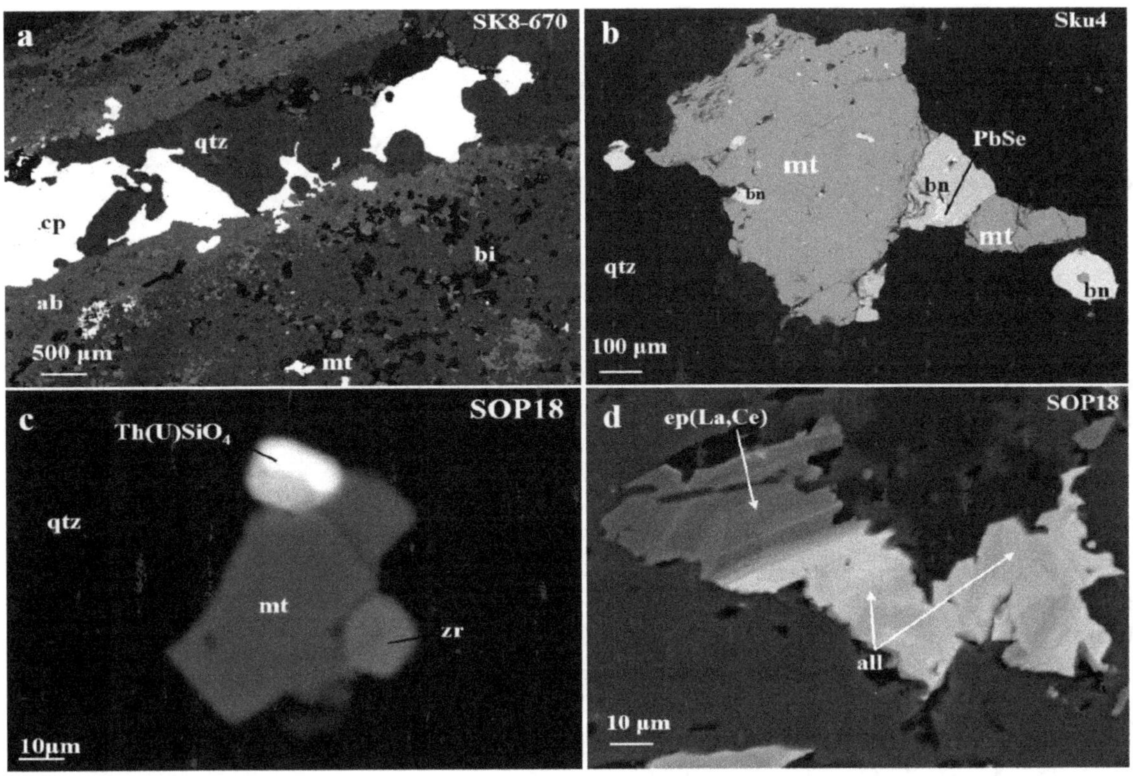

Figure 3. Back-scattered electron (BSE) images from the Skouries porphyry deposit, showing textural relationships between co-existing minerals. Quartz–chalcopyrite vein crosscutting biotite and magnetite-rich potassic altered porphyry (**a**); a close intergrowth of disseminated Ti–magnetite, chalcopyrite, bornite, and clausthalite (PbSe) (**b**); magnetite associated with thorite and zircon hosted by quartz (**c**); rare earth element (REE)-enriched epidote overgrown by allanite (**d**). Their composition is given (Table 2). Abbreviations: qtz = quartz; ab = albite; bi = biotite; mt = magnetite; cp = chalcopyrite; bn = bornite; PbSe = clausthalite; zr = zircon; ep = epidote; all = allanite.

3.1.3. Massive Magnetite Associated with Fe–Cu–Ni–Co Sulphides in Ophiolites (Pindos)

Sulfide mineralization in the Pindos ophiolite complex is located in the Smolicas Mountains, near the Aspropotamos dismembered ophiolite unit. This sequence, belonging to the ophiolitic unit of the Pindos complex, includes: (1) An intrusive section composed of dunites, lherzolites, olivine-websterites, olivine-gabbros, anorthosite gabbros, gabbros, and occasionally gabbronorites; and (2) a volcanic and subvolcanic sequence composed mainly by basalts and basaltic andesite pillow lavas ranging from high- to low-Ti affinity [22,23]. At the Perivoli (Tsoumes Hill) of the Aspropotamos unit, within gabbro close to its tectonic contact with serpentinized harzburgite, small irregular to lens-like occurrences (4 m × 1.5 m) of massive sulfide mineralization of Fe–Cu–Ni–Co type have been described [24]. Magnetite, forming often a network texture, is associated with sulphides, either as massive ore with inclusions of sulfides (chalcopyrite, pyrite, and pyrrhotite), or as individual grains dispersed within sulfide ore. A salient feature of magnetite is a network texture with occasionally deformed crystals (Figure 4). The composition of magnetite separates is characterized by very low Al, Ti, Cr, Mn, Ni, and V contents (Table 1 and Table S1).

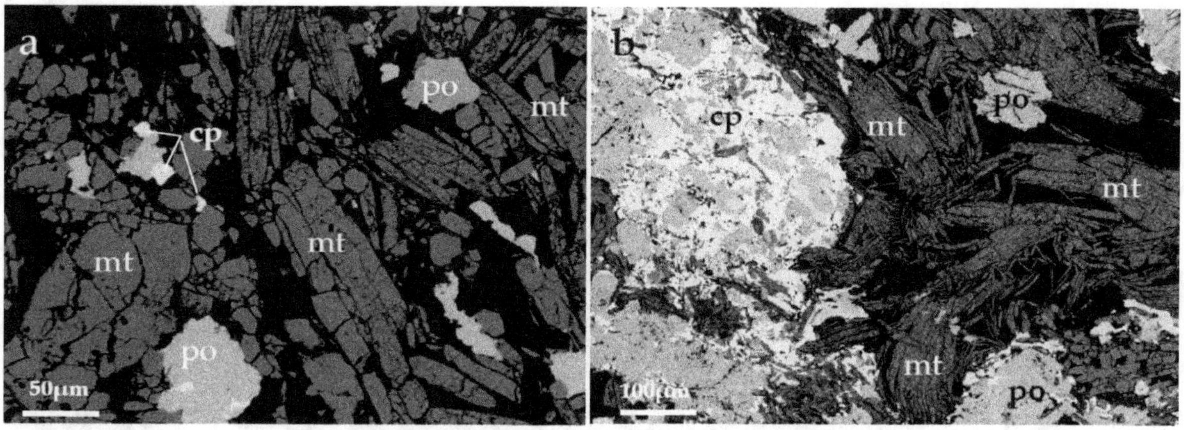

Figure 4. Back-scattered electron (BSE) images showing elongated curved magnetite associated with sulphides (**a**,**b**) from the Perivoli area, Pindos ophiolite complex. Abbreviations as in Figure 3.

3.1.4. Massive Magnetite in Mafic Ophiolitic Rocks Edessa (Samari)

The isolated ophiolite masses of Eastern Edessa belong to the Axios zone (Almopias subzone) [25–27]. At the area of Samari, Edessa, with dominant rock types diabase, gabbros, and serpentinized peridotites, ore bodies consisting of magnetite separates and sulphides have been described [27]. Magnetite occurs (1) as disseminated euhedral and/or subhedral crystals, along with sulphides (pyrite, chalcopyrite) within gabbros and diabase, accompanied by epidote, chlorite, quartz, sphene and tremolite; and (2) as massive ore hosted within gabbros, close to its contact with peridotite, with a gangue silicate mass dominant by greenalite, while sulphide veinlet crosscutting the magnetite separate ore are common (Figure 5) [27]. The magnetite is pure, characterized by the presence of only negligible contents of other elements (Table 1).

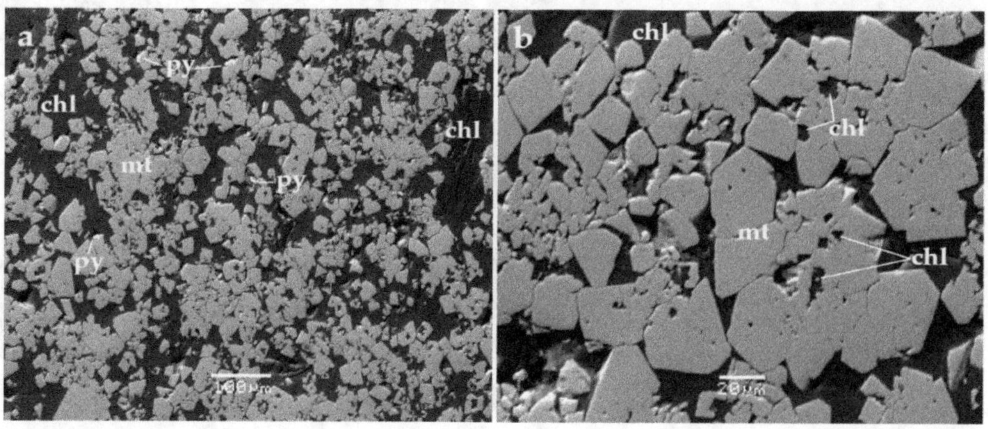

Figure 5. Back-scattered electron (BSE) images showing massive magnetite associated with little amounts of pyrite in a matrix of chlorite (greenalite) (**a**) and inclusions of chlorite in magnetite (**b**) from the Samari area (E. Edessa). Abbreviations as in Figures 2 and 3.

3.1.5. Massive Magnetite Associated with Apatite in Ophiolites

Major Mesozoic ophiolite complexes in Greece, including the Othris, Pindos, and Vourinos suites, were all obducted from the Middle-to-Late Jurassic and have been interpreted as parts of the same oceanic slab, although fore-arc and back-arc ridges have been tectonically fragmented and separated by overlapping sediments [28,29]. The Othrys ophiolite complex consists of a stack of thrust sheets in which overlapping stratigraphic successions, consisting mainly of harzburgites, lherzolites, gabbros, mafic dikes, and pillow lavas, have been recognized, while small irregularly shaped bodies of gabbro and gabbroic dikes or veins intrude the plagioclase lherzolite of the mantle sequence [30]. Irregular to lens-like occurrences (0.5 m × 1 m) of massive pure magnetite associated with large (up to 3 cm long) well-formed crystals of chlor–hydroxyapatite apatite, and lesser amounts of serpentine, tremolite, and Ni silicates (nepouite, pimelite), and by Ni sulfides (pentlandite, violarite, heazlewoodite) have been described in the Agoriani area of the Othrys ophiolite complex, central Greece [31]. The analysis of magnetite separates showed very little Al, Mg, Ni, Ti, and REE content (Table S1).

3.2. Magmatic Magnetite Separates

3.2.1. Disseminated Magnetite in Ultramafic Lavas and Wherlites

The Agrilia formation, Othrys ophiolite complex, about 6 km Northwestern of Lamia, is an unusual ultramafic lava characterized by high Mg content (31–33 wt.% MgO) [32]. Olivines from the Agrilia ultramafic lavas display high forsterite (Fo) contents, and calculated parental magma (in equilibrium with $Fo_{90.5}$) approximately 17 wt.% MgO [33]. Disseminated chromite occurs within groundmass (devitrified glass) and as inclusions in olivine and clinopyroxene [32–35]. The chromite spinel within groundmass is commonly rimmed by Ti–magnetite with sharp contacts (Figure 6a). Chromite is Cr-rich, exhibiting a limited range of variation (Table 3). Amphibole (tschermakite after [34], rhönite, chromite, and Ti–magnetite all associated with groundmass (Figure 6)). Rhönite in association with Ti–magnetite within devitrified glass (Table 3), and inter-grown with amphibole, may be of genetic significance [33,34]. The above ultramafic lavas are associated with medium-to-coarse grained wehrlite. They are mainly composed by olivine, clinopyroxene, hornblende, and in lesser amounts, spinel, orthopyroxene, phlogopite, and magnetites.

Table 3. Representative electron SEM/EDS analyses of rhönite (Rho) and amphibole (amph) within devitrified glass (Figure 6) from the Agrilia ultramafic lavas [35]. Symbol n.d. = no detected.

Host	**Devitrified Glass (d.G.)**							
	Rhonite (Rho)			d.G.		Amphibole (amph)		
SiO_2	26.3	26.6	25.8	33.1	45.6	39.8	40.4	55.2
MgO	8.8	8.7	8.8	14.5	6.5	9.2	10.6	11.1
Al_2O_3	15.7	14.9	15.2	12.1	9.9	16.5	17.5	10.6
Cr_2O_3	n.d.	n.d.	n.d.	0.8	1.1	n.d.	n.d.	1.3
CaO	10.7	10.6	10.6	4.3	3.1	10.1	9.6	10.4
TiO_2	4.9	3.9	4.7	1.6	0.2	1.9	1.2	0.2
FeO	28.9	29.8	30.3	27.1	21.8	17.7	17.1	6.2
MnO	0.2	0.2	0.3	0.2	0.2	n.d.	0.3	n.d.
Na_2O	0.5	0.5	1.3	1.6	6.9	2.5	1.7	2.7
K_2O	n.d.	n.d.	n.d.	0.2	0.9	0.3	0.2	0.3
NiO	n.d.	n.d.	n.d.	n.d.	n.d.	n.d.	n.d.	n.d.
Cl	n.d.	n.d.	n.d.	n.d.	n.d.	0.2	0.2	n.d.
Total	96.0	95.2	97.0	95.5	96.2	98.2	98.8	98.0

3.2.2. Disseminated Magnetite Separates in Norite Gabbros (Central Vourinos, Krapa Hills)

The Vourinos ophiolite complex located in NW Greece constitutes a complete but tectonically disrupted ophiolite sequence. Two magmatic series have been preserved at the central part of the complex, which constitute the earlier Krapa sequence, consisting of magnetite-bearing gabbro-norite and leuco-gabbro-norite (minor magnetite-bearing) and the younger Asprokambo sequence [36]. The Fe–Ti mineralization occurs as disseminated grains in a proportion of 5–20 vol.% between coarse-grained clinopyroxene and plagioclase, while sulphides (pyrrhotite, pyrite, chalcopyrite) are present in lesser amounts. Ti–magnetite occurs as interstitial to the major phases as small disseminated euhedral and irregular grains (Figure 7). Ti-bearing magnetite separates exhibits low Si, Mg, and high Al (1.7–2.3 wt.% Al_2O_3), Ti (3.5–6.9 wt.% TiO_2), and V (1.4–2.1 wt.% V_2O_5 contents) (Table 1).

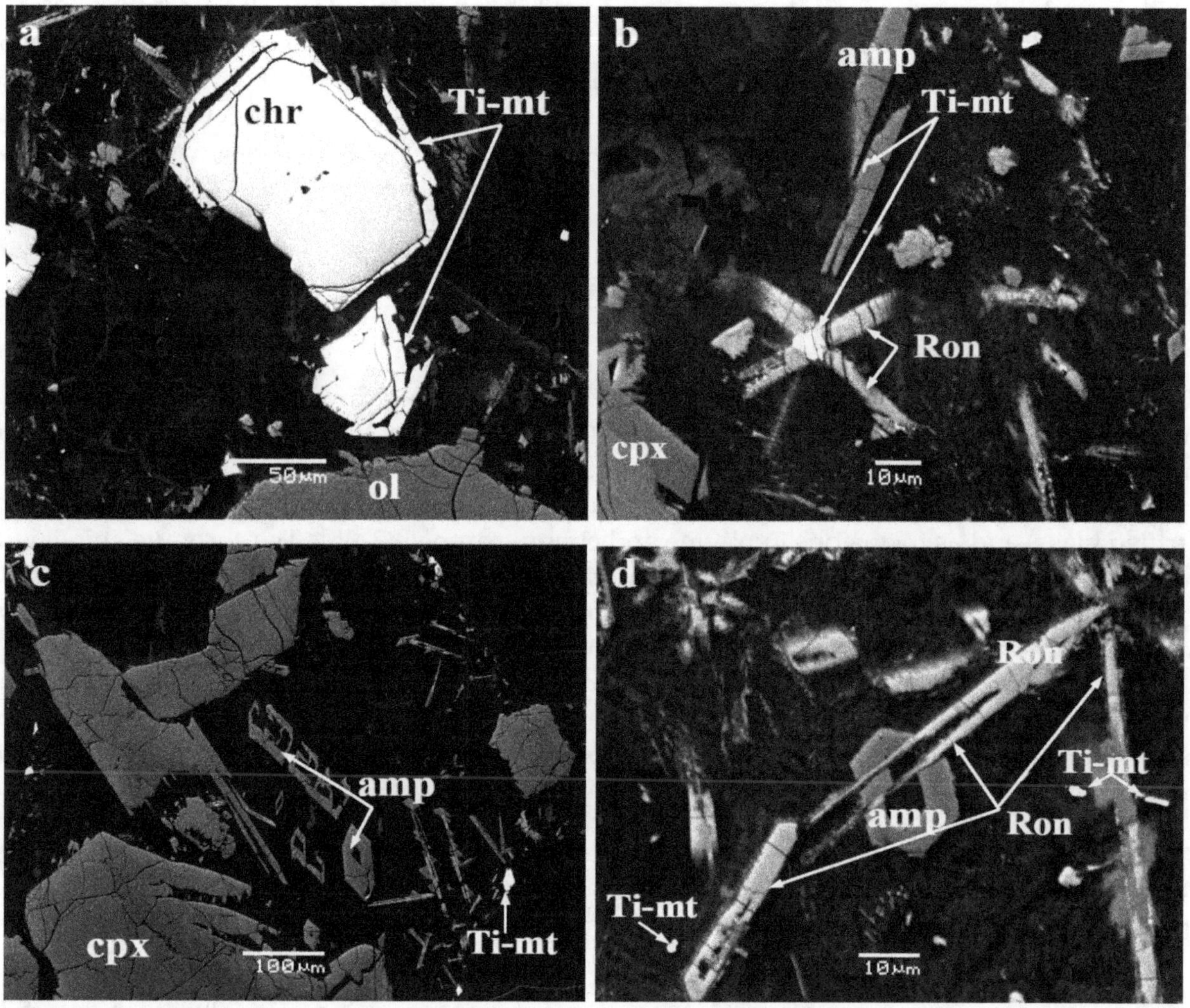

Figure 6. Back-scattered electron (BSE) images from ultramafic lavas of the Agrilia Formation, showing the association of Ti–magnetite (Ti–Mt) with chromite (chr) (**a**), microcrystals of rhönite (Ron) (**b**) associated, and randomly oriented amphibole (Amp) (**c**,**d**).

3.3. Magnetite in Coastal Black Sand

The volcanic rocks of the South Aegean arc form a chain from the Saronikos Gulf in the west, to Kos, Nisyros, and the east, through the Milos and Santorini islands [37]. Multiple eruptions of calc-alkaline to tholeiitic composition and Plio-Pleistocene age are located from the Santorini toward the Nisyros–Kos volcanic rocks (Figure 1a) [37–39]. Those volcanic rocks are part of the volcanic system related to the northward subduction of the last remnant of the oceanic crust of the African plate beneath the southern edge of the active margin of the European plate [38–41]. Magnetite and other heavy minerals have been found in all volcanic types in small quantities (1–2 modal %).

The investigated coastal black sand samples (black sea sand) collected from the Northern part of Kos Island revealed the presence of abundant Ti–magnetite with inclusions of apatite, ilmenites, Fe–Mn-oxides and monazite [(Ce, La, Th, Nd, Y) PO_4] (Figure 8). Feldspars and quartz are common too.

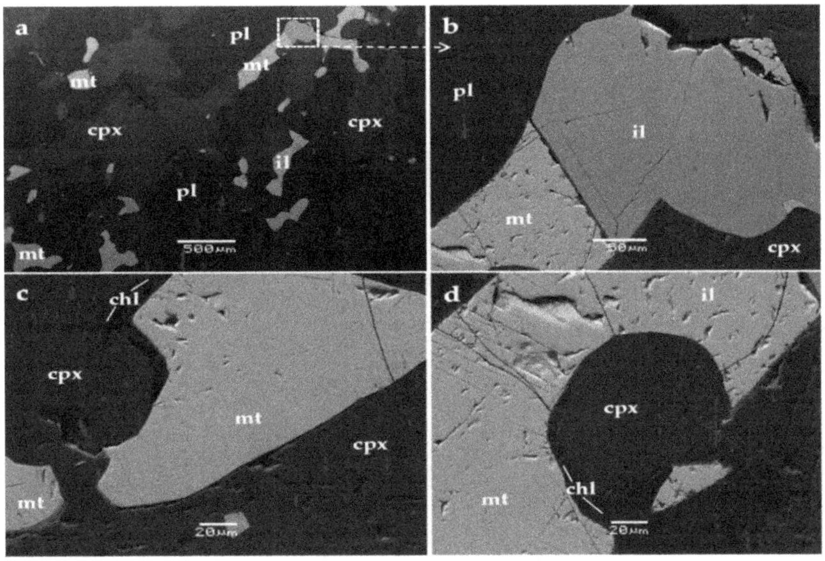

Figure 7. Back-scattered electron (BSE) images from the gabbro-norite sequence of Central Vourinos (Krapa area) showing disseminated Ti–magnetite as small euhedral and irregular phase associated with ilmenite (il) interstitially to the major phases of clinopyroxene (cpx) anf plagioclase (pl) (**a**–**d**).

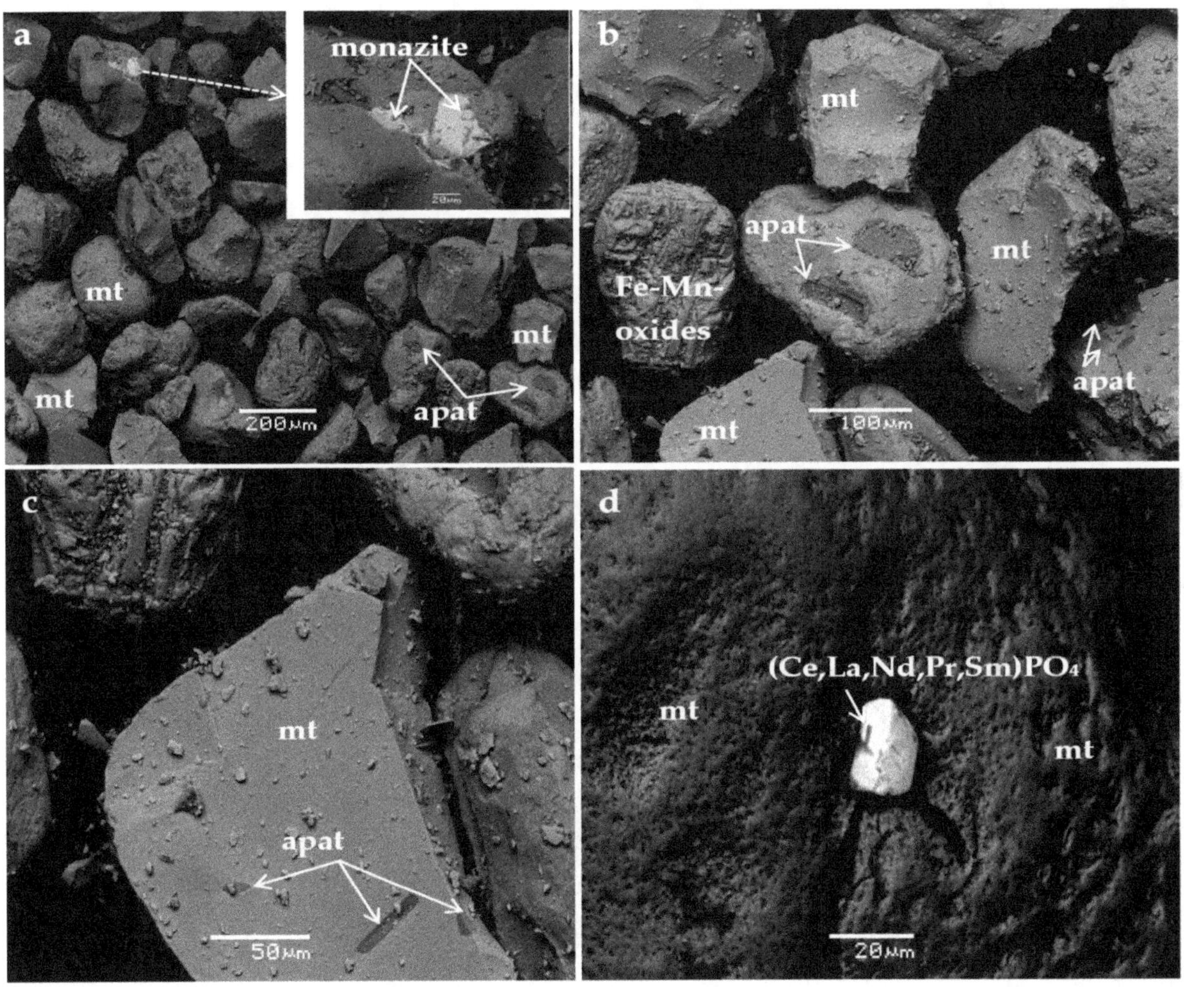

Figure 8. Back-scattered electron (BSE) images of coastal black sand collected from the northern part of Kos Island, showing that Ti–magnetite is the dominant oxide, while apatite (apat) forms inclusions of approximately 100-μm-long grains (**a**–**c**); Fe–Mn-oxides (**b**), monazite [(Ce, La, Th, Nd, Y) PO_4 and (Ce, La, Nd, Pr, Sm) PO_4] are common (**a**,**d**).

The sample from the Nisyros caldera (neighboring Kos Island) comes from whitish extinct fumaroles co-existing with native sulphur, that is dominant by opal (Figure 9). The analyzed sample of the formation from the Nisyros caldera is opal–CT and quartz [41], which it is associated with abundant native sulphur (Figure 9).

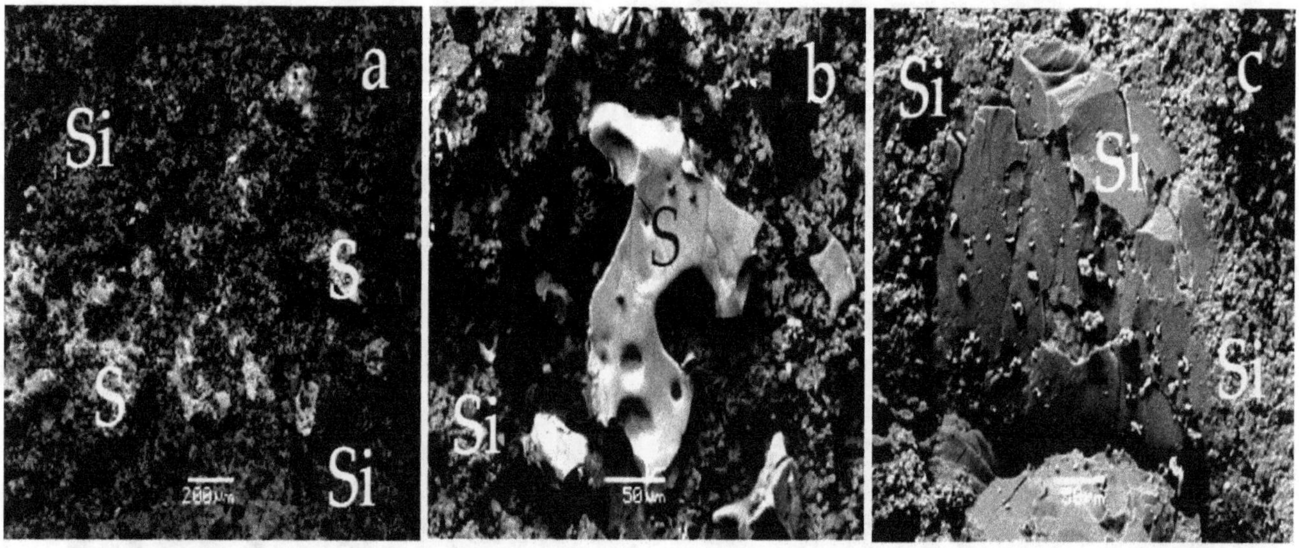

Figure 9. Back-scattered electron (BSE) images (unpolished sections) of extinct fumarole formation from the hydrothermal crater of the Nisyros caldera, showing native sulphur (**a**,**b**) and the association of native sulphur with whitish Si material (opal) (**c**), having low REE content (Table S1).

3.4. Magnetite Separates from Metamorphosed Fe–Ni-Laterites (Olympos, Edessa, and Vermion)

The dismembered ophiolite masses of the Olympos, Edessa, and Vermio mountains (Upper Jurassic–Lower Cretaceous), composed mainly by serpentinized harzburgite and, to a lesser extentm crustal magmatic rocks (pyroxenites and gabbros), outcrop along the eastern margin of the Pelagonian massif. They are considered to have been derived from the Almopias subzone of the Vardar zone and were overthrust westward onto the Pelagonian massif [22]. They are mostly composed of a mantle sequence (harzburgites) and crustal magmatic rocks, both characterized by supra-subduction zone (SSZ) features [23]. Small laterite bodies are located in the Olympos, Edessa, and E. Vermio Mountains, at the contact of serpentinized harzburgites with Upper Cretaceous limestones or within the serpentinites themselves. Due to intense tectonism, the Fe–Ni-laterite occurrences are often entirely enclosed within serpentinized harzburgites [1,25]. Apart from fine-grained magnetite separates in the matrix of the Fe–Ni ore, magnetite separates are associated with chromite, as the product of a gradual alteration of chromite or epitaxial as rims on chromite (Figure 10). A salient feature of zoned chromite grains from the Olympos, Edessa, and Vermion laterite occurrences is a Cr, Al, and Mg gradual decrease as Fe increases outward from the core; Mn, Zn, and Co increases gradually outwards, attaining greatest values at the periphery of chromite cores and in the Fe–chromite, up to 13.0, 2.1, and 4.1 wt.%, respectively, and drops off to negligible values in magnetite separates (Table 4). At the area of Vermio, there is abundant garnet (grossularite) and calcite, while Ni is mainly hosted in chlorite, serpentine, and theophrastite containing 80 wt.% NiO (Figure 10; Table 4) [42]. A (Co, Mn, Ni)-hydroxide with a wide compositional range occurs in a spatial association with theophrastite (Figure 11; Table 4) [43]. Several SEM-EDS analyses of hydroxides can be classified into two groups (Figure 11): Group A includes phase mineral dominated by cobalt, and group B those dominated by Mn, towards the peripheral parts of the concentrated development (Figure 10).

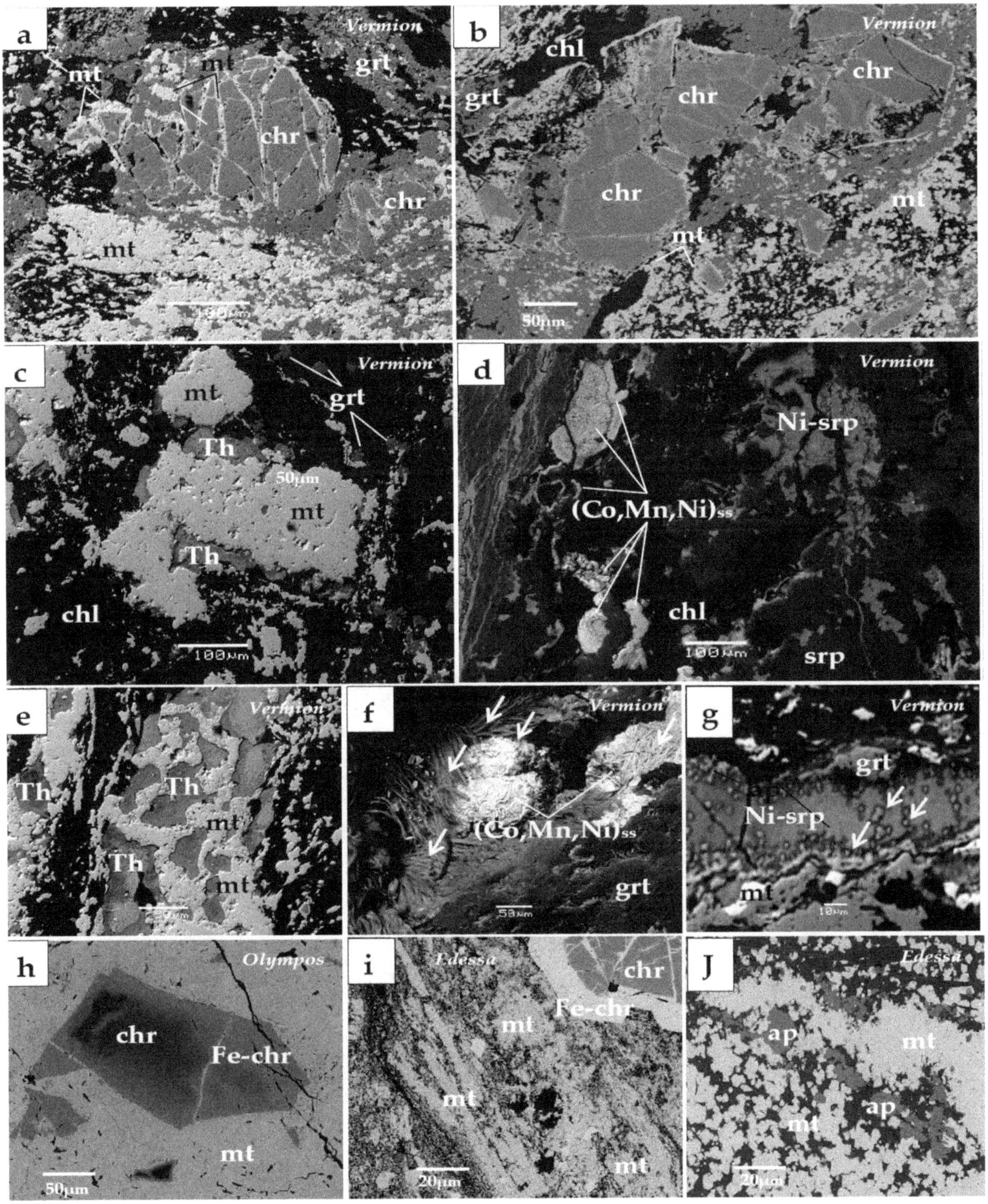

Figure 10. Back-scattered electron (BSE) images of metamorphosed Ni-laterites from northern Greece. Magnetite (mt) surrounding chromite (chr) grains and occurring along cracks (**a**,**b**,**e**,**f**); various morphological forms and textural relationship between magnetite–theophrastite (Th) and garnet (grt) (**c**,**f**,**g**); magnetite crosscutting veinlets (**e**,**i**,**j**); magnetite–(Co–Mn–Ni)$(OH)_{ss}$ showing successive thin layers, composed by fine fibrous crystals (**d**); microtextures resembling fossilized microorganisms coated by (Co–Mn–Ni)$(OH)_{ss}$ (white arrows) (**f**) or isolated and aggregates of bacterio-morphic Fe–Ni–serpentine (white arrows) within a veinlet of Ni–serpentine (Ni–srp) (**g**); chromite–Fe–chromite (**h**); chromite–magnetite (**i**); magnetite–apatite (ap) (**j**).

Table 4. Representative electron SEM/EDS analyses of chromite cores, magnetite (mt) and Fe–chromite (Fe–chr), Theophrastite and (Co, Mn, Ni)-hydroxides from Ni-laterites of the Vermion mountain (Northern Greece; Figure 10). Symbol n.d. = no detected.

		Olympos				**Edessa**			**E. Vermion**				
Location		Paliambela				Nissi		Mavrolivado		Stournari			
Mineral	Core Chromite	Fe-chr	Rim mt	Core Chromite	Fe-chr	Rim mt	Core Chromite	Fe-chr	Rim mt	Theophastit	(Co,Mn,Ni)-hydroxides		
sample	L1	L1	L1	Ed10	Ed10	EdP10	V.AK	V.AK	V.AK				
wt.%													
SiO_2	0.2	n.d.	0.4	n.d.	n.d.	0.5	0.3	0.7	1.2	n.d.	n.d.	n.d.	n.d.
Al_2O_3	8.2	8.6	0.4	9.7	6.5	0.2	24.4	7.7	0.5	n.d.	n.d.	n.d.	n.d.
Cr_2O_3	62.1	53.7	5.9	60.2	58.1	1.6	43.3	42.3	1.6	n.d.	n.d.	n.d.	n.d.
Fe_2O_3	1.8	3.5	62.2	2.7	3.3	67.7	2.2	17.6	66.5	n.d.	n.d.	n.d.	n.d.
FeO	13.9	27.6	28.4	15.5	8.3	29.3	17.3	26.6	28.9	1.0	n.d.	n.d.	n.d.
MgO	11.3	0.6	0.4	11.2	3.5	0.3	11.1	2.2	0.3	n.d.	n.d.	n.d.	n.d.
MnO	1.9	4.1	0.5	1.6	13.1	n.d.	n.d.	2.8	0.3	n.d.	25.0	13.9	19.2
ZnO	n.d	0.4	n.d.	n.d.	2.1	n.d.	0.5	0.7	0.3	n.d.	n.d.	n.d.	n.d.
CoO	n.d.	n.d.	0.4	n.d.	4.1	n.d.	n.d.	n.d.	n.d.	n.d.	38.0	54.3	46.9
TiO_2	n.d.	n.d.	n.d.	n.d.	n.d.	n.d.	n.d.	n,d.	n.d.	n.d.	n.d.	n.d.	n.d.
NiO	n.d	n.d.	1.6	n.d.	n.d.	1.4	n.d.	0.2	n.d.	79.2	17.4	11.2	13.6
Total	99.4	98.5	100.2	100.9	99.0	101	99.1	100.8	99.6	80.2	80.4	79.44	79.7
Cr/(Cr + Al)	0.83	0.81	0.81	0.78	0.86	0.87	0.54	0.79	0.94				
$Mg/(Mg + Fe^{2+})$	0.55	0.03	0.13	0.47	0.17	0.01	0.51	0.12	0.01				

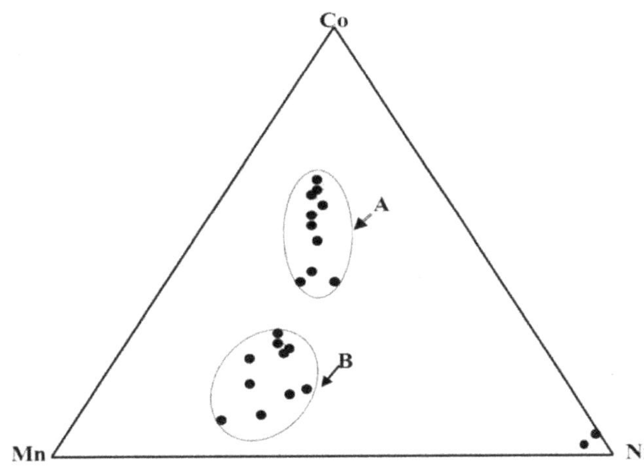

Figure 11. Triangular plot of Co, Mn, and Ni (atom%) showing the compositional variation of the (Co, Mn, Ni)$_{\text{solid solutions}}$ [43].

The (Co, Mn, Ni)$_{\text{solid-solution}}$ phases are associated with theophrastite, pyrochroite, Ni–serpentine (garnierite), garnet (grossularite), and magnetite, all having a common origin, subsequently of the strong late tectonic evolution, overprinting earlier deformation events (Figure 10c–g). They occur as successive thin (a few to tens of μm layers, composed by fine fibrous crystals up to a few μm (Figure 10d), microtextures resembling fossilized microorganisms coated by (Co–Mn–Ni) (OH)$_{ss}$ (white arrows) (Figure 10f), or isolated and aggregates of bacterio-morphic Fe–Ni–serpentine (white arrows) within a veinlet of Ni–serpentine (Ni–srp) (Figure 10g).

4. Geochemical Characteristics of Magnetite Separates

Although a good separation of magnetite from very fine inter-grown silicates and/or sulphides (Figures 2–10) was not achieved, it is clear that the highest Cr, Mn, and Co contents were recorded in Fe–Ni-laterite occurrences of northern Greece, ranging from 1.1 wt.% Cr in Vermio to 9.0 wt.% Cr in Olympos, from 0.38 wt.% Mn in Vermio to 0.69 wt.% Mn in Olympos, and from 460 ppm Co in Olympos to 1060 ppm Co in the Edessa laterites (Table S1).

The highest Ti and V contents were determined in disseminated magnetite separates from the magmatic sequence (norite gabbros) of the Vourinos (Krapa) ophiolite complex (2.56 wt.% Ti, 6330 ppm V), the Agrilia ultramafic rocks from the Othrys complex (7.48 wt.% Ti, 1760 ppm V), and the coastal black sands (or placer deposits) from Kos Island. In addition, magnetite separates from the Agrilia ultramafic rocks is richer in Al, Mg, Ti, and Cr compared to that from the Krapa norite-gabbros. Magnetite separates from the Skouries deposit shows relatively significant contents of Ni (up to 640 ppm Ni) and Cr (up to 1060 ppm Cr) (Table S1).

Hydrothermal magnetite separates from the ancient Lavrion (Plaka) mine, that is of skarn-type, and the Skouries porphyry deposits, that is disseminated, both associated with calc-alkaline rocks differ in terms of the higher Ba, Bi, Cr, Ni, Co, V, Sr, Sc, Rb, ΣREE, Y, Hf, Th, U, and Zr contents in the latter (Table S1). Gallium is higher in magnetite separates associated with cal-alkaline rocks (38–59 ppm) compared to that associated with ophiolites (2–21 ppm Ga) (Table S1). The highest Ga contents were recorded in magnetite separates from the Skouries porphyry-Cu deposit (40–59 ppm), skarn-type from Lavrion mine (38–39 ppm), and coastal sand from Kos Island (42 ppm), all related with calc-alkaline rocks. A potential order of the average Ga content in magnetite separates, from low to high is from Ni-laterites, hydrothermal (ophiolites) associated with sulphides, ultramafic lavas (ophiolites), hydrothermal (ophiolites) associated with apatite, magmatic (norite-gabbros), skarn-type (calc-alkaline), coastal sands (calc-alkaline volcanic), and porphyry-Cu deposits (hydrothermal calc-alkaline), resulting into a discrimination using a plot of (V + Ti) versus Ga content (Figure 12).

The rare earth element (REE) content of the majority of hydrothermal magnetite separates analyzed, such as those from the Lavrion mine (skarn-type), the Skouries porphyry-Cu deposit, the Pindos

(Perivoli) and Edessa (Samari) is generally low (a few decades ppm and REEs exhibit similar trends. Although the REE content, as well as actinides (Th, U) and Zr, in the porphyry (Skouries) deposit are higher compared to those in skarn-type (Lavrion), the REE patterns are similar in terms of the negative Eu, Tm, and Ho anomalies, but the former display occasionally positive Eu anomalies as well, and slight negative anomalies in Sm and Tm (Figure 13a,b). Moreover, considering the magnetite separate samples from the Skouries deposit, it seems likely that there is a small increase of the REE_{total} with the increasing depth from 43 to 363 m. In addition, the magnetite separate sample from the lowest depth (43 m) exhibits a positive Eu anomaly, in contrast to the samples from deeper parts of the deposit (Figure 13; Table S1). The hydrothermal magnetite separates associated with apatite from the Othrys (Agoriani) complex show higher REE content (54 and 80 ppm), that is, increasing with the presence of apatite, reaching values to more than 1300 ppm REE in apatite [31].

Calculations of Ce, Eu, Pr, and Gd anomalies were defined [44] by equations such as: Ce/Ce* = (2Ce/Ce chondrite)/(La/La chondrite + Pr/Pr chondrite) and Eu/Eu* = (2Eu/Eu chondrite)/(Sm/Sm chondrite + Gd/Gd chondrite). Assuming that values > 1 and < 1 are called positive and negative anomalies, respectively, chondrite normalized REE patterns for Ni-laterites from the Vermio mountains showed negative Ce, Eu anomalies, which is in agreement with the Ce/Ce* (0.56–0.24) and Eu/Eu* (0.40–0.24) values (Table S1). The highest REE contents were recorded in Ti–magnetite separates from the ultramafic lavas (130 ppm) and from the coastal sands, Kos Island (220 ppm). Sulphur occurs to be impregnated with particles of rocks, including opal–CT, quartz, alunite, anhydrite, and kaolinite (Figure 9) [41]. There is enrichment for light rare earth elements (LREE) from La to Sm and decrease for the heavier REE in magnetite separates compared to andesitic volcanic rock from Santorini Island (Figure 13f), and enrichment for all REE compared to the recent formation, associated with abundant native sulphur in Nisyros Island (Figure 9a,b). A distinct negative Eu anomaly is a common feature of the magnetite separates from the coastal sand of Kos and the andesitic rock from Santorini Island.

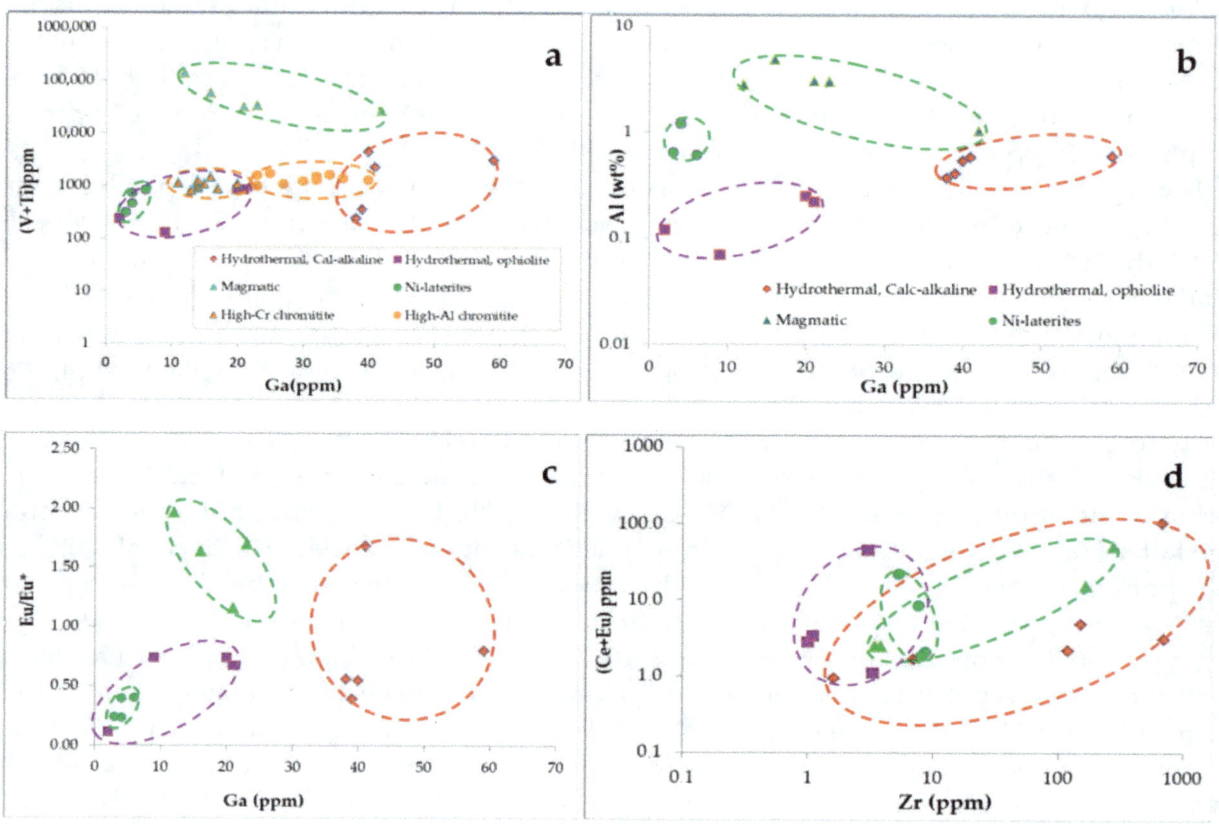

Figure 12. *Cont.*

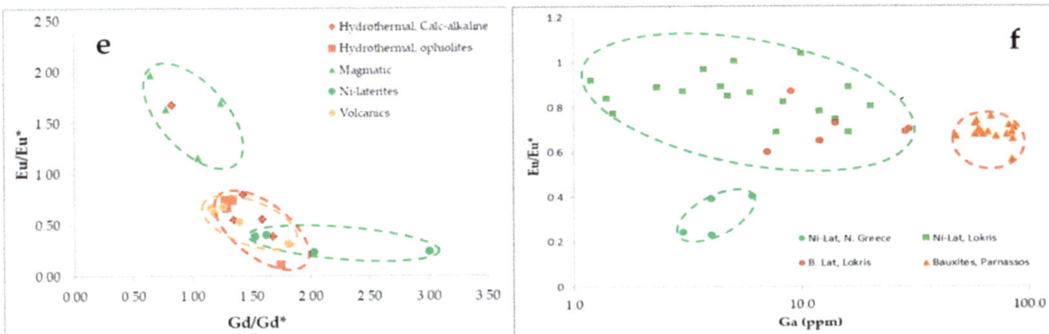

Figure 12. Plots of (V + Ti) and Al versus Ga contents (**a**,**b**), Eu/Eu* ratio versus Ga content (**c**); (Ce + Eu) versus Zr content (**d**), Eu/Eu* versus Gd/Gd* ratios (**e**), all for magnetite separates (data from Table S1) and plot of the Eu/Eu* ratio versus Ga content for Fe–Ni-laterites from Northern and Central Greece (Ni-lat), bauxite latererite (B-lat), and bauxites from Parnassos-Giona zone (**f**). Data from Table S1 and literature [45,46].

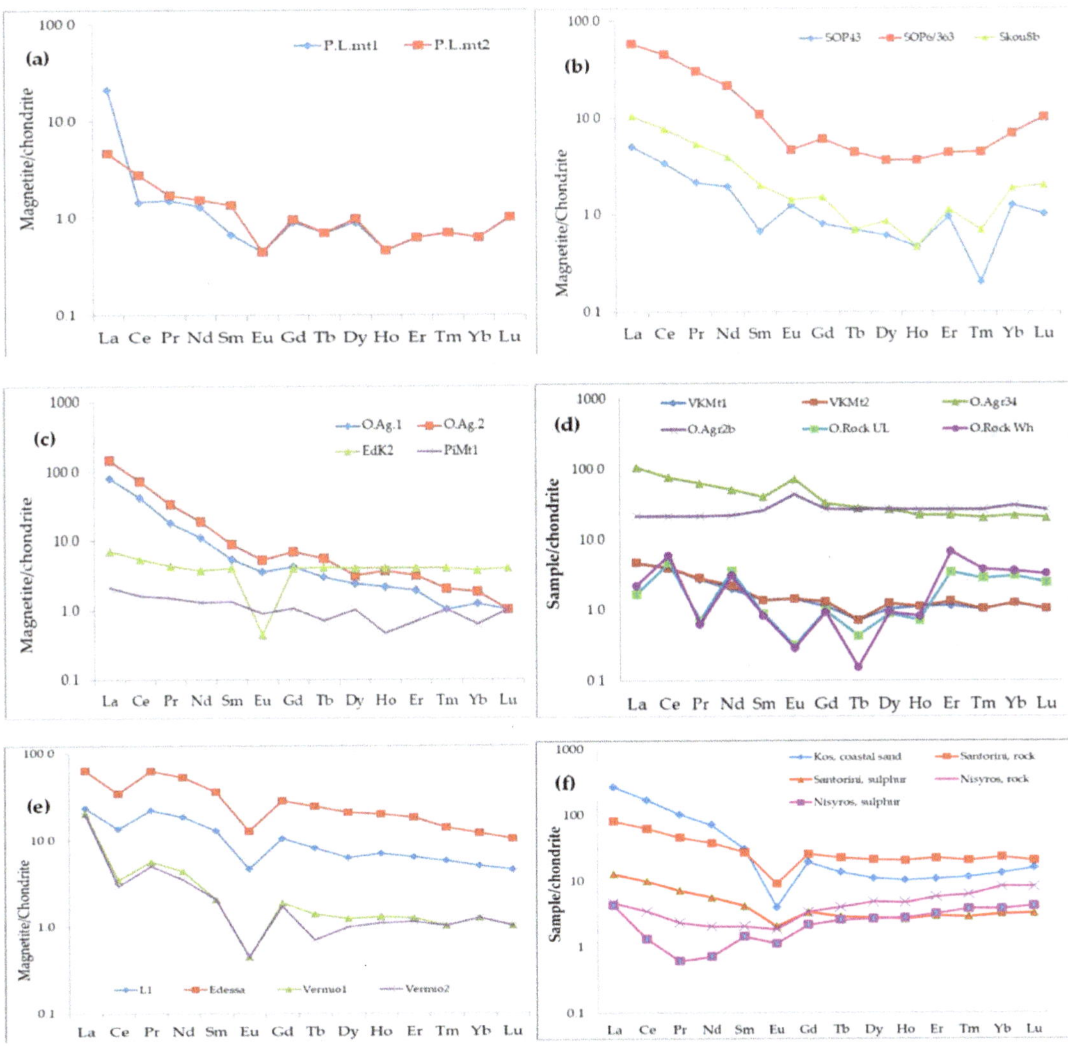

Figure 13. Chondrite normalized diagrams for magnetite separates (**a**–**f**) and volcanic rock (**d**,**f**). Data from Table S1 and [47] for ultramafic lavas (sample labeled as Rock UL, $n = 3$, and wehrlite (Wh), $n = 7$) from the Othrys ophiolite complex. Chondrite values from McDonough and Sun [48].

Moreover, the same trace elements were analyzed in two samples representing extinct fumaroles of andesitic and rhyolite composition, co-existing with native sulphur from the Aegean active volcanic arc, collected from Santorini (andesite) and Nisyros Island (Figures 9 and 13f). There is enrichment of LREE from La to Sm and decrease of the heavier REE in magnetite separates from costal sands (Kos), that is comparable to andesitic material from Santorini Island (Figure 13f). In addition, they exhibit elevated REE content compared to the extinct fumarole from Nisyros Island (Table S1; Figure 12f). Furthermore, REE patterns of both extinct fumarole material and native sulphur from Nisyros differ in terms of the low LREE content, the relative HREE enrichment compared to magnetite separates and andesitic material from Santorini and the negative Ce, Pr, and Eu anomalies (Figure 13f; Table S1).

Relatively high REE contents were recorded in Ti–magnetite separates from the ultramafic lavas from the Agrilia (99 ppm), while in wehrlite it is only 23 ppm. However, both are clearly higher compared to ultramafic lavas and wehrlite rocks (literature data), ranging from 14 to 17 ppm REE_{total}, respectively [47]. Chondrite normalized REE for magnetite separates from ultramafic lavas and wehrlite display sub-parallel patterns, with slight LREE enrichment in ultramafic lava, and an inverse relationship in terms of heavy rare earth elements (HREE) (Figure 13d). The Eu content in Ti–magnetite separates from the ultramafic lavas of Agrilia (Othrys) is enriched, showing a positive anomaly; in contrast, it is depleted in the magmatic sequence (norite gabbros) of the Vourinos (Krapa) complex (Figure 13d). In addition, the REE patterns for ultramafic lavas and wehrlites (literature data [47]) show a slight HREE enrichment for wehrlites compared to ultramafic lavas too (Figure 13d), while both rock types exhibit clear positive Ce, Nd, Gd, and Er anomalies and negative Pr, Eu, and Tb anomalies (Figure 13d).

In summary, the analyzed samples of magnetite, both massive ore and separates, covering a wide range of associated rock type, reveal wide compositional variations (Table S1 and Table 5). Although much compositional overlaps were recorded, a potential classification of magnetite separates is: (a) Magmatic: (a1) Associated with ophiolites and (a2) associated with calc-alkaline volcanic rocks (coastal sands); (b) hydrothermal: (b1) Associated with ophiolites and (b2) associated with calc-alkaline rocks (skarn-type and porphyry-Cu–Au deposits); and (c) epigenetic origin of magnetite separates in laterites (Figures 2–13). Average values show the highest ΣREE contents in disseminated magnetite separates from ultramafic lavas (Othrys) and coastal black sands (Kos Island) (Table 5).

Available data from previous publications have shown that an enrichment in Cr and Ni is a common feature of porphyry-Cu–Au–Pd–Pt, such as the Skouries, Elatsite, and other porphyry deposits worldwide, characterized by significant Pd and Pt contents, in contrast to the porphyry-Cu ± Mo deposits of Russia and Mongolia with lower Pd–Pt content (Figure 14) [4,12–14,49].

Table 5. Summary of the certain trace element contents (ICP-MS analyses) in magnetite separates and/or host ores.

MAGMATIC	Location	Rock-Type	Texture	ΣREE	Ce/Ce*	Eu/Eu*	Pr/Pr*	Gd/Gd*	Ga	V	Ti	Cr	Ni	Reference
Ophiolites	Vourinos (Krapa)	Norite gabbro	Disseminated	6	1.06	1.43	0.92	1.14	22	6520	2.6	1070	100	Present study
	Othrys	Ultramafic lava	Disseminated	132	0.9	1.97	0.98	0.65	12	1820	13.4	1600	160	Present study
Calc-alkaline	Kos	Coastal sand	Placer	224	0.92	0.32	0.85	1.81	42	1480	2.5	1150	11	Present study
HYDROTHERMAL														
Ophiolites	Edessa (Samari)	Peridotite	Massive	10	0.94	0.11	0.95	1.75	2	34	0.02	560	56	Present study
	Othrys (Agoriani)	Association with apatite	Massive	76	0.82	0.70	0.71	1.28	20	640	0.02	1100	280	Present study
	Pindos (Perivoli)	Irregular	Massive	3.1	0.90	0.74	1.02	1.34	9	80	<0.01	12	6	Present study
Calk-alkaline	Skouries	Porphyry-Cu	Disseminated	26	0.98	1.01	0.88	1.20	46	760	0.25	630	340	Present study
	Lavrion (Plaka)	Skarn	Massive	3	0.81	0.48	0.94	1.64	3.9	44	0.02	160	30	Present study
METAMORPHOSED	Olympos,Vermio, Edessa	Ni-laterites (n = 4)	Major, matrix	37	0.41	0.32	1.45	0.93	4.2	280	380	33,980	9800	Present study
Fe-Ni-laterite	Lokris	Ni-laterites (n = 13)	Rare, matrix	120	1.02	0.85	0.96	1.14	7.6	75	2900	1900	2200	[46]
Bauxite laterite	Lokris	Bauxitic laterites (n = 6)	Rare, matrix	350	1.01	0.71	0.97	1.2	17	340	7200	3500	1400	[46]
Bauxite	Parnassos-Giona	Bauxites (n = 17)	Rare, matrix	470	2.64	0.69	0.58	0.8	70	460	16,000	820	150	[45]

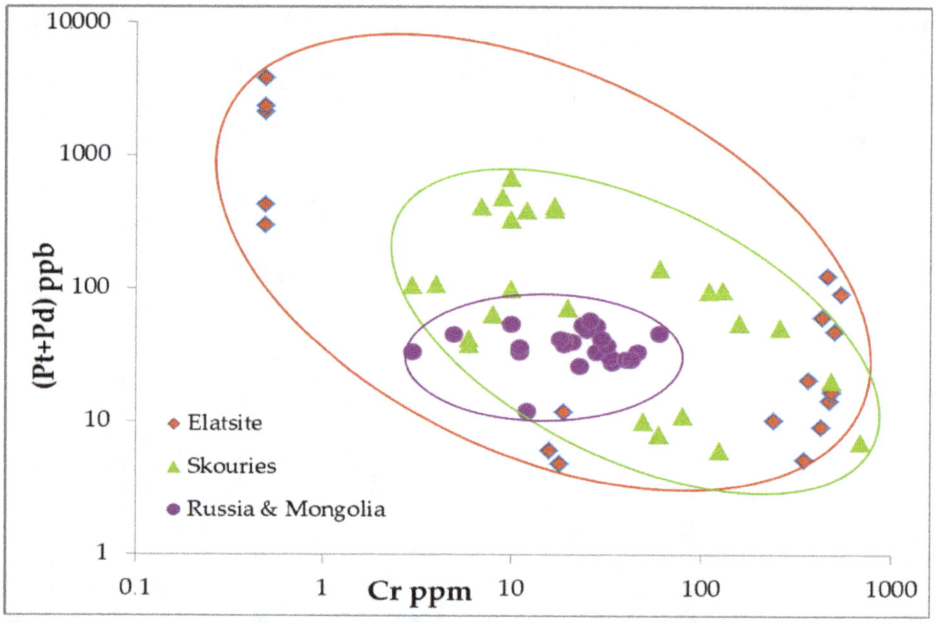

Figure 14. Plot of the (Pt + Pd) versus Cr content in porphyry-Cu–Au–Pd–Pt and porphyry-Cu–Mo deposits. Data from literature [12,14,49].

5. Discussion

The trace element content in magnetite separates, coupled with mineral chemistry and texture relationships, may provide evidence on (1) their parent magma or hydrothermal ore forming system, (2) partition coefficient of trace elements into magnetite separates, and (3) physico/chemical conditions (temperature, fO_2) during deposition and/or re-deposition of magnetite separates, and can be used as a prospective tool for many types of ore deposits [5–10,50]. The recorded wide variations in the trace element contents of massive and concentrates of disseminated magnetite separates from Greece may be used to discriminate between magmatic and hydrothermal magnetite separates, associated either with calc-alkaline rocks and ophiolites, between porphyry-Cu + platinum-group elements (PGEs) and porphyry-Cu lacking precious metals deposits, and between laterite types.

5.1. Trace Element Distribution in Magnetite Structure and Associated Minerals

The geochemistry of magnetite separates is affected by a variety of cations accommodated in the spinel-type structure [3,14,15] and the co-precipitation of other minerals [5–10]. The high-resolution studies of minerals applying TEM techniques provide a new and very important way to elucidate valuable information on the physico/chemical processes and mechanisms during formation of ore deposits and post-depositional changes of ore minerals [51]. The investigation of hydrothermal magnetite separates from the Los Colorados iron oxide–apatite deposit in Chile showed three generations of magnetite, which vary significantly in terms of their Si, Al, Ca, Mg, Ti, Mn, Na, and K minor and trace element contents, in contrast to the V content (remarkably constant), and concluded a complex incorporation of trace elements during growth of magnetite [51]. Although, there is a discrepancy between experimentally determined values and those derived from natural occurrences, the partition coefficients for spinel–liquid pairs ($D^{sp/lq}$), for high field strength elements, such as Ga, V, Zr, Zn, and Co, are extremely low [50,52]. Considering the presented data (Table S1), potential components of sulphide (Cu, Zn) and/or silicate (Si, Mg, K) inclusions in magnetite separates are used with caution. Selected elements, such as V, Ti, and Ga, in individual types of magnetite separates are plotted, showing different fields for magmatic and hydrothermal magnetite, and furthermore, between hydrothermal magnetite associated with ophiolites and calc-alkaline rocks (Figure 12).

Hydrothermal apatite from the Othrys (Agoriani) ophiolite complex contain over 1300 ppm ΣREE [31], whereas the associated magnetite separates contain less than 100 ppm ΣREE, due probably to incomplete separation of apatite (low P and Ca contents) (Table S1) [31]. Similarly, magnetite separates from the coastal black sand from Kos, containing inclusions of apatite (Figure 8b,c), exhibit relatively high (over 220 ppm ΣREE) (Table S1) and monazite of varying composition [(Ce, La, Th, Nd, Y) PO_4 and (Ce, La, Nd, Pr, Sm)PO_4] (Figure 8a,d). Magnetite separates from black coastal sands can be linked back to a volcanic source, and may be a potential REE source [41]. On a global scale, in many placer deposits, such as the river placers in Malaysia, paleo-placers in South Africa, Elliot Lake, Ontario, and beach sand deposits in India, monazite is an important REE mineral and many monazite ore deposits are economically important REE resources, although small contents of Th and U make monazite a restricted mineral in India [53]. Also, hydroxyl–bastnaesites and a (La, Nd, Y) (CO_3) F member of the bastnaesite group, which are among the most economically important as well, have been identified at the lowest part of karst-type laterites in Greece, close to their contact with carbonates [54–57].

The first data on magnetite separates from black coastal sand from Kos Island are compared with potential parental rocks of the neighboring active Nisyros Island and Santorini, all belonging to the Aegean volcanic arc, showing REE, Ti, V, Cr, Mn, Zn, and Ga contents in magnetite separates are higher than those in the volcanic rocks of basaltic composition from Santorini Island [50] and Nisyros Island (Table S1). The similarity between the REE patterns, including the clear Eu negative anomaly of magnetite separates from coastal black sand of Kos Island and that of andesitic material from Santorini Island (Figure 13f), may provide evidence of an andesitic source of the placer magnetite separates in Kos. Also, the composition of native sulfur has been used to qualitatively estimate elemental abundances in low-temperature fumarole gases. At low temperatures (<250 °C), oxygen in the atmosphere interacts with H_2S, resulting to precipitation of native sulphur [58].

5.2. Re-Mobilization and Re-Deposition of Magnetite

Magnetite separates as a late magmatic or early hydrothermal phase in both calc-alkaline systems and ophiolite complexes may be dissolved and re-precipitated, due to changes in temperature, salinity, pressure, and fO_2 [51,59–61]. An increase in temperature would enhance the solubility and, consequently, the undersaturation of iron in the fluids [61], leading to dissolution of the primary magnetite separates and formation of the secondary varieties. Thus, primary magnetite may have been re-mobilized during late brittle deformation and motion along the detachments contacts [60]. In general, textural and compositional data for magnetite from nine iron skarn deposits in Canada, Romania, and China have shown that most samples have been re-equilibrated by dissolution and re-precipitation and/or recrystallization [16,51]. These authors concluded that the applicability and reliability of trace element discrimination diagrams may be used with caution where magnetite has been subjected to multiple re-equilibration processes, and that detailed textural characterization must be undertaken in order to evaluate the effects of hydrothermal overprinting events and other re-equilibration processes on the trace element compositions of magnetite.

A peculiarity of the above Fe–Ni-laterite occurrences is the relatively high Cr, Mn, Co, and Ni content in Fe–chromite zones of spinels. The epigenetic mobilization of Fe, Mn, Co, and Ni and their re-deposition is suggested by a significant Mn, Zn, and Co enrichment in Fe–chromite (Table 1) and the abundance of Ni–serpentine (Figure 10; Table 4). The substitution for Mg^{2+} and Fe^{2+} by Mn, Zn, and Co in the chromite lattice resulted to a gradual transformation of chromite to Fe–chromite, while in the magnetite, that is considered to be the final product of this alteration process [1,62] and the above elements drop to negligible amounts (Table 4). Calculations for spinel structures [14,15,63,64] have shown that in normal spinels, there is a tendency for the electrostatic energy to favor the larger ions ($Fe^{2+} > Mg^{2+}$ and $Fe^{3+} > Al^{3+}$) into the tetrahedral site. Thus, the resulting small expansion of the tetrahedral and octahedral sites hampers the entry of other trace elements in the magnetite separate lattice at low temperatures [65,66]. Besides, the re-mobilization of Fe and re-deposition of magnetite at low temperature is indicated by its association with (Ni, Fe, Co, Mn)-hydroxides. Specifically, the texture relationships between magnetite, theophastite [$Ni(OH)_2$], and (Co, Fe, Mn, Ni) $(OH)_2$ (Figure 10a–h) suggest a common mineral forming system, along shear zones, subsequently of the strong late tectonic evolution, which overprints earlier deformation events, probably at low temperature, approximately 100 °C [67] and alkaline conditions, as it is suggested by the presence of calcite and the high Ca content in the magnetite separates concentrates (Table S1). Moreover, the association of the (Co, Mn, Ni) $(OH)_2$ solid-solution phases with microtextures resembling fossilized microorganisms coated by (Co–Mn–Ni) $(OH)_2$ (Figure 10f, white arrows) and the occurrence of isolated and aggregates of bacterio-morphic Fe–Ni–serpentine (Figure 10g) suggest the potential effect of microorganisms in the Vermio mineral-forming processes. Currently, much research interest has focused on a consortium of microorganisms of varying morphological forms, producing enzymes, which are considered to be a powerful factor to catalyze redox reactions, and act as nucleation sites for the precipitation of secondary minerals [68–70]. In general, the presence of organic matter and microorganism traces in highly tectonized zones of the metamorphosed Fe–Ni-laterites may suggest direct involvement of microorganisms, and have created appropriate conditions for metal bio-leaching and bio-mineralization. In general karst-type Fe–Ni-laterites, where dominant Fe minerals are goethite ± hematite [54,55,71], may display REE contents ranging from a few decades to thousands of more than 5000 ppm (Table S1) [54], depending on the site of the Ni-laterite samples along vertical profiles and the post-deposition processes affected their composition. The epigenetic mobilization of REE, along with Fe, Mn, Ni, Co, under reducing and acidic conditions and, subsequently, re-deposition under alkaline conditions at the lowest parts of contacts between laterites of Central Greece (Lokris) and carbonate rocks has been well established as well [54–56,71]. Thus, the potential depletion of Ga and Al, and elevated Mn, Co, Zn, and Fe ± REE contents has been recorded in the above large Fe–Ni-laterite deposits, but in small Fe–Ni-laterite occurrences in Northern Greece, strongly transformed during post-magmatic metamorphism, the dominant Fe mineral is depleted magnetite, due probably to their small size, facilitating potential changes during sub-solidus reactions. Therefore, the dominance of magnetite in Fe–Ni-laterite ores is not an encouraging factor for the exploration of Ni sources of laterite-type.

5.3. *Implications of the Trace Element Distribution for the Origin of Magnetite Mineralization*

Despite the potential re-distribution of trace elements in magnetite separate-bearing deposits, disseminated magnetite separates from mafic–ultramafic magmatic rocks of ophiolite complexes exhibit much higher Ti and V contents compared to hydrothermal magnetite, and they are comparable to magnetite separates from coastal sands derived from the weathering of calc-alkaline volcanic rocks (Table 5). The massive hydrothermal magnetite separates of skarn-type differs from the disseminated magnetite separates from porphyry deposits, in terms of the lower Ba, Bi, Cr, Ni, Co, V, Sr, Sc, Rb, ΣREE, Y, Hf, Th, U, and Zr contents (Table S1). Also, Ga is higher in magnetite separates associated with calc-alkaline rocks (38–59 ppm) compared to that related to ophiolites (2–21 ppm Ga) (Table S1). The highest Ga contents were recorded in magnetite separates from the Skouries porphyry-Cu deposit (40–59 ppm), skarn-type from Lavrion mine (38–39 ppm), and coastal sand from Kos Island (42 ppm), all related with calc-alkaline rocks. A potential order of the average Ga content in magnetite separates concentrates, from low to high, is from Ni-laterites, hydrothermal (ophiolites) associated with sulphides, ultramafic lavas (ophiolites), hydrothermal (ophiolites) associated with apatite, magmatic (norite gabbros), skarn-type (calc-alkaline), coastal sands (calc-alkaline volcanic), and porphyry-Cu deposits (hydrothermal calc-alkaline), resulting in a discrimination using a plot of (V + Ti) versus Ga content (Figure 12).

Magnetite separates associated with Fe–Cu–Ni–Co sulphides with dominant minerals pyrrhotite, chalcopyrite, and pentlandite exhibit relatively high Cu and Zn, due to incomplete separation from the associated sulphides, but it is REE-, Ti-, and V-depleted. Although an originally primary magmatic origin for these sulphides is not precluded, characteristics of the highly transformed pure magnetite separates (Table 1 and Table S1) at Pindos may indicate that potential original magmatic features have been overprinted by a low-level hydrothermal circulation process.

The plot of the Eu/Eu* ratio versus Ga content shows a less negative anomaly Eu/Eu* (average 0.85) for large Fe–Ni-laterite deposits [56,72,73], bauxite laterites and bauxites (63 and 0.69, respectively) [74], and the most negative Eu/Eu* anomaly (average 0.32), for the magnetite separates from Fe–Ni-laterites of Northern Greece (Figure 12e; Table 5). Assuming that the presence of Eu anomaly, caused by the ability of Eu to exist in either Eu^{2+} or Eu^{3+} states, the Eu/Eu* anomalies seem to be effected by physico/chemical conditions, such as a redox potential [74]. Diagenetic re-mobilization of Eu is possible under conditions of reduction to Eu^{2+} at low oxidation potential, as is suggested by the presence of organic matter [73–76]. In addition, the increasing order of the average Ga content from the Fe–Ni-laterites to bauxite laterites and bauxites (Table 5; Figure 12e) is in agreement with the preference of Ga to high-Al ores. The average Ga content for karst bauxite deposits is 58 ppm with a range from less than 10 to 812 ppm, because the close geochemical affinity of Ga to Al enables Ga to substitute easily in rock-forming alumino-silicates [77]. Such a preference Ga with high-Al chromitites, as well and the structural incorporation of Ga into chromite lattice, is evidenced by the progressive and linear increase of Al or decreasing Cr/(Cr + Al) atomic ratio [78,79].

5.4. Implications of Magnetite Separates for Exploration of Precious Metals in Porphyry-Cu Deposits

Alkaline or K-rich calc-alkaline porphyry deposits worldwide represent significant gold resources, owing to their large sizes. They are derived from hydrous calc-alkaline magmas and are genetically linked to magmatic and hydrothermal processes. Despite their common origin in a supra-subduction environment, platinum-group elements (PGEs) have been recorded only in certain porphyry-Cu deposits. During the last decades, significant Pd and Pt contents were described in the Cordillera of British Columbia, in the Balkan–Karpathian system, including the Skouries deposit, in the Philippines and elsewhere [4,12,13,80,81]. The estimated Pd, Pt, and Au potential for porphyry deposits, combined with the association of merenskyite [$(Pd, Pt)(Te, Bi)_2$)] with major Cu minerals (bornite and chalcopyrite), are considered to be encouraging economic factors for Pd and Pt, as by-products, the Au being the main precious metal product [13,80,81].

It has been suggested that first-stage arc magmas tend to generate Cu-rich, relatively Au-poor porphyry systems, while relatively Au-rich residue is left in the lower crust and lithospheric mantle after first-stage arc magmas [82]. Subsequently, a second stage of melting, due probably to lithospheric thickening, thermal rebound, mantle lithosphere delamination, or lithospheric extension, may generate magmas with relatively high Au/Cu ratios [82–84] and PGE enrichments [84–87]. Critical requirements for a significant base/precious metal potential in porphyry Cu + Au + Pd ± Pt deposits are considered to be the geotectonic environment, controlling the precious/base metal endowment in the parent magma (metasomatized asthenospheric mantle wedge) [86,87]. It is indicated by significant Cr, Co, Ni, and Re contents, the oxidized nature of parent magmas, that facilitate the capacity for transporting sufficient Au and PGE, and the degree of evolution of the mineralized system, as is exemplified by the enrichment of (Pd + Pt) versus (Os + Ir + Ru) and LREE versus HREE [4,86,87]. Arc magmas with high potential to produce hydrothermal systems with ideal chemistry for transporting precious metals (Cu + Au + Pd ± Pt deposits) are characterized by fO_2 more than two log units above fayalite-magnetite-quartz (FMQ), where fO_2 is oxygen fugacity and FMQ is the fayalite–magnetite–quartz oxygen buffer [88]. Since the oxidized nature of parent magma is connected with the ability to produce a magmatic–hydrothermal system with ideal chemistry that facilitates the capacity for transporting sufficient precious metals, abundant magnetite (reaching up to 10 vol.% in the Skouries deposit) linked to pervasive potassic and propylitic alteration type [13], in contrast to "reduced" porphyry Cu–Au deposits, lacking primary hematite, magnetite separates, and sulphate minerals [89], is considered to be a characteristic feature of the Pd-bearing porphyry-Cu deposits. Thus, the abundance of magnetite separates is a characteristic feature of all porphyry-Cu–Au–Pd deposits, as indicated by unexpected high-grade Cu–(Pd + Pt) (up to 6 ppm) mineralization at the Elatsite porphyry deposit, which is found in a spatial association with the Chelopech epithermal deposit (Bulgaria) and the Skouries porphyry deposit, Greece [4,12]. Disseminated hydrothermal magnetite separate is often associated with fine (less than 10 μm) crystals of U-thorite and zircon within veinlets of quartz (Figure 3c), which is consistent with the elevated Th, U, and Zr contents (Table S1).

The enrichment in Cr, Co ± Ni is a common feature, recorded in the Skouries, Elatsite, and other porphyry deposits of the Balkan Peninsula, characterized by significant Pd and Pt contents, in contrast to the porphyry-Cu ± Mo deposits of Russia and Mongolia with lower (Pd + Pt) content (less than 25 ppm Cr [49,90]) supports the origin of their parent magma from an enriched mantle source [91,92]. In addition, a salient feature is the evolved geochemical signature of the Skouries and Elatsite porphyry-Cu deposits, as indicated by the negative trend between (Pd + Pt) and Cr contents (Figure 14). The application of stable isotope analyses of oxygen ($\delta^{18}O$ = 4.33–9.45‰) and hydrogen (δD = −110‰ to −73‰) on quartz veins from various drill holes and depths may provide evidence to constrain the origin of fluids trapped in quartz veins [85]. Although any systematic variation of the isotopic data with depth is uncertain, it seems likely that the samples with higher Cr contents correspond to lower $\delta^{18}O$ values, while those with lower Cr contents correspond to higher $\delta^{18}O$ values. This isotope variation, coupled with the higher (Pd + Pt) content Pd/Pt ratios corresponding to higher $\delta^{18}O$ values and lower Cr contents, may point toward the more evolved mineralizing system (Figure 14) [86].

6. Conclusions

Texture, mineral chemistry and geochemical data on massive and disseminated magnetite separates from a wide range of geotectonic settings and rock types led us to the following conclusions:

- Magnetite separates of hydrothermal origin associated with calc-alkaline rocks can be distinguished from that associated with ophiolites in terms of their higher Ga content, while magnetite separates of magmatic origin are characterized by the highest (V + Ti) content.

- Hydrothermal magnetite separates from skarn-type (Lavrion mine) differs compared to the disseminated separates from porphyry-Cu deposit (Skouries) in terms of the higher Ba, Bi, Cr, Ni, Co, V, Sr, Sc, Rb, ΣREE, Y, Hf, Th, U, and Zr contents in the latter.

- Although an initial primary magmatic origin for the Fe–Cu–Ni–Co sulphides in the Pindos ophiolite is not precluded, geochemical characteristics (REE-, Ti-, and V-depleted) of the highly transformed magnetite separates may indicate that potential original magmatic features have been overprinted by low-level hydrothermal circulation processes.

- Similarity between the REE patterns, including the clear Eu negative anomaly of magnetite separates from coastal black sand of Kos Island, may provide evidence of an andesitic source.

- A major requirement controlling the Pd and Pt potential of porphyry-Cu deposits is probably the oxidized nature of parent magmas that facilitate the capacity for transporting sufficient Pd and Pt, and the crystallization of abundant magnetite.

- Abundance of magnetite in porphyry-type deposits may be used to discriminate between porphyry-Cu–Au–Pd–Pt and porphyry-Cu–Mo deposits lacking precious metals, providing evidence for the exploration of precious metals in porphyry systems.

Author Contributions: D.G.E. collected the samples, provided the field information, and performed all types of analyses. Both authors (D.G.E. and M.E.-E.) contributed to the elaboration and interpretation of the data. The writing of the paper was completed by M.E.-E.

Acknowledgments: Many thanks are expressed to the two anonymous reviewers for their constructive criticism and suggestions on an earlier draft of the manuscript. Vassilis Skounakis, University of Athens, is thanked for his assistance with the SEM/EDS analyses.

References

1. Paraskevopoulos, G.M.; Economou, M.I. Zoned Mn-rich chromite from podiform type chromite ore in serpentinites of northern Greece. *Am. Miner.* **1981**, *66*, 1013–1019.
2. Economou, M.; Skounakis, S.; Papathanasiou, C. Magnetite deposits of skarn type from the Plaka area of Laurium. *Greece* **1981**, *40*, 241–252.
3. Lindsley, D.H. Experimental studies of oxide minerals: Reviews Mineral. *Geochemistry* **1991**, *25*, 69–106.
4. Eliopoulos, D.G.; Economou-Eliopoulos, M. Platinum-group element and gold contents in the Skouries porphyry-copper deposit, Chalkidiki Peninsula, northern Greece. *Econ. Geol.* **1991**, *86*, 740–749. [CrossRef]
5. Dupuis, C.; Beaudoin, G. Discriminant diagrams for iron oxide trace element fingerprinting of mineral deposit types. *Miner. Depos.* **2011**, *46*, 319–335. [CrossRef]
6. Dare, S.A.S.; Barnes, S.-J.; Beaudoin, G. Variation in trace element content of magnetite crystallized from a fractionating sulfide liquid, Sudbury, Canada: Implications for provenance discrimination. *Geochim. Cosmoch. Acta* **2012**, *88*, 27–50. [CrossRef]
7. Nadoll, P.; Mauk, J.L.; Hayes, T.S.; Koenig, A.E.; Box, S.E. Geochemistry of magnetite from hydrothermal ore deposits and host rocks of the Mesoproterozoic Belt Supergroup, United States. *Econ. Geol.* **2012**, *107*, 1275–1292. [CrossRef]
8. Nadoll, P.; Angerer, T.; Mauk, J.L.; French, D.; Walshe, J. The chemistry of hydrothermal magnetite: A review. *Ore. Geol. Rev.* **2014**, *61*, 1–32. [CrossRef]
9. Dare, S.A.; Barnes, S.-J.; Beaudoin, G.; Mıric, J.; Boutroy, E.; Potvin-Doucet, C. Trace elements in magnetite as petrogenetic indicators. *Miner. Depos.* **2012**, *49*, 785–796. [CrossRef]
10. Rapp, J.F.; Klemme, S.; Butler, I.B.; Harley, S.L. Extremely high solubility of rutile in chloride and fluoride-bearing metamorphic fluids: An experimental investigation. *Geology* **2010**, *38*, 323–326. [CrossRef]
11. Economou-Eliopoulos, M. Apatite and Mn, Zn, Co-enriched chromite in Ni-laterites of northern Greece and their genetic significance. *J. Geochem. Exp.* **2003**, *80*, 41–54. [CrossRef]
12. Augé, T.; Petrunov, R.; Bailly, L. On the mineralization of the PGE mineralization in the Elastite porphyry Cu–Au deposit, Bulgaria: Comparison with the Baula-Nuasahi Complex, India, and other alkaline PGE-rich porphyries. *Can. Miner.* **2005**, *43*, 1355–1372. [CrossRef]
13. Economou-Eliopoulos, M. Platinum-Group Element Potential of Porphyry Deposits. In *Mineralogical Association of Canada Short Course 35*; Mineralogical Association of Canada: Quebec, QC, Canada, 2005; pp. 203–245.
14. Navrotsky, A.; Kleppa, O.J. The thermodynamics of cation distributions in simple spinels. *J. Inorg. Nucl. Chem.* **1967**, *29*, 2701–2714. [CrossRef]
15. O'Neill, H.C.; Navrotsky, A. Simple spinels: Crystallographic parameters, cation radii, lattice energies, and cation distribution. *Am. Miner.* **1983**, *68*, 181–194.
16. Hu, H.; Lentz, D.; Li, J.W.; McCarron, T.; Zhao, X.F.; Hall, D. Reequilibration processes in magnetite from iron skarn deposits. *Econ. Geol.* **2015**, *110*, 1–8. [CrossRef]
17. Marinos, G.P.; Petrascheck, W.E. Laurium. *Geol. Geophys. Res.* **1956**, *4*, 1–247.
18. Skarpelis, N.; Tsikouris, B.; Pe-Piper, G. The Miocene igneous rocks in the Basal Unit of Lavrion (SE Attica, Greece): Petrology and geodynamic implications. *Geol. Mag.* **2007**, *145*, 1–15. [CrossRef]
19. Voudouris, P.; Melfos, V.; Spry, P.; Bonsall, T.; Tarkian, M.; Economou-Eliopoulos, M. Mineralogical and fluid inclusion constraints on the evolution of the Plaka intrusion-related ore system, Lavrion, Greece. *Miner. Petrol.* **2008**, *93*, 79–110. [CrossRef]
20. Ducoux, M.; Branquet, Y.; Jolivet, L.; Arbaret, L.; Grasemann, B.; Rabillard, A.; Gumiaux, C.; Drufin, S. Synkinematic skarns and fluid drainage along detachments: The West Cycladic Detachment System on Serifos Island (Cyclades, Greece) and its related mineralization. *Tectonophysics* **2017**, *695*, 1–26. [CrossRef]
21. Kockel, F.; Mollat, H.; Gundlach, H. Hydrothermally altered and (copper) mineralizedp orphyritic intrusionsin the ServomacedoniaMn assif s (Greece). *Miner. Depos.* **1975**, *10*, 195–204. [CrossRef]

22. Pearce, J.A.; Lippard, S.J.; Roberts, S. Characteristics and tectonic significance of supra-subduction zone ophiolites. *Geol. Soc. Lond. Spec. Pub.* **1984**, *16*, 77–94. [CrossRef]
23. Saccani, E.; Photiades, A.; Santato, A.; Zeda, O. New evidence for supra-subduction zone ophiolites in the Vardar Zone from the Vermion Massif (northern Greece): Implication for the tectono-magmatic evolution of the Vardar oceanic basin. *Ofioliti* **2008**, *33*, 17–37.
24. Economou-Eliopoulos, M.; Eliopoulos, D.; Chryssoulis, S. A comparison of high-Au massive sulfide ores hosted in ophiolite complexes of the Balkan Peninsula with modern analogues: Genetic significance. *Ore Geol. Rev.* **2008**, *33*, 81–100. [CrossRef]
25. Paraskevopoulos, G.M.; Economou, M.I. Genesis of magnetite ore occurrences by metasomatism of chromite ores in Greece. *Neues Jahrb. Fur Miner. Abh.* **1980**, *140*, 29–53.
26. Mercier, J. Édute géologique des zones internes des Hellénides en Macédoine centrale (Grèce). Ire Thèse (1966). *Ann. Géol. Des. Pays Hell.* **1968**, *20*, 596.
27. Sideris, C.; Skounakis, S.; Economou, M. The ophiolite complex of Edessa area and the associated mineralization. International Symposium on metallogeny of mafic and ultramafic complexes. *I.G.C.P. 169 Athens* **1980**, *2*, 142–153.
28. Hynes, A.J. The Geology of Part of the Western Othrys Mountains, Greece. Ph.D. Thesis, University of Cambridge, Cambridge, UK, 1972.
29. Rassios, A.; Dilek, Y. Rotational deformation in the Jurassic Mesohellenic Ophiolites, Greece, and its tectonic significance. *Lithos* **2009**, *108*, 207–223. [CrossRef]
30. Rassios, A.; Konstantopoulou, G. Emplacement tectonism and the position of chrome ores in the Mega Isoma peridotites, SW Othris, Greece. *Bull. Geol. Soc. Greece* **1993**, *28*, 463–474.
31. Mitsis, I.; Economou-Eliopoulos, M. Occurrence of apatite with magnetite in an ophiolite complex (Othrys), Greece. *Am. Miner.* **2001**, *86*, 1143–1150. [CrossRef]
32. Paraskevopoulos, G.; Economou, M. Komatiite-type ultramafic lavas from the Agrilia Formation, Othrys ophiolite complex, Greece. *Ofioliti* **1986**, *11*, 293–304.
33. Economou-Eliopoulos, M.; Paraskevopoulos, G. Platinum -group elements and gold in komatiitic rocks from the Agrilia Formation, Othrys ophiolite complex, Greece. *Chem. Geol.* **1989**, *77*, 149–158. [CrossRef]
34. Koutsovitis, P.; Magganas, A.; Ntaflos, T. Rift and intra-oceanic subduction signatures in the Western Tethys during the Triassic: The case of ultramafic lavas as part of an unusual ultramafic–mafic–felsic suite in Othris, Greece. *Lithos* **2012**, *144*, 177–193. [CrossRef]
35. Baziotis, I.; Economou-Eliopoulos, M.; Asimow, P.D. Ultramafic lavas and high-Mg basaltic dykes from the Othris ophiolite complex, Greece. *Lithos* **2017**, *288*, 231–247. [CrossRef]
36. Rassios, A. Geology and Evolution of the Vourinos Complex, Northern Greece. Ph.D. Thesis, University of California (Davis), Davis, CA, USA, 1981.
37. Francalanci, L.; Varekamp, J.C.; Vougioukalakis, G.; Defant, M.J.; Innocenti, F.; Manetti, P. Crystal retention, fractionation and crustal assimilation in aconvecting magma chamber Nisyros Volcano, Greece. *Bull. Volcanol.* **1995**, *56*, 601–620. [CrossRef]
38. Papanikolaou, D. Geotectonic evolution of the Aegean. *Bull. Geol. Soc. Greece* **1993**, *27*, 33–48.
39. Royden, L.H.; Papanikolaou, D.J. Slab segmentation and late Cenozoicdisruption of the Hellenic arc. *Geochem. Geophys. Geosyst.* **2011**, *12*, Q03010. [CrossRef]
40. Mitropoulos, P.; Tarney, J.; Saunders, A.D.; Marsh, N.G. Petrogenesis of Cenozoic rocks from the aegean island arc. *J. Volcanol. Geotherm. Res.* **1987**, *32*, 177–193. [CrossRef]
41. Tzifas, I.T.; Misaelides, P.; Godelitsas, A.; Gamaletsos, P.N.; Nomikou, P.; Karydas, A.G.; Kantarelou, V.; Papadopoulos, A. Geochemistry of coastal sands of Eastern Mediterranean: The case of Nisyros volcanic materials. *Chem. Der Erde-Geochem.* **2017**, *77*, 487–501. [CrossRef]
42. Marcopoulos, T.; Economou, M. Theophrastite, $Ni(OH)_2$, a new mineral from Northern Greece. *Am. Miner.* **1981**, *66*, 1020–1021.
43. Economou-Eliopoulos, M.; Eliopoulos, D. *A New Solid Solution [(Co,Mn,Nn)(OH)*$_2$*], in the Vermion Mt (Greece) and its Genetic Significance for the Mineral Group of Hydroxides*; McLaughlin, E.D., Braux, L.A., Eds.; Chemical Mineralogy, Smelting and Metallization; Nova Science Publishers: New York, NY, USA, 2009; pp. 1–18.
44. Taylor, S.R.; McLennan, S.M. *The Continental Crust: Its Composition and Evolution*; Blackwell Scientific: Oxford, UK, 1985.

45. Gamaletsos, P. Mineralogy and Geochemistry of Bauxites from Parnassos-Ghiona Mines and the Impact on the Origin of the Deposits. Ph.D. Thesis, University of Athens, Athens, Greece, 2014.
46. Kalatha, S. Metallogenesis of Bauxite Laterites and Fe-Ni-laterites-Enrichnment in Rare Earth Elements. Ph.D. Thesis, University of Athens, Athens, Greece, 2017.
47. Barth, M.; Gluhak, T. Geochemistry and tectonic setting of mafic rocks from the Othris Ophiolite, Greece. *Contrib. Miner. Petrol.* **2009**, *157*, 23–40. [CrossRef]
48. McDonough, W.F.; Sun, S.S. Chemical evolution of the mantle. The composition of the Earth. *Chem. Geol.* **1995**, *120*, 223–253. [CrossRef]
49. Sotnikov, V.I.; Berzina, A.N.; Economou-Eliopoulos, M.; Eliopoulos, D.G. Palladium, platinum and gold distribution in porphyry Cu _ Mo deposits of Russia and Mongolia. *Ore Geol. Rev.* **2001**, *18*, 95–111. [CrossRef]
50. Schock, H.H. Distribution of rare-earth and other trace elements in magnetite separates. *Chem. Geol.* **1979**, *26*, 119–133. [CrossRef]
51. Deditius, A.P.; Reich, M.; Simon, A.C.; Suvorova, A.; Knipping, J.; Roberts, M.P.; Rubanov, S.; Dodd, A.; Saunders, M. Nanogeochemistry of hydrothermal magnetite. *Contrib. Miner. Petrol.* **2018**, *173*, 46. [CrossRef]
52. Horn, I.; Foley, S.F.; Jackson, S.E.; Jenner, G.A. Experimental determined partitioning of high field strength and selected transition elements between spinel and basaltic melt. *Chem. Geol.* **1994**, *117*, 193–318. [CrossRef]
53. Balaram, V. Rare earth elements: A review of applications, occurrence, exploration, analysis, recycling, and environmental impact. *Geosci. Front.* **2019**, *10*, 1285–1303. [CrossRef]
54. Maksimovic, Z.; Skarpelis, N.; Panto, G. Mineralogy and geochemistry of the rare earth elements in the karstic nickel deposit of Lokris area, Greece. *Acta Geol. Hung.* **1993**, *36*, 331–342.
55. Economou-Eliopoulos, M.; Eliopoulos, D.G.; Apostolikas, A.; Maglaras, K. Precious and rare earth element distribution in Ni-laterite deposits from Lokris area, Central Greece. In Proceedings of the Fourth Biennial SGA Meeting, Turku, Finland, 11–13 August 1997; pp. 411–413.
56. Kalatha, S.; Perraki, M.; Economou-Eliopoulos, M.; Mitsis, I. On the origin of bastnaesite-(La,Nd,Y) in the Nissi (Patitira) bauxite laterite deposit, Lokris, Greece. *Minerals* **2017**, *7*, 45. [CrossRef]
57. Goodenough, K.M.; Schilling, J.; Jonsson, E.; Kalvig, P.; Charles, N.; Tuduri, J.; Deady, E.A.; Sadeghi, M.; Schiellerup, H.; Muller, A.; et al. Europe's rare earth element resource potential: An overview of REE metallogenetic provinces and their geodynamic setting. *Ore Geol. Rev.* **2016**, *72*, 838–856. [CrossRef]
58. Colony, W.E.; Nordlie, B.E. Liquid sulfur at Volcan Azufre, Galapagos Islands. *Econ. Geol.* **1973**, *68*, 371–380. [CrossRef]
59. GustafsonL, B.; Hunt, J.P. The porphyry copper deposit at E1 Salvador, Chile. *Econ. Geol.* **1975**, *70*, 857–912. [CrossRef]
60. Hemley, J.J.; Hunt, J.P. Hydrothermal ore-forming processes in the light of studies in rock-buffered systems. II. Some general geologic applications. *Econ. Geol.* **1992**, *87*, 23–43.
61. Hemley, J.J.; Cygan, G.L.; Fein, J.B.; Robinson, G.R.; d'Angelo, W.M. Hydrothermal ore-forming processes in the light of studies in rock-buffered systems I.: Iron-copper-zinc-lead sulfide solubility relations. *Econ. Geol.* **1992**, *87*, 1–22. [CrossRef]
62. Bliss, N.W.; MacLean, W.H. The paragenesis of zoned chromite from central Manitoba. *Geochim. Cosmochim. Acta* **1975**, *39*, 973–990. [CrossRef]
63. Bosi, F.; Andreozzi, G.B.; Hålenius, U.; Skogby, H. Zn-O tetrahedral bond length variations in normal spinel oxides. *Am. Mineral.* **2011**, *96*, 594–598. [CrossRef]
64. Martignago, F.; Dal Negro, A.; Carbonin, S. How Cr^{3+} and Fe^{3+} affect Mg-Al order disorder transformation at high temperature in natural spinels. *Phys. Chem. Miner.* **2003**, *30*, 401–408. [CrossRef]
65. Gervilla, F.; Padrón-Navarta, J.A.; Kerestedjian, T.; Sergeeva, I.; González-Jiménez, J.M.; Fanlo, I. Formation of ferrian chromite in podiform chromitites from the Golyamo Kamenyane serpentinte, Eastern Rhodopes,SE Bulgaria: A two-stage process. *Contrib. Mineral. Petrol.* **2012**, *164*, 643–657. [CrossRef]
66. Colás, V.; González-Jiménez, J.M.; Griffin, W.L.; Fanlo, I.; Gervilla, F.; O'Reilly, S.Y.; Pearson, N.J.; Kerestedjian, T.; Proenza, J.A. Fingerprints of metamorphism in chromite: New insights from minor and trace elements. *Chem. Geol.* **2014**, *389*, 137–152. [CrossRef]
67. Glemser, O.; Einerhand, J. Uber hohere Nickelhydroxide. *Z. Anorg. Chem.* **1950**, *261*, 26–42. [CrossRef]
68. Baskar, S.; Baskar, R.; Kaushik, A. Role of microorganisms in weathering of the Konkan-Goa laterite formation. *Curr. Sci.* **2003**, *85*, 1129–1134.

69. Russell, M.J.; Hall, A.J.; Boyce, A.J.; Fallick, A.E. On hydrothermal convection systems and the emergence of life. *Econ. Geol.* **2005**, *100*, 418–438.
70. Southam, G.; Saunders, J. The geomicrobiology of ore deposits. *Econ. Geol.* **2005**, *100*, 1067–1084. [CrossRef]
71. Valeton, I.; Biermann, M.; Reche, R.; Rosenberg, F. Genesis of Nickel laterites and bauxites in Greece during the Jurassic and Cretaceous, and their relation to ultrabasic parent rocks. *Ore Geol. Rev.* **1987**, *2*, 359–404. [CrossRef]
72. Eliopoulos, D.; Economou-Eliopoulos, M. Geochemical and mineralogical characteristics of Fe–Ni and bauxite–laterite deposits of Greece. *Ore Geol. Rev.* **2000**, *16*, 41–58. [CrossRef]
73. Kalatha, S.; Economou-Eliopoulos, M. Framboidal pyrite and bacteriomorphic goethite at transitional zones between Fe–Ni–laterites and limestones: Evidence from Lokris, Greece. *Ore Geol. Rev.* **2015**, *65*, 413–425. [CrossRef]
74. Norton, S. Laterite and bauxite formation. *Econ. Geol.* **1973**, *63*, 353–361. [CrossRef]
75. Laskou, M.; Economou-Eliopoulos, M. The role of microorganisms on the mineralogical and geochemical characteristics of the Parnassos-Ghiona bauxite deposits, Greece. *J. Geochem. Explor.* **2007**, *93*, 67–77. [CrossRef]
76. Laskou, M.; Economou-Eliopoulos, M. Bio-mineralization and potential biogeochemical processes in bauxite deposits: Genetic and ore quality significance. *Miner. Petrol.* **2013**, *107*, 471–486. [CrossRef]
77. Schulte, R.F.; Foley, N.K. *Compilation of Gallium Resource Data for Bauxite Deposits*; U.S. Geological Survey: Reston, VA, USA, 2014.
78. Zhou, M.F.; Robinson, P.T.; Su, B.X.; Gao, J.F.; Li, J.W.; Yang, J.S.; Malpas, J. Compositions of chromite, associated minerals, and parental magmas of podiform chromite deposits: The role of slab contamination of asthenospheric melts in suprasubduction zone environments. *Gondwana Res.* **2014**, *26*, 262–283. [CrossRef]
79. Eliopoulos, I.P.D.; Eliopoulos, G.D. Factors Controlling the Gallium Preference in High-Al Chromitites. *Minerals* **2019**, *9*, 623. [CrossRef]
80. Tarkian, M.; Koopmann, G. Platinum-group minerals in the Santo Tomas II (Philex) porphyry copper–gold deposit, Luzon Island, Philippines. *Miner. Depos.* **1995**, *30*, 39–47. [CrossRef]
81. Thompson, J.F.H.; Lang, J.R.; Stanley, C.R. Platinum-group elements in alkaline porphyry deposits, British Columbia. *Explor. Min. B. C. Mines Branch Part B* **2001**, *2001*, 57–64.
82. Richards, J.P. Tectono-magmatic precursors for porphyry Cu–(Mo–Au) deposit formation. *Econ. Geol.* **2003**, *98*, 1515–1533. [CrossRef]
83. Richards, J.P. Postsubduction porphyry Cu–Au and epithermal Au deposits: Products of remelting of subduction-modified lithosphere. *Geology* **2009**, *37*, 247–250. [CrossRef]
84. Richards, J. Tectonic, magmatic, metallogenic evolution of the Tethyan orogeny: From subduction to collision. *Ore Geol. Rev.* **2015**, *70*, 323–345. [CrossRef]
85. Tarkian, M.; Hünken, U.; Tokmakchieva, M.; Bogdanov, K. Precious-metal distribution and fluid-inclusion petrography of the Elatsite porphyry copper deposit, Bulgaria. *Miner. Depos.* **2003**, *38*, 261–281. [CrossRef]
86. Eliopoulos, D.G.; Economou-Eliopoulos, M.; Zelyaskova-Panayiotova, M. Critical factors controlling Pd and Pt potential in porphyry Cu–Au deposits: Evidence from the Balkan Peninsula. *Geosciences* **2014**, *4*, 31–49. [CrossRef]
87. Holwell, D.; Fiorentini, M.; McDonald, I.; Lu, Y.; Giuliani, A.; Smith, D.; Keith, M.; Locmelis, M. A metasomatized lithospheric mantle control on the metallogenic signature of post-subduction magmatism. *Nat. Commun.* **2019**, *10*, 1–10. [CrossRef]
88. Mungall, J.E.; Andrews, D.R.A.; Cabri, L.J.; Sylvester, P.J.; Tubrett, M. Partitioning of Cu, Ni, Au, and platinum-group elements between monosulfide solid solution and sulfide melt under controlled oxygen and sulphur fugacities. *Geochim. Cosmochim. Acta* **2005**, *69*, 4349–4360. [CrossRef]
89. Sillitoe, R.H. Porphyry copper systems. *Econ. Geol.* **2010**, *105*, 3–41. [CrossRef]
90. Berzina, A.P.; Berzina, A.N.; Gimon, V.O. Paleozoic–Mesozoic porphyry Cu (Mo) and Mo (Cu) deposits within the southern margin of the Siberian Craton: Geochemistry, geochronology, and petrogenesis (a review). *Minerals* **2016**, *6*, 125. [CrossRef]
91. McInnes, B.I.A.; Cameron, E.M. Carbonated, alkaline hybridizing melts from a sub-arc environment: Mantle wedge samples from the Tabar-Lihir-Tanga-Feni arc, Papua New Guinea. *Earth Planet. Sci. Lett.* **1994**, *122*, 125–141. [CrossRef]
92. Moritz, R.; Kouzmanov, K.; Petrunov, R. Upper Cretaceous Cu–Au epithermal deposits of the Panagyurishte district, Srednogorie zone, Bulgaria. *Swiss Bull. Miner. Petrol.* **2004**, *84*, 79–99.

6

Characteristics of Precipitation of Rare Earth Elements with Various Precipitants

Kenneth N. Han

South Dakota School of Mines and Technology, Rapid City, SD 57701-3995, USA; kennethhydro@gmail.com

Abstract: The effective and selective leaching of rare earth elements (REEs) from various sources is frequently possible in practice by adopting a carefully coordinated strategy incorporating a selective precipitation of these elements from undesired ones in solution. In this study, the behavior of chemical precipitation of REEs with commonly used precipitants such as sulfate, carbonate, fluoride, phosphate, and oxalate was examined using thermodynamic principles and calculations. It was found that the pH of the system has a profound effect on determining particular chemical species of precipitants, which are subsequently responsible for the precipitation of REEs. The role of various anions such as Cl^-, NO_3^-, and SO_4^{2-} derived from the acid used in the leaching process on the precipitation behavior of REEs was examined. These anions form complexes with REEs and display a very positive effect on the precipitation behavior. The nitrate environment exhibits most conducive to precipitation followed by sulfate and then chloride.

Keywords: precipitation; leaching; complexation; anion effect; equilibrium calculation; rare earth elements

1. Introduction

Rare earth elements (REEs) are becoming very important in modern society. REEs are essential ingredients in the transition to green technology and there have been numerous recent articles describing their extraction processes from various sources [1,2].

Most rare earth bearing minerals are refractory in nature and therefore, a pretreatment or strong acid treatment of ores bearing these minerals is usually necessary [3–6] before leaching is affected. Such pretreated ores are then subjected to the leaching process of REEs. The leach liquor is frequently subjected to the removal of impurities such as iron, calcium, uranium, and thorium by going through stage-wise precipitation using various compounds such as hydroxide, carbonate, sulfate, and oxalate [7–13]. After the removal of these impurities by chemical precipitation, REEs in solution are then subjected to precipitation using various precipitants, which in general takes place at pH above 3–4, since most of these impurities are precipitated at around these pHs. However, during the precipitation of impurities, some REEs are also removed from the solution by coprecipitation or adsorption. Therefore, it is desirable to precipitate REEs before removal of the impurities. It is the hope of this study to identify the precipitants and chemical conditions to achieve this objective.

The chemical complexity of the aqueous medium introduced by various chemical treatment of leach liquor has introduced a complex nature to the system making the analysis of the subsequent chemical precipitation a challenge. In addition to metal ions in the leach liquor extracted from ores, many anions such as Cl^-, NO_3^-, and SO_4^{2-} mostly derived from acids used in the leaching operation also play an important role in leaching as well as precipitation.

The effect of various anions on the leaching behavior has been discussed in earlier studies [14,15]. For example, the leaching of Re-phosphate in the presence of sulfuric acid can be simply described by

Equation (1). (It is noted that Re is the elemental symbol for rhenium. However, Re(III) refers to only REE(III) in this paper.)

$$<RePO_4> + \{H_2SO_4\} = \{Re^{3+}\} + \{PO_4{}^{3-}\} + \{SO_4{}^{2-}\} + 2\ \{H^+\} \quad (1)$$

Here the symbols, < > and { } stand for solid and liquid form, respectively. The oxidation state of REEs in solution and solids is assumed to be Re(III) throughout this presentation. It should be noted however, that some REEs can be oxidized to Re(IV) or reduced to Re(II) in aqueous media in very rare cases [16]. The concentration of the free REEs at equilibrium can be estimated if the concentration of all species involved in Equation (1) is known. However, such calculated values are far from real values since the chemicals produced from the leaching process undergo hydrolysis and complexation [14,15]. As a result, the correct evaluation of the product concentration requires a very complicated process involving calculations of equilibrium equations simultaneously in some cases.

The chemical precipitation of REEs is similarly complicated. For example, a simple chemical precipitation of a free Re-ion, Re^{3+} with phosphate ion, $PO_4{}^{3-}$ can be given by Equation (2).

$$\{Re^{3+}\} + \{PO_4{}^{3-}\} = <RePO_4> \quad (2)$$

It should be noted that REEs precipitate with the phosphate ion as a free Re-ion, Re^{3+} at relatively high pHs such as 4 and 6. However, when the pH of the solution changes to a lower value, for example 1 or 2, by adding acids of HCl, HNO_3, or H_2SO_4, the anions such as Cl^-, $NO_3{}^-$, and $SO_4{}^{2-}$ will form various complexes with REEs [3,14,15], and then these complexes are subjected to precipitation. For example, in the presence of sulfate ions, there will be at least Re^{3+}, $Re(SO_4)^+$, $Re(SO_4)_2{}^-$, and $Re_2(SO_4)_3$, present in solution, and for most of the sulfate concentration, $Re(SO_4)_2{}^-$ is the predominant species when the concentration of sulfate ions is especially more than 0.1 M [14,15]. As a result, in such cases, $Re(SO_4)_2{}^-$ should be used instead of the free ion and the resulting concentration of rare earth elements after precipitation is an order of magnitude different from when the free ion is used.

In addition, the phosphate ion also may not exist depending upon the pH of the system. It should be noted that the values of pKa are 2.0 for $H_3PO_4/H_2PO_4{}^-$, 7.2 for $H_2PO_4{}^-/HPO_4{}^{2-}$, and 12.0 for $HPO_4{}^{2-}/PO_4{}^{3-}$ (see Table 1) [17]. This means that H_3PO_4 should be used instead of the phosphate ion in Equation (2) for the system whose pH lies below 2 and likewise, $H_2PO_4{}^-$ for pH between 2 and 7.2 and so on.

Table 1. The pKa values for various precipitants [17].

Precipitants	pKa
$H_3PO_4/H_2PO_4{}^-$	2.1
$H_2PO_4{}^-/HPO_4{}^{2-}$	7.2
$HPO_4{}^{2-}/PO_4{}^{3-}$	12.0
$H_2CO_3/HCO_3{}^-$	6.38
$HCO_3{}^-/CO_3{}^{2-}$	10.34
HF/F^-	3.0
$H_2C_2O_4/HC_2O_4{}^-$	1.0
$HC_2O_4{}^-/C_2O_4{}^{2-}$	4.2
$HSO_4{}^-/SO_4{}^{2-}$	2.0

In this paper, factors such as pH, complexation, and hydration of precipitants affecting the chemical precipitation with various precipitants will be examined and their practical implication in the analysis of such precipitation processes will be discussed. The precipitation behavior of REEs with carbonate, fluoride, phosphate, sulfate, and oxalate was considered in this study, since these precipitants are most widely used in precipitation of REEs in practice.

2. Acquisition of Thermodynamic Information

Table 1 presents the values of pKa for the precipitants considered in this study. It should be noted that there are some variations in the numerical values of pKa based on various literature sources. For example, the values for oxalic acid given by Pourbaix [17] and Dean [18] are different from those by HSC [19] and Wagman [20]. However, the effect of such differences on the final outcome of the behavior of chemical precipitation is not very significant.

The thermodynamic data used in this study are listed in Table 2. Most data given in the table were taken from the literature information [21–26] and those unavailable were estimated using the parallel plot method as described earlier [14].

Table 2. The Gibbs standard free energy formation of various compounds in kCal/mol. Information taken from [14,15] with some modifications.

	Carbonate	Phosphate	Sulfate	Fluoride	Oxalate	Oct-Sulfate
La	−768.5	−455.3	−892.6	−399.8	−872.2	−1399.9
Ce	−761.4	−455.2	−884.7	−396.8	−868.3	−1340.2
Pr	−764.5	−451.0	−879.4	−394.2	−860.6	−1343.5
Nd	−761.6	−448.7	−877.2	−391.5	−861.8	−1335.0
Sm	−759.6	−447.6	−878.8	−392.3	−860.1	−1330.5
Eu	−759.4	−446.7	−875.3	−391.0	−859.1	−1275.3
Gd	−758.6	−445.5	−874.2	−390.5	−857.7	−1328.1
Tb	−757.0	−444.7	−873.7	−388.7	−856.9	−1313.2
Dy	−750.6	−442.2	−868.9	−385.9	−856.0	−1323.3
Ho	−744.2	−440.4	−861.3	−383.0	−849.6	−1314.3
Er	−743.9	−439.1	−855.9	−380.9	−842.0	−1309.5
Tm	−736.9	−438.2	−862.9	−379.6	−846.6	−1308.1
Yb	−735.5	−437.6	−861.5	−378.4	−839.5	−1293.4
Lu	−734.5	−435.7	−860.1	−376.8	−841.3	−1319.1
Y	−749.9	−443.0	−866.8	−389.4	−851.5	−1320.6
Sc	−714.9	−418.2	−845.1	−361.5	−828.4	−1300.0

Oct-sulfate: $Re_2(SO_4)_3 \cdot 8H_2O$ (here Re represents REEs).

Some of the standard Gibbs free energy formation values, such as Y- and Sc-oxalates were not available in the literature and, therefore, were estimated using the parallel method as described by earlier studies [14,15]. It should be noted that there were typographical errors in the values for Re-octa-hydrated sulfates in earlier publications [14,15]. It should also be noted that subsequent relevant calculations in the same articles were carried out using correct values, however.

3. Carbonate System

It is well known that rare earth elements are readily precipitated with carbonate at high pHs. Rare earth elements are readily subjected to precipitation with carbonate [27,28] especially at high pHs. Re-carbonates are ideal precipitates, since carbonates are readily available, relatively less expensive than other precipitants, and precipitates can be easily dissolved in a mild acid for further treatments of REEs in subsequent processes of purification. However, it is not easily precipitated in acidic media. As a result, the leach liquor is usually subjected to a neutralization step before the precipitation of REEs with carbonate is affected. During the neutralization process, however, impurities such as Fe^{3+}, Ca^{2+} are precipitated first resulting in loss of REEs due to adsorption and coprecipitation of REEs with oxides or sulfates of impurities.

A simple stoichiometric equation describing the precipitation of REEs with carbonate can be written in Equation (3), which is analogous to Equation (2).

$$2\,\{Re^{3+}\} + 3\,\{CO_3^{2-}\} = <Re_2(CO_3)_3> \quad (3)$$

As discussed earlier, the concentration of CO_3^{2-} at pH 1–6 that was considered in this study, which is consistent with the range of pH practiced in industry, is practically zero, since the pK value for HCO_3^-/CO_3^{2-} is 10.34. The stable species at pH less than 6, is H_2CO_3, which is the dominant species as seen in Table 1 and, therefore, H_2CO_3 should be used in Equation (3). Furthermore, when pH of the solution is less than 6, Re^{3+} is subjected to complexation to become either $ReCl^{2+}$, $Re(NO_3)^{2+}$, or $Re(SO_4)_2^-$ depending upon the kind of acid used [14,15]. Therefore, for the working pH range (pH 1–6) in this study, Equations (4)–(6) should be used for chloride, nitrate, and sulfate systems instead of Equation (3).

$$2\,\{ReCl^{2+}\} + 3\,\{H_2CO_3\} = <Re_2(CO_3)_3> + 2\,\{Cl^-\} + 6\,\{H^+\} \tag{4}$$

$$2\,\{Re(NO_3)^{2+}\} + 3\,\{H_2CO_3\} = <Re_2(CO_3)_3> + 2\,\{NO_3^-\} + 6\,\{H^+\} \tag{5}$$

$$2\,\{Re(SO_4)_2^-\} + 3\,\{H_2CO_3\} = <Re_2(CO_3)_3> + 4\,\{SO_4^{2-}\} + 6\,\{H^+\} \tag{6}$$

In order to illustrate the steps involved in calculating the equilibrium concentration of REEs in solution, let us take Equation (4) as an example. The equilibrium constant for Equation (4), K_{eq} can be found knowing the Gibbs standard free energy formation of each component involved in the reaction. Such found equilibrium constant is given in Equation (7) below:

$$K_{eq} = \frac{\{Cl^-\}^2\{H^+\}^6}{\{ReCl^{2+}\}^2\{H_2CO_3\}^3} \tag{7}$$

Here $\{H^+\}$ is known if the pH of the system is known, $\{Cl^-\}$ is known since the pH of the system is adjusted by HCl in this reaction, (see Table 3) and $\{H_2CO_3\}$ was set as 10^{-4} M. Therefore, $\{ReCl^{2+}\}$ can be determined.

It should be noted that Equation (6) is valid above pH 2 but SO_4^{2-} should be replaced by HSO_4^- when the pH of the system is less than 2 (See Table 3). When Equations (4)–(6) are solved to calculate the equilibrium concentrations of REEs in solution, it was assumed that the final concentrations of Cl^-, NO_3^-, and SO_4^{2-} are those set by respective acid concentration to adjust the pH of the solution as discussed above. This assumption is reasonable for low pHs but the accurate estimation of the final equilibrium concentrations of REEs at pH 6 should be carried out taking into account all of practical conditions of the system.

It should also be noted that most of precipitants considered here are mostly weak acids and, therefore, they contribute in setting the pH of the solution (see Table 4). This can be a cause of inaccuracy in the final results especially at relatively high pHs, say pH greater than 4. The correction of such effects can be made but no attempt on such corrections was made in this study. This is because the objective of the study is to observe the precipitation behavior at low pHs.

Table 3. Concentrations of various anions used in the study at different pH's [17].

	HCl	**HNO_3**	**H_2SO_4**	
pH	Cl^- (M)	NO_3^- (M)	SO_4^{2-}(M)	HSO_4^- (M)
6	1.0×10^{-6}	1.0×10^{-6}	1.0×10^{-6}	9.9×10^{-11}
5	1.0×10^{-5}	1.0×10^{-5}	1.0×10^{-5}	9.9×10^{-9}
4	1.0×10^{-4}	1.0×10^{-4}	9.9×10^{-5}	9.9×10^{-7}
3	1.0×10^{-3}	1.0×10^{-3}	9.1×10^{-4}	9.0×10^{-5}
2	1.0×10^{-2}	1.0×10^{-2}	5.0×10^{-3}	5.0×10^{-3}
1	1.0×10^{-1}	1.0×10^{-1}	9.0×10^{-3}	9.1×10^{-2}
0	1.0	1.0	1.0×10^{-2}	9.9×10^{-1}

Table 4. The pH values of some weak acids for various concentrations.

Concentration, M.	H_3PO_4	HF	H_2CO_3	$H_2C_2O_4$
1.0×10^{-6}	5.97	6	6.31	6
1.0×10^{-4}	4	4.05	5.2	4
1.0×10^{-3}	3	3.26	4.69	3
1.0×10^{-2}	2.3	2.64	4.18	2.09
1.0×10^{-1}	1.63	2.1	3.68	1.31

It is important to note that any calculated value exceeding the highest concentration of REEs in practice should be treated with caution. The highest concentration of REEs possible in the aqueous system is based on the grade of REEs in the ore being leached. Assuming the highest grade of REEs of the ore being treated is 50%, the highest concentration of REEs possible in the aqueous media could be about 1.5 M [15]. It is safe to assume that when the calculated value is more than 1–2 M, the precipitation is impossible under the conditions considered. On the other hand, when the calculated value is less than the target value, 1 ppm, then the precipitation could be favorable.

Figure 1 presents the equilibrium concentration of dissolved REEs in solution when REEs are subjected to precipitation with carbonate. In this study, the concentration of the precipitant, was kept constant at 10^{-4} M throughout the paper as stated earlier. The choice of the precipitant concentration being 10^{-4} M, which is the final concentration after the precipitation is completed, is arbitrary and deemed reasonable in view of practical application. The horizontal dashed line across the figure represents the concentration of REEs in solution being 3×10^{-6} M, which is about 1 ppm of REEs. This is an arbitrary concentration representing an acceptable target level of precipitation of REEs after subjecting to the carbonate precipitation.

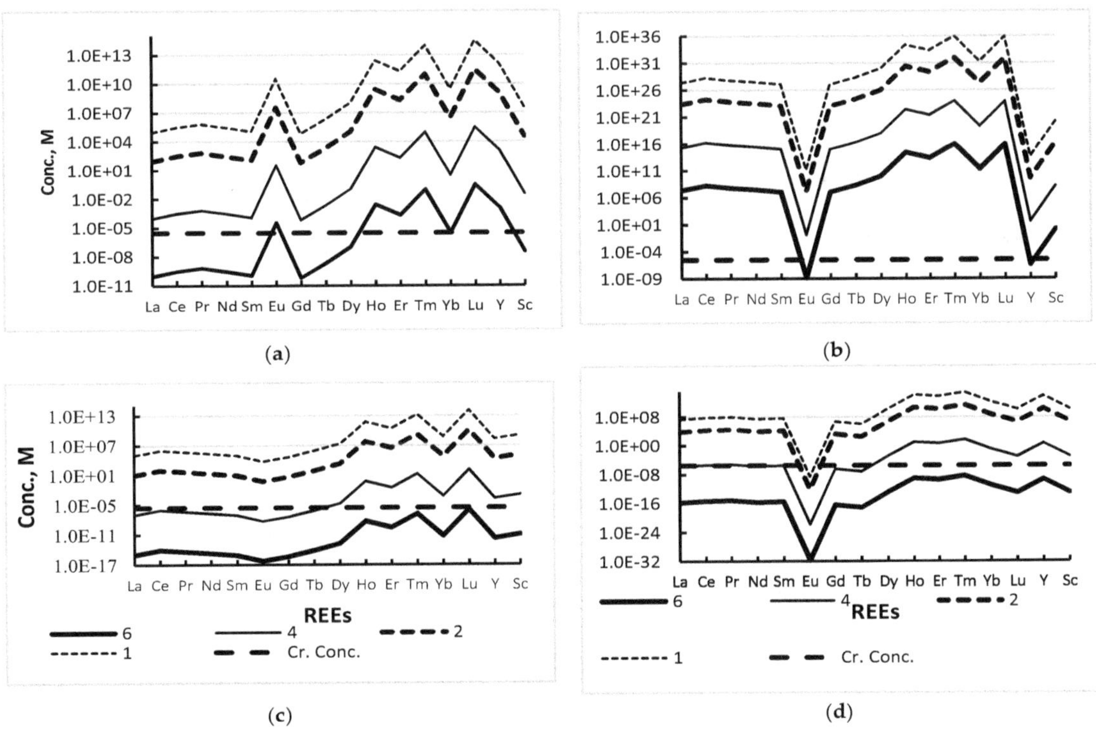

Figure 1. Concentrations of (**a**) Re^{3+}, (**b**) $ReCl_2^+$, (**c**) $Re(NO_3)_2^+$, and (**d**) $Re(SO_4)_2^-$ in equilibrium with carbonate precipitate with 10^{-4} M of total concentration of carbonate in solution at pH 6, 4, 2, and 1. Cr.concentration: 1 ppm line as a target concentration.

It is a reasonable assumption that the initial concentration of REEs in the leach liquor would be in the range of 3×10^{-1} to 3×10^{-3} M [15]. It is a usual practice that the initial concentration of the precipitant is a little more than the stoichiometric amount. It is obvious that carbonate is incapable of precipitating most of REEs, except a few light REEs at pH 6. It is interesting to note that the precipitation of nitrate and sulfate complexes is significantly better than free REE ions. The anomaly behavior by Eu seems to be due to inaccuracy in the standard Gibbs free energy data given in the literature.

4. Fluoride System

It has been well understood that fluoride readily complexes with rare earth elements [6,29]. The pKa value of HF/F^- is 3 as seen in Table 1 and, therefore, the relevant equations governing the precipitation of REEs with fluoride should be formulated in such a way that HF is the precipitant at pH less than 3 and F^- should be used at higher pHs. For example, the precipitation of $ReCl_2^+$ with fluoride is described by Equation (8) when the pH of the solution is 6 or 4, while Equation (9) is for pH 2 and 1. Here too, the concentration of the precipitant is assumed to be 10^{-4} M. The precipitation of the complexed REEs is remarkable as in the case with the carbonate system. Most of REEs is easily complexed with chloride, nitrate, and sulfate as long as the concentration of these anions is more than 0.1 M, below which they remain as Re^{3+} [14,15].

$$\{Re^{3+}\} + 3\,\{F^-\} = <ReF_3> \tag{8}$$

$$\{Re^{3+}\} + 3\,\{HF\} = <ReF_3> + 3\,\{H^+\}9 \tag{9}$$

Figure 2a–d present the precipitation of Re^{3+}, $ReCl_2^+$, $Re(NO_3)_2^-$, and $Re(SO_4)_2^-$ with fluoride, respectively. With regard to Figure 2a, there appears to be only one line representing pH 6 and 4. This is because when pH is above 3, F^- is responsible for the precipitation reaction as seen in Equation (7), in which the precipitation becomes independent of pH. As seen in these figures, fluoride is a more powerful precipitant for REEs than carbonate in the pH range considered in this study.

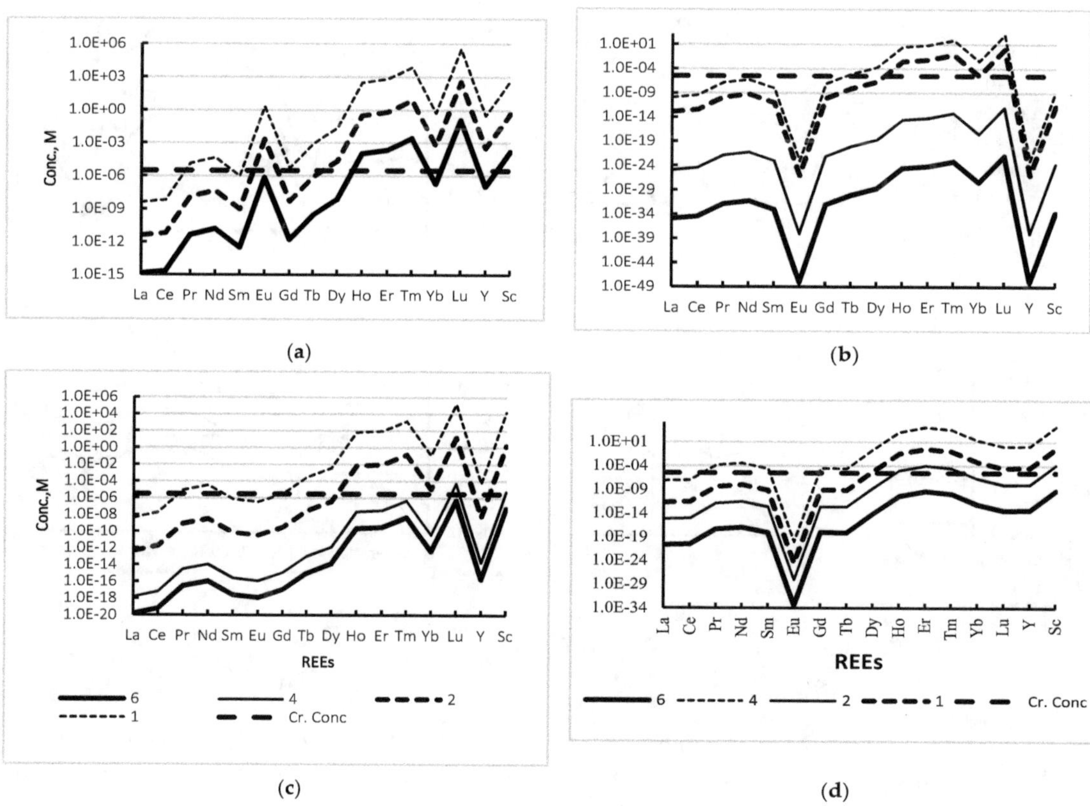

Figure 2. Concentrations of (**a**) Re^{3+}, (**b**) $ReCl_2^+$, (**c**) $Re(NO_3)_2^+$, and (**d**) $Re(SO_4)_2^-$ in equilibrium with fluoride precipitate with 10^{-4} M of total concentration of fluoride in solution at pH 6, 4, 2, and 1. Cr. Concentration: 1 ppm line as a target concentration.

5. Phosphate System

When phosphoric acid, H_3PO_4 is placed in water, it dissociates into dihydrogen phosphate, $H_2(PO_4)^-$, hydrogen phosphate, $H(PO_4)^{2-}$, or phosphate ion, PO_4^{3-} depending upon the pH of the solution. The pKa values of these conjugated species are 2.1 for $H_3PO_4/H_2PO_4^-$, 7.2 for $H_2PO_4^-/HPO_4^{2-}$, and 12.0 for HPO_4^{2-}/PO_4^{3-} as noted in Table 1. Therefore, the relevant equilibrium equations for the precipitation of REEs with phosphate should involve phosphoric acid for pH 1 and 2, and dihydrogen phosphate for pH 4 and 6. These are shown in Equations (10) and (11), respectively, for precipitation of $ReCl_2^+$.

$$\{ReCl_2^+\} + H_2PO_4^- = <RePO_4> + 2\{Cl^-\} + 2\{H^+\} \quad (10)$$

$$\{ReCl_2^+\} + \{H_3PO_4\} = <RePO_4> + 2\{Cl^-\} + 3\{H^+\} \quad (11)$$

Figure 3 illustrates the precipitation of Re^{3+}, $ReCl_2^+$, $Re(NO_3)_2^-$ and $Re(SO_4)_2^-$ with phosphate. Unlike the carbonate but similar to fluoride systems, the phosphate system shows a powerful precipitation behavior even for free REEs. REEs are readily precipitated for all forms of rare earth element ions in solution except Re-chloride complexes. Here again, the unusual behavior shown by europium is most likely due to erratic thermodynamic data given in the literature. It should be noted that industrially important REE-bearing minerals are monazite and xenotime, both of which are phosphate compounds. Chemical interaction between REEs and phosphate has been investigated in numerous studies [30,31].

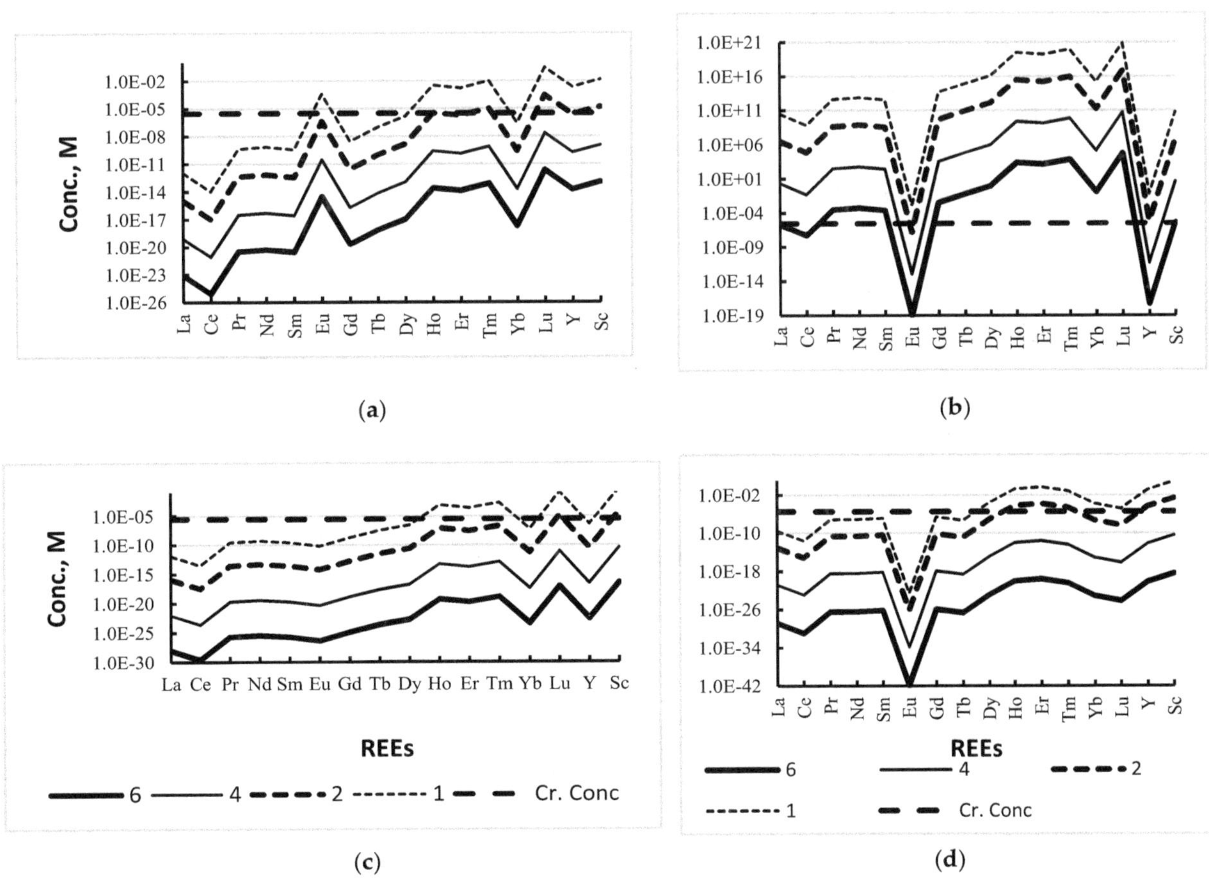

Figure 3. Concentrations of **(a)** Re^{3+}, **(b)** $ReCl_2^+$, **(c)** $Re(NO_3)_2^+$, and **(d)** $Re(SO_4)_2^-$ in equilibrium with phosphate precipitate with 10^{-4} M of total concentration of phosphate in solution at pH 6, 4, 2, and 1. Cr. Concentration: 1 ppm line as a target concentration.

6. Sulfate System

6.1. Sulfate (SO_4^{2-})

The precipitation of REEs with sulfate is given in Figure 4. In Figure 4a, the concentration of free REEs in solution in equilibrium with Re-sulfate precipitates is shown. In the construction of this figure, Equations (12) and (13) were used. Equation (12) was used for pH 6 and 4, while Equation (13) was used for pH 2 and 1. As mentioned earlier, the concentration of precipitants was kept constant at 10^{-4} M. At pH 6, the concentration of sulfate ion is 10^{-4} M plus 10^{-6} M, in which the first was added by the rule and the latter was from sulfuric acid to adjust the pH. In addition, at pH 4, on the other hand, the concentration of sulfate ion is 2×10^{-4} M, because of the added sulfate plus sulfate provided by sulfuric acid to adjust the pH of the solution. The same is also applicable for Figure 4d, in which case, Equations (14) and (15) were used instead of Equations (12) and (13).

It is noted that the precipitation is better at low pHs, namely pH 2 and 1, compared to those at pH 6 and 4. The reason for this is because there is more sulfate present at low pHs due to the supply of sulfate via sulfuric acid to adjust the pH of the solution.

$$2\,\{Re^{3+}\} + 3\,\{SO_4^{2-}\} = <Re_2(SO_4)_3> \quad (12)$$

$$2\,\{Re(SO_4)_2^{-}\} = <Re_2(SO_4)_3> + \{SO_4^{2-}\} \quad (13)$$

$$2\,\{Re(SO_4)_2^{-}\} = <Re_2(SO_4)_3> + \{SO_4^{2-}\} \quad (14)$$

$$2\,\{Re(SO_4)_2^{-}\} + \{H^{+}\} = <Re_2(SO_4)_3> + \{HSO_4^{-}\} \quad (15)$$

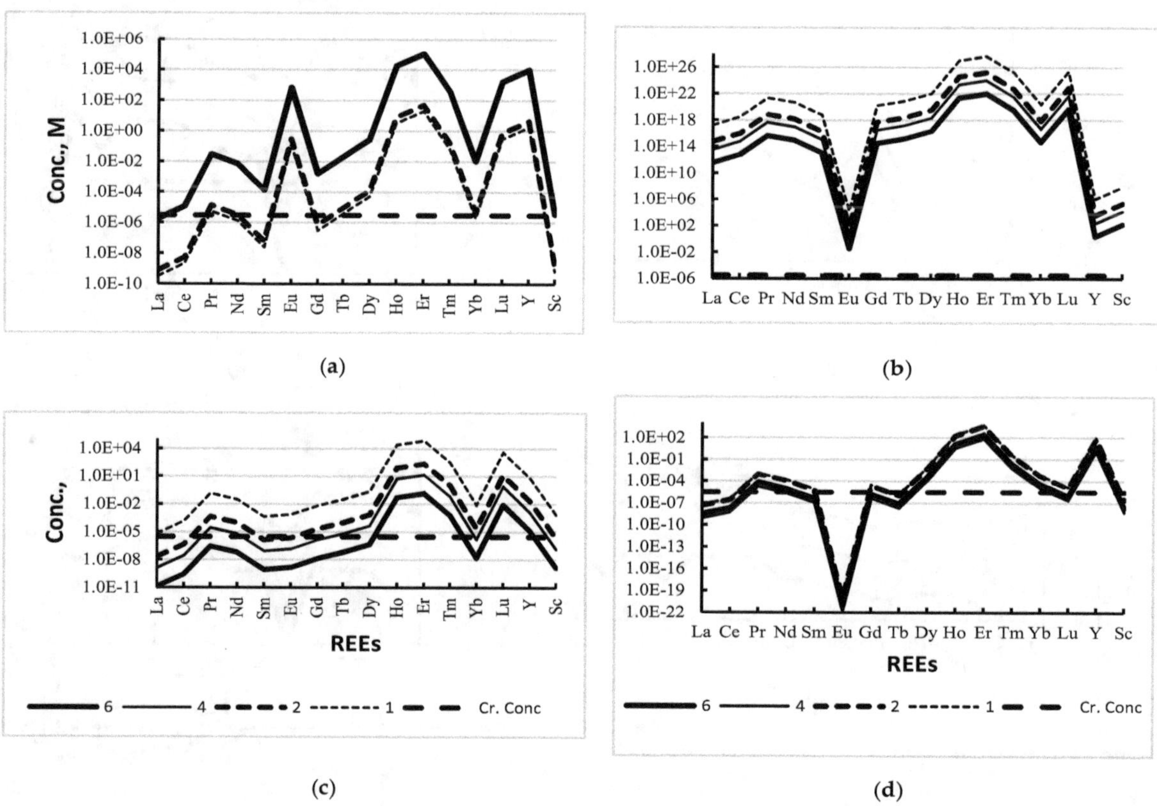

Figure 4. Concentrations of (**a**) Re^{3+}, (**b**) $ReCl_2^{+}$, (**c**) $Re(NO_3)_2^{+}$, and (**d**) $Re(SO_4)_2^{-}$ in equilibrium with sulfate precipitate with 10^{-4} M of total concentration of sulfate in solution at pH 6, 4, 2, and 1. Cr. Concentration: 1 ppm line as a target concentration. Additional sulfate is provided by sulfuric acid in (**a**) and (**d**).

For the precipitation of $Re(SO_4)_2^-$ using sulfate, the precipitation calculations were done using Equations (14) and (15). The precipitation of $Re(SO_4)_2^-$ with sulfate is unusual in that the species does not need any precipitant but it has to release one sulfate or bisulfate to form the precipitate $<Re_2(SO_4)_3>$ as seen in Equations (14) and (15). Such precipitation can be referred to as a decomposition precipitation.

It should be noted that in practice, REEs are usually precipitated with double salt sulfate [5,32], since the precipitation is more effective with double sodium sulfate. However, in this study, precipitation of REEs with only sulfate was carried out. Although double salt sulfate precipitation is more powerful than simple sulfate precipitation, the precipitation behavior of the two systems is expected to be very similar.

6.2. Octa-Hydrated Sulfate ($SO_4{\cdot}8H_2O$)

The chemical precipitation of REEs with sulfate to form octa-hydrated sulfate is very similar to the precipitation behavior of sulfate. The relevant equations used in this analysis are given in Equations (16)–(19).

$$2\,\{Re^{3+}\} + 3\,\{SO_4^{2-}\} + 8\,\{H_2O\} = <Re_2(SO_4)_3{\cdot}8H_2O> \quad (16)$$

$$2\,\{Re^{3+}\} + 3\,\{HSO_4^-\} + 8\,\{H_2O\} = <Re_2(SO_4)_3{\cdot}8H_2O> + 3\,\{H^+\} \quad (17)$$

$$2\,\{Re(SO_4)_2^-\} + 8\{H_2O\} = <Re_2(SO_4)_3{\cdot}8H_2O> + \{SO_4^{2-}\} \quad (18)$$

$$2\,\{Re(SO_4)_2^-\} + \{H^+\} + 8\,\{H_2O\} = <Re_2(SO_4)_3{\cdot}8H_2O> + \{HSO_4^-\} \quad (19)$$

As in the case with the sulfate system, the appearance of sulfate or bisulfate is based on the pH of the solution. In other words, when the pH of the system is greater than 2, sulfate is the predominant species and at low pHs, bisulfate should be used as shown in Table 3. It was expected that the octa-hydrated system is favored in the chemical precipitation compared to simple sulfate precipitation [14,15]. However, the results shown in Figure 4; Figure 5 show that the two systems behave very similarly.

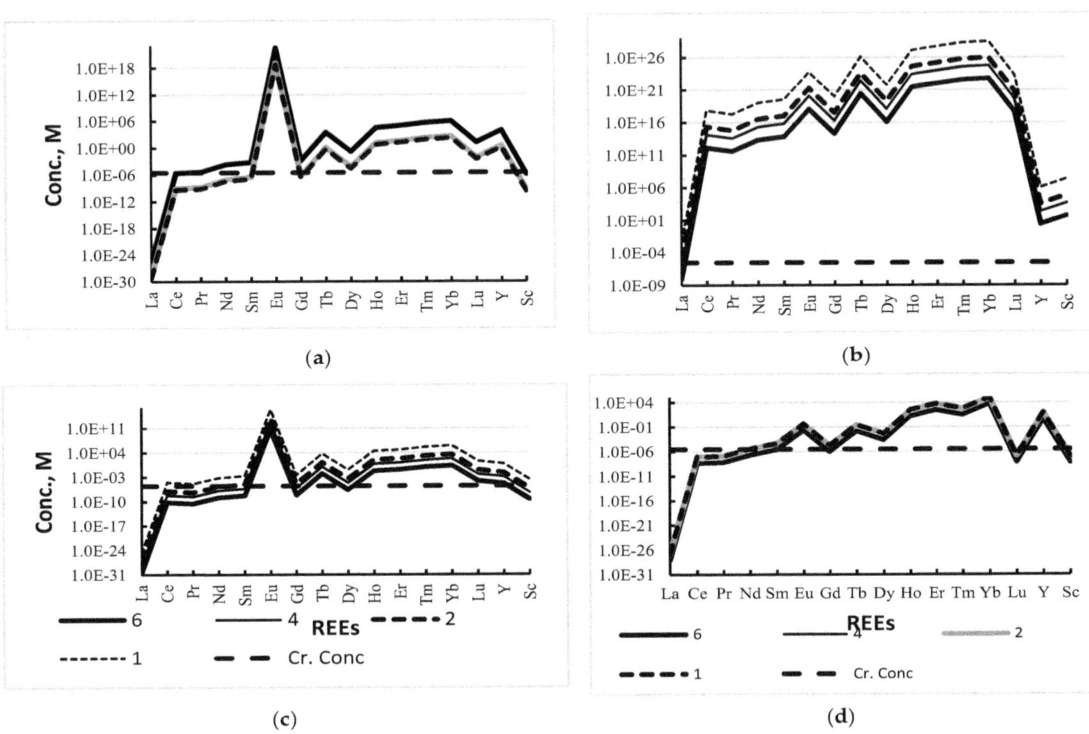

Figure 5. Concentrations of (**a**) Re^{3+}, (**b**) $ReCl_2^+$, (**c**) $Re(NO_3)_2^+$, and (**d**) $Re(SO_4)_2^-$ in equilibrium with octa-hydrated sulfate precipitate with 10^{-4} M of total concentration of sulfate in solution at pH 6, 4, 2, and 1. Cr. Concentration: 1 ppm line as a target concentration. Additional sulfate is provided by sulfuric acid in (**a**) and (**d**).

7. Oxalic Acid

Oxalic acid is the most commonly used precipitant for REEs from solution and the majority of investigators have used this precipitant to separate REEs in the solution from other dissolved ions. This is because oxalic acid has in general a very favorable affinity with REEs in general [4,6,9,13]. As seen in Table 1, the pKa values are 1 and 4.2 for $H_2C_2O_4/HC_2O_4^-$ and $HC_2O_4^-/C_2O_4^{2-}$, respectively. As a result, the relevant equations to be used to solve the equilibrium concentrations of REEs in solution after precipitation are Equations (20) and (21) for pH 6 and 4, and Equation (22) for pH 2 and 1, respectively.

$$2\,\{Re(SO_4)_2\} + 3\,\{C_2O_4{}^{2-}\} = \langle Re_2(C_2O_4)_3\rangle + 4\,\{(SO_4)^{2-}\} \tag{20}$$

$$2\,\{Re(SO_4)_2{}^-\} + 3\,\{HC_2O_4{}^-\} = \langle Re_2(C_2O_4)_3\rangle + 4\,\{(SO_4)^{2-}\} + 3\,\{H^+\} \tag{21}$$

$$2\,\{Re(SO_4)_2{}^-\} + 3\,(HC_2O_4{}^-) + \{H^+\} = \langle Re_2(C_2O_4)_3\rangle + 4\,\{(HSO_4)^-\} \tag{22}$$

The concentrations of (a) Re^{3+}, (b) $ReCl_2{}^+$, (c) $Re(NO_3)_2{}^+$, and (d) $Re(SO_4)_2{}^-$ in equilibrium with oxalate precipitate with 10^{-4} M of total concentration of oxalate in solution at pH 6, 4, 2, and 1 is plotted in Figure 6.

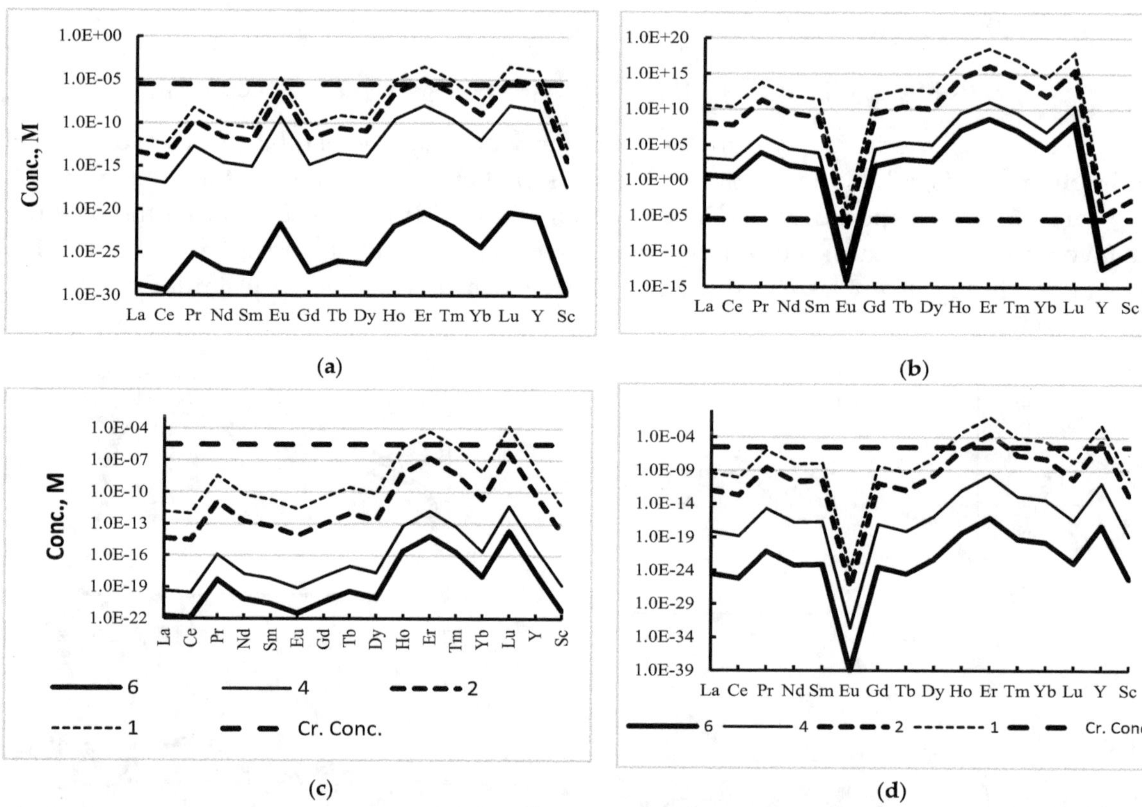

Figure 6. Concentrations of (**a**) Re^{3+}, (**b**) $ReCl_2{}^+$, (**c**) $Re(NO_3)_2{}^+$, and (**d**) $Re(SO_4)_2{}^-$ in equilibrium with oxalate precipitate with 10^{-4} M of total concentration of oxalate in solution at pH 6, 4, 2, and 1. Cr. Concentration: 1 ppm line as a target concentration.

As seen in this figure, oxalic acid is proven to be the most effective precipitant for REEs in solution. The degree of precipitation is very favorable for all pH ranges.

8. Comparison Among Precipitants

As expected, all of the precipitants considered in this study have shown good precipitation characteristics. In general, the most effective precipitant is oxalate, which has shown an outstanding precipitation behavior for REEs even at low pHs. On the other hand, most of precipitants are not effective unless the pH of the system is relatively high. As seen in Figure 7, the precipitation of free Re-ions, Re^{3+} is relatively poor even at pH 6 with the exception of oxalate and phosphate. When the pH of the system is low, such as pH 1 (Figure 7b), even oxalate loses its effectiveness. As seen in Figure 7, it is in general that the order of precipitation power follows: oxalate > phosphate > fluoride > sulfate > carbonate.

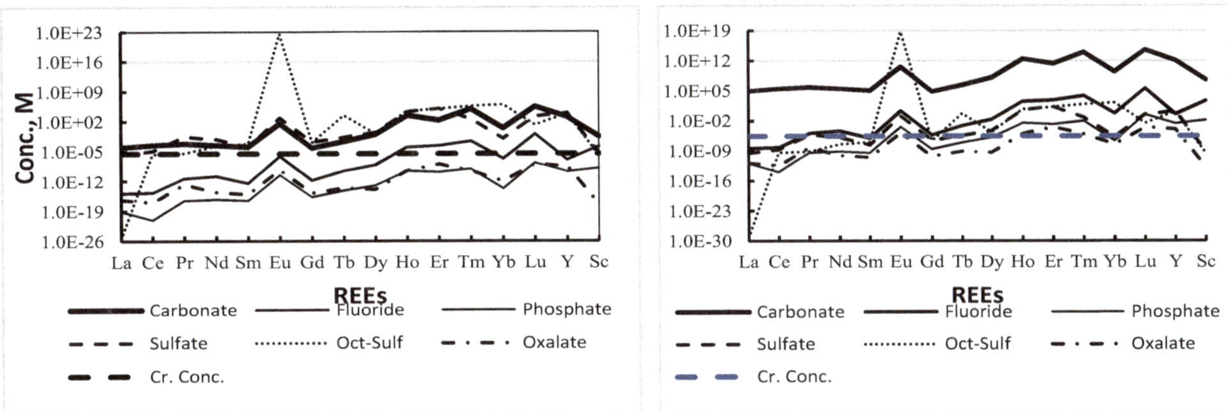

(a) (b)

Figure 7. Concentration of Re^{3+} in equilibrium with various precipitants at 10^{-4} M and at (**a**) pH 4 and (**b**) pH 1.

However, when REEs complex with the anions such as Cl^-, NO_3^-, and SO_4^{2-} or HSO_4^-, which almost always exist, especially at low pHs, since acids frequently used in dissolving Re-bearing minerals in the leaching operation are either HCl, HNO_3, or H_2SO_4. As discussed in the earlier studies [8,9], when such acids are present in the aqueous system, REEs immediately form complexes with these anions. In the presence of chloride, $ReCl_2^+$ is the most dominating species, $Re(NO_3)_2^+$ for the nitrate system and $Re(SO_4)_2^-$ is for the sulfate system. As a result, the calculations involved to find the equilibrium concentrations of REEs should be carried out using these chemical moieties in place of Re^{3+}. Consequently, the precipitation behavior of REEs will also change and therefore, Figure 8 will be the relevant figure in place of Figure 7 for the nitric system as an example.

It is very notable in the case of the nitrate system, the precipitation power of carbonate has overtaken that of fluoride and sulfate. Similar plots were made for the chloride system as well as the sulfate system and the results were quite similar but the intensity of precipitation was not as strong as the nitric system.

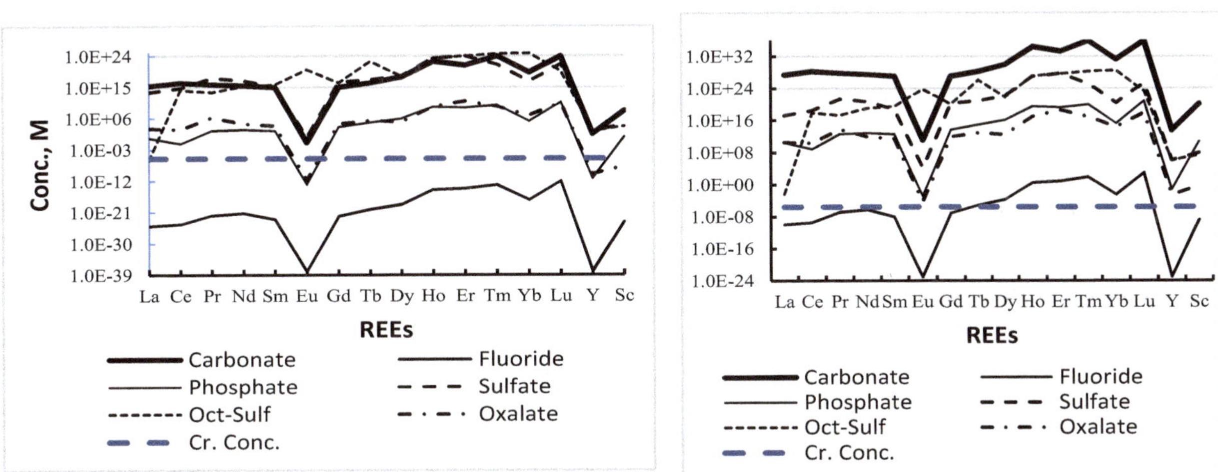

(a) (b)

Figure 8. Concentration of $ReCl_2^+$ in equilibrium with various precipitants at 10^{-4} M and at (**a**) pH 4 and (**b**) pH 1.

9. Conclusions

The chemical precipitation of REEs with precipitants including carbonate, fluoride, phosphate, sulfate, and oxalate was studied over a wide range of pH such as 1, 2, 4, and 6. The concentration of precipitants was fixed at 10^{-4} M chosen as the standard concentration for comparison. Such precipitation occurs in practice after primary leaching as a part of the purification process.

With all of the five precipitants studied, Ligjt REEs tend to precipitate much easier than Heavy REEs. The effect of anions associated with acids used in the leaching process was found to have had a significant impact on the overall precipitation process. In general, the complexed REEs with these anions tend to provide a conducive environment in precipitation resulting in an order of magnitude difference in the precipitation extent. The nitrate system was found to be most effective followed by sulfate and chloride systems. The pH of the system has another significant factor in determining the precipitation in that the difference in pKa values of the precipitants introduces precipitant moiety of the precipitants, which has a pronounced effect on the overall precipitation.

Acknowledgments: This paper was written in preparation of a lecture to be given at the 7th International Conference on Hydrometallurgy to be held in Ganzhou, China in May of 2020.

References

1. Amato, A.; Becci, I.; De Michelis, B.; Innocenzi, F.V.; Ippolito, N.M.; Gomez, P.J.; Beolchini, V.F. Sustainability analysis of innovative technologies for the rare earth elements recovery. *Renew. Sustain. Energy Rev.* **2019**, *106*, 41–53. [CrossRef]
2. Binnemans, K.; Jones, P.T.; Blanpain, B.; Gerven, T.V.; Yang, Y.; Walton, A.; Buchert, M. Recycling of rare earths: A critical review. *J. Clean. Prod.* **2013**, *51*, 1–22. [CrossRef]
3. Han, K.N.; Kellar, J.J.; Cross, W.M.; Safarzadeh, S. Opportunities and challenges for treating rare-earth elements. *Geosyst. Eng.* **2014**, *17*, 178–194. [CrossRef]
4. Gupta, C.K. Krishnamurthy. Extractive metallurgy of rare earth. *Int. Mater. Rev.* **1992**, *37*, 197–248. [CrossRef]
5. Vijayalakshmi, R.; Mishra, S.L.; Singh, H.; Gupta, C. Processing of xenotime concentrate by sulphuric acid digestion and selective thorium precipitation for separation of rare earths. *Hydrometallurgy* **2001**, *61*, 75–80. [CrossRef]
6. Yan, C.; Jia, J.; Liao, C.; Wu, S. Rare earth separation in China. *Tsinghua Sci. Technol.* **2006**, *11*, 241–247. [CrossRef]
7. Ahmed, S.H.; Helaly, O.S.; Ghany, E. Evaluation of rare earth double sulphate precipitation from monazite leach solutions. *Int. J Inorg. Bioinorg. Chem.* **2015**, *5*, 1–8.
8. Helaly, O.S.; Abd, M.S.; Ghany, E.; Borai, E.H.; Aly, H.F.; Abded Fattan, T.M. Separation of cerium, light and heavy rare earth concentrates from Egyptian crude monazite. *Chem. Technol.* **2015**, *10*, 184–192.
9. Chi, R.; Xu, Z. A solution chemistry approach to the study of rare earth element precipitation by oxalic acid. *Metall. Mater. Trans. B* **1999**, *30*, 189–195. [CrossRef]
10. Konishi, Y.; Noda, Y. Precipitation stripping of rare-earth carbonate powders from rare-earth-loaded carboxylate solutions using carbon dioxide and water. *Ind. Eng. Chem. Res.* **2001**, *40*, 1793–1797. [CrossRef]
11. Cantrell, K.J.; Byrne, R.H. Rare earth element complexation by carbonate and oxalate ions. *Geochim. Cosmochim. Acta* **1987**, *51*, 597–605. [CrossRef]
12. Vasconcellos, M.E.; da Queiroz, C.A.; Abrao, A. Sequential separation of the yttrium—Heavy rare earths by fractional hydroxide precipitation. *J. Alloys Compd.* **2004**, *374*, 405–407. [CrossRef]
13. Khawassek, Y.M.; Eliwa, A.A.; Gawad, E.A.; Abdo, S.M. Recovery of rare earth elements from El-Sela effluent solutions. *J. Radiat. Res. Appl. Sci.* **2015**, *8*, 583–589. [CrossRef]
14. Han, K.N. Effect of anions on the solubility of rare earth element-bearing minerals in acids. In *Mining, Metallurgy & Exploration*; Springer: Berlin/Heidelberg, Germany, 2019; Volume 36, pp. 215–225. [CrossRef]
15. Han, K.N. Effect of metal complexation on the solubility of rare earth compounds. In *Critical and Rare Earth Elements/Recovery from Secondary Resources*; Abhilash Akcil, A., Ed.; CRC Press: Boca Raton, FL, USA, 2019; pp. 59–84.

16. Aide, M.T.; Aide, C. Rare Earth Elements: Their Importance in Understanding Soil Genesis. *Int. Sch. Res. Not.* **2012**. [CrossRef]
17. Pourbaix, M. *Atlas of Electrochemical Equilibria*; National Association of Corrosion Eng: Houston, TX, USA, 1974; p. 644.
18. Dean, J.A. *Lange's Handbook of Chemistry*, 15th ed.; McGraw-Hill: New York, NY, USA, 1999; p. 51.
19. HSC Chemistry 5.11. *Chemical Reaction and Equilibrium Software with Extensive Thermochemical Database*; Version 5.0; Outokumpu Research Oy: Piori, Finland, 2002.
20. Wagman, D.D. The NBS tables of chemical thermodynamic properties: Selected values for inorganic and C_1 and C_2 organic substances in SI units. American chemical society and the American institute of physics for the national bureau of standards. *J. Phys. Chem. Ref. Data* **1982**, *11*, 407.
21. Kim, E.; Osseo-Asare, K. Aqueous stability of thorium and rare earth metals in monazite hydrometallurgy:Eh-pH diagrams for the systems Th-, La-, Nd-, (PO_4)-(SO_4)-H_2O at 25 °C. *Hydrometallurgy* **2012**, *113*, 67–78. [CrossRef]
22. Firsching, F.H.; Mohammadzadel, J. Solubillty Products of the Rare-Earth Carbonates. *Am. Chem. Soc.* **1986,** *31*, 40–42.
23. Firsching, F.H.; Brune, S.N. Solubility Products of the Trivalent Rare-Earth Phosphates. *J. Chem. Eng. Data* **1991**, *36*, 93–95. [CrossRef]
24. Speddin, F.H.; Jaffe, G.S. Conductances, Solubilities and Ionization Constants of Some Rare Earth Sulfates in Aqueous Solutions at 25°. *Inst At. Res. Dep. Chem.* **1953**, *76*, 882–884.
25. Migdiscov, A.; Williams-Jones, A.E.; Wagner, T. An experimental study of the solubility and speciation of the rare earth elements (III) in fluoride- and chloride-bearing aqueous solutions at temperatures up to 300 °C. *Geochim. Cosmochim. Acta* **2009**, *73*, 7087–7109. [CrossRef]
26. Chung, D.Y.; Kim, E.H.; Lee, E.H.; Yoo, J.Y. Solubility of rare earth oxalate in oxalic and nitric acid media. *J. Ind. Eng. Chem.* **1998**, *4*, 277–284.
27. De Vasconcellos, M.E.; da Rocha, S.M.R.; Pedreira, W.R.; da Queiroz, C.A.; Abrao, A. Solubility behavior of rare earths with ammonium carbonate and ammonium carbonate plus ammonium hydroxide: Precipitation of their peroxicarbonates. *J. Alloys Compd.* **2008**, *451*, 426–428. [CrossRef]
28. Kim, P.; Anderko, A.; Navrotsky, A.; Riman, R.E. Trends in structure and thermodynamic properties of normal rare earth carbonates and rare earth hydroxycarbonates. *Minerals* **2018**, *8*, 106. [CrossRef]
29. Tokunaga, S.; Haron, M.J.; Wasay, S.A.; Wong, K.F.; Laosangthum, K.; Uchiumi, A. Removal of fluoride ions from aqueous solutions by multivalent metal compounds. *Int. J. Environ. Stud.* **1995**, *48*, 17–28. [CrossRef]
30. Wu, S.; Zhao, L.; Wang, X.; Huang, X.; Zhang, Y.; Feng, Z.; Cui, D. Simultaneous recovery of rare earth elements and phosphorus from phosphate rock by phosphoric acid leaching and selective precipitation: Towards green process. *J. Rare Earths* **2019**, *37*, 652–658. [CrossRef]
31. Zakharova, B.; Komissarova, L.; Trasking, V.; Naumov, S.; Melnikov, P. Precipitation of Rare Earth Phosphates from H_3PO_4 Solutions. *Phosphorus* **1996**, *111*, 1–4. [CrossRef]
32. Moeller, T.; Kremers, H.E. Observations on Rare Earths Double Sodium Sulfate Precipitation for Separation of the Terbium and Yttrium Earths. *Ind. Eng. Chem. Anal.* **1945**, *17*, 44–45. [CrossRef]

7

A Comparative Study of Porphyry-Type Copper Deposit Mineralogies by Portable X-ray Fluorescence and Optical Petrography

Connor A. Gray and Adrian D. Van Rythoven *

Department of Environmental, Geographical, and Geological Sciences, Bloomsburg University of Pennsylvania, Bloomsburg, PA 17815, USA; connorgray@mymail.mines.edu

* Correspondence: avanrythov@bloomu.edu

Abstract: Porphyry-type deposits are crucial reserves of Cu and Mo. They are associated with large haloes of hydrothermal alteration that host particular mineral assemblages. Portable X-ray fluorescence analysis (pXRF) is an increasingly common tool used by mineral prospectors to make judgments in the field during mapping or core logging. A total of 31 samples from 13 porphyry copper deposits of the Western Cordillera were examined. Whole-rock composition was estimated over three points of analysis by pXRF. This approach attempts to capture the rapid and sometimes haphazard application of pXRF in mineral exploration. Modes determined by optical petrography were converted into bulk rock compositions and compared with those determined by pXRF. The elements S, Si, Ca, and K all were underestimated by optical mineralogy, and the elements Cu, Mo, Al, Fe, Mg, and Ti were overestimated by optical mineralogy when compared with pXRF results. Most of these porphyry samples occur in veined porphyritic quartz monzonite that is characteristic of these deposits. Sulfide and silicate vein stockworks are pervasive in most of the samples as well as dissemination of sulfides outwards from veinlets. Ore minerals present include chalcopyrite and molybdenite with lesser bornite. Chalcocite, digenite, and covellite are secondary. Potential sources of analytical bias are discussed.

Keywords: Western Cordillera; pXRF; North America; modal mineralogy; optical mineralogy; porphyry deposit; copper; molybdenum; major element; handheld X-ray fluorescence

---

1. Introduction

1.1. Porphyry-Type Deposits

Porphyry-type deposits, often termed just "porphyry deposits", represent one of the most studied geological systems on Earth, and the economic fortunes that entail from the proper extraction of the metals found in these deposits have pushed the academic and industrial communities to study them. This deposit model that came to popularity in the mid-late 1960s [1,2], was summarized by Sillitoe [3] and then later expanded upon [4–8]. Porphyry-type deposits represent 60% of world Cu production annually, and 65% of known world Cu resources [6]. These deposits also represent half of the world's Mo production and may also contain economic grades of Ag, Zn, W, Sn, Re, and Au [7]. Low-grade resources of Pt-group elements are also possible [9,10]. Tonnages commonly exceed one billion [8]. An example of a deposit of this scale is Bingham Canyon (Utah), a deposit analyzed in this study, at 2.6 Gt of ore [11]. These high tonnages are offset by lower grades of metals extracted, typically between 0.5–1.0 wt% Cu [7]. The deposits are typically found along convergent plate margins where the subduction of an oceanic plate below the adjacent continental or oceanic plate drives partial melting, thus creating arc magmatism [3]. Although similar types of deposits occur in rarer continental-continental orogens [12]. These magmatic arc systems allow metals found in greater concentrations deeper in the earth to rise buoyantly as magma and emplace themselves as intrusive bodies capable of metal concentration. These systems cool, and depressurization processes occur such that metals are highly concentrated in an immiscible hydrothermal fluid located at the top of such intrusive bodies. These metals are then injected into the surrounding rock by hydraulic fracturing forming large stockworks of cross-cutting veins that contain concentrated metals [8]. The particular samples analyzed in this study are from the Western Cordillera of the Americas. Deposits of this type are essential to the growth of economies in developing countries, such as Chile where copper mining has contributed an average of 10% of GDP over the past 20 years [13]. Deposits found along the Western Cordillera of North America are, or have been influential regarding the economies of the regions and countries these deposits are found in.

1.2. Portable X-ray Fluorescence

X-ray fluorescence analysis produces compositional data by exciting the sample in question with X-rays. Those X-rays will ionize the sample's atoms by knocking out core-shell electrons. Outer-shell electrons will then fall/relax into the subsequent core-shell vacancy. As this involves a loss of energy, a photon is emitted with energy equal to the difference in electron orbitals. The energy quantum of the photon is unique to the element in question and is also in the X-ray portion of the electromagnetic spectrum. Thus, the X-ray's energy is characteristic of the element and the number of photons with that quantum is proportional to the element's concentration in the sample. Engineering advances over the last few decades have allowed this laboratory-scale instrumentation to be miniaturized into a portable format. A variety of commercial portable and handheld X-ray fluorescence analyzers are now available [14]. Portable X-ray fluorescence (pXRF) is an increasingly common method for on-site and rapid materials analysis. For exploration and development of mineral deposits, it has particularly been applied to geochemical exploration [15–17], assessments of soil contamination [18], core logging [19], and mineralization type [20,21]. This is a method that requires little training, is portable, fast, fairly affordable, and can provide useful data. However, compared to other lab methods (including more sophisticated, lab-based, and non-portable XRF analyzers), analyses can be very susceptible to measurement bias, insensitivity to elements lighter than Mg or Al, and matrix effects [22,23]. Whereas "portable" can refer to a variety of XRF configurations, for this study it refers to handheld instruments, typically resembling a pistol.

As pXRF is often used as a tool by field or core logging geologists, it is typically one of the first analytical methods applied to samples. The absence of typical sample preparation in the field, the small sample point size, and human bias in selecting what points to analyze introduce additional uncertainties to the data. Although they may be collected under less than optimal analytical conditions, results from pXRF can influence decisions regarding geologic mapping, cross-sections, and selection of samples for more detailed assay/investigation. Another investigative method that can follow pXRF is optical petrography. In fact, data from pXRF may influence a geologist to select a sample for thin sectioning and petrographic examination. A geologist can later re-interpret and reconcile the pXRF data in the context of sample mineralogy. Likewise, pXRF data can be useful in interpreting the observed mineralogy in thin sections. The combined use of complementary but independent methods such as pXRF and optical petrography has the potential to augment the abilities of the investigating geologist in a fairly fast and economical manner.

In addition to its portability and ease of use, pXRF is typically non-destructive. Outcrop, drill core, grab samples, soil, and thin section billets (this study) can be analyzed with no crushing, grinding, fusion, dissolution, etc., that lab-based XRF or ICP-MS require. Similarly, polished thin sections that can be made from a modest amount of sample, are not expended when examined, preserve mineral textures, and can later be analyzed by other methods such as electron microscopy or other microbeam methods.

We present that pXRF used in a rapid fashion with no standardization, in the manner of many field applications, can still provide semi-quantitative data that are useful to answer geological questions in the field. We simulate these haphazard conditions using point analysis on thin-section billets of rock samples from porphyry-type deposits. Sources of analytical uncertainty arising from these conditions are examined and discussed. The results are compared with those of optical petrography, another common, albeit much older, standby in non-destructive geological sample characterization.

2. Materials and Methods

2.1. Sample Locations

The 31 hand samples used in this comparative study are from 13 different deposits found along the oceanic-continental subduction zone of the Western Cordillera of North America (Table 1, Figure 1). These are Cu ± Mo porphyry-type deposits and represent different portions within each deposit (for those deposits with more than one sample). The samples include specimens from the hypogene and supergene zones of these deposits. The zone of interest for this study was primarily the hypogene. These samples were collected and donated by Peter H. Kirwin and John R. Ray, formerly of the American Copper and Nickel Company.

Table 1. List of samples investigated in this study and their origins.

Sample	Locality	Location	Sample	Locality	Location
Be-1	Bethlehem	British Columbia	L-1	Lornex	British Columbia
Be-4	Bethlehem	British Columbia	Mo-2	Morenci	Arizona
Be-6	Bethlehem	British Columbia	Mo-4	Morenci	Arizona
Bi-2	Bingham Canyon	Utah	Mo-6	Morenci	Arizona
Bi-3	Bingham Canyon	Utah	Mo-8	Morenci	Arizona
Bu-2	Butte	Montana	OH-1	Orange Hill	Alaska
Bu-6	Butte	Montana	S-1	Sierrita	Arizona
Bu-7	Butte	Montana	S-2	Sierrita	Arizona
C-3	Cananea	Mexico	S-4	Sierrita	Arizona
C-4	Cananea	Mexico	SM-3	San Manuel	Arizona
CC-1	Copper Canyon	Nevada	SM-4	San Manuel	Arizona
F-1	Florence	Arizona	Y-3	Yerington	Nevada
F-2	Florence	Arizona			
F-3	Florence	Arizona			
IP-1	Ithaca Peak	Arizona			
IP-2	Ithaca Peak	Arizona			
IP-3	Ithaca Peak	Arizona			
IP-10	Ithaca Peak	Arizona			
IP-12	Ithaca Peak	Arizona			

2.2. Analytical Methods

In this comparative study two methods were used to determine the overall mineralogy of the samples. These methods include using pXRF giving elemental weight percent readings for each sample, as well as using polarized transmitted and reflected light microscopy to estimate mineral abundances and related textures. Limiting the study to these two methods allows the investigation to be conducted in an essentially non-destructive manner at minimal cost.

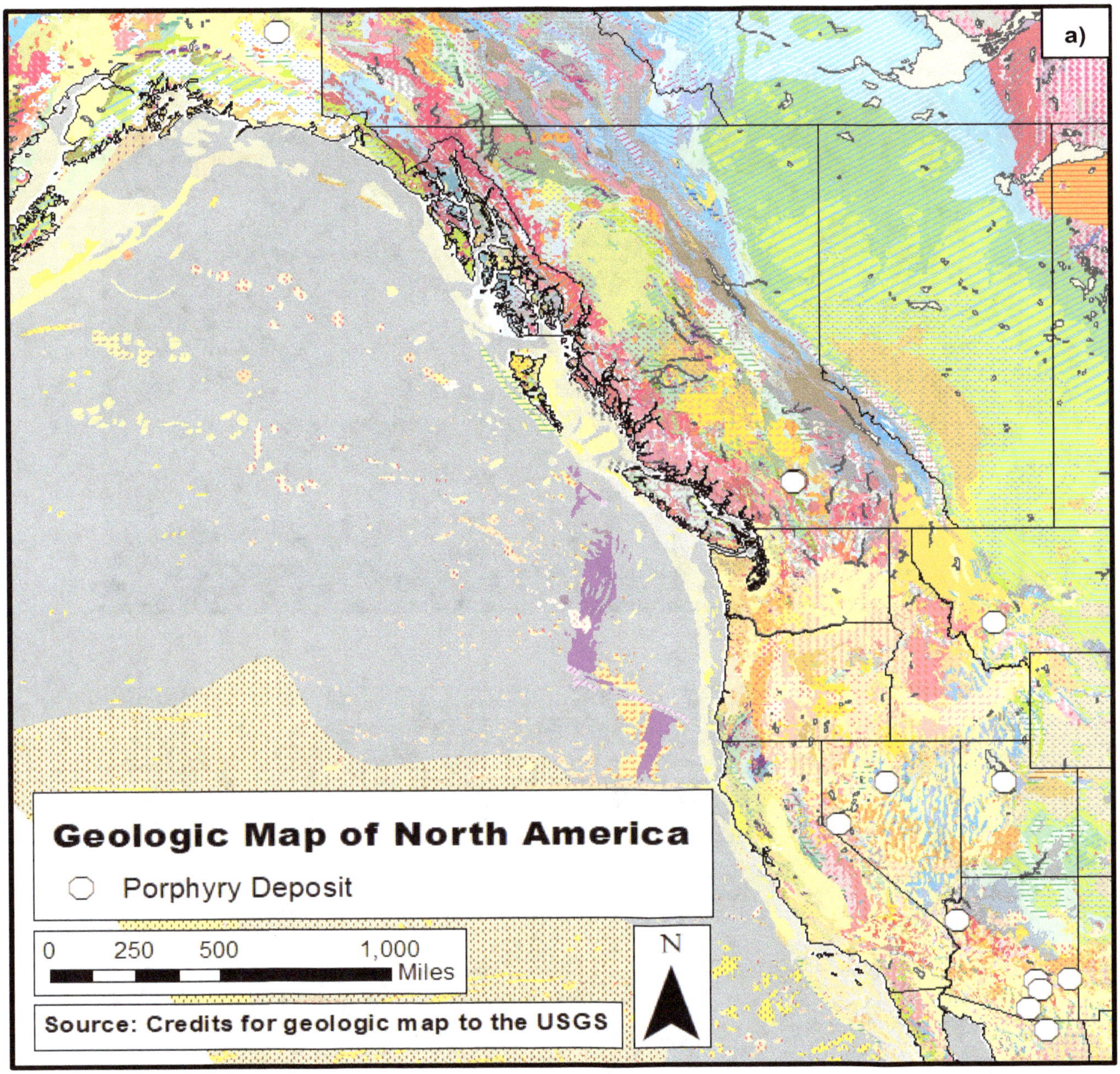

Figure 1. *Cont.*

Figure 1. *Cont.*

Figure 1. Maps of deposit locations in western North America that were sampled. (**a**) Overall geologic map of the Western Cordillera (map constructed in ArcGIS). (**b**) Satellite image of the Orange Hill deposit in Alaska. (**c**) Satellite image of the British Columbian, Nevadan, Montanan, and Utahan deposits. (**d**) Satellite image of Arizonan and Mexican deposits.

2.2.1. Optical Petrography

The 31 samples were selected from a larger collection on the basis of interesting mineralogy or textures visible in hand samples, geographic location, and budget limitations (for thin sectioning). They were cut into rectangular billets by a rock saw (e.g., Figure 2). The billets were then used to shave off 30 μm × 26 mm × 46 mm polished thin sections for further analysis under a petrographic microscope. The thin sectioned samples were then examined under a Leica DM2700 P microscope (Leica Microsystems Inc., Buffalo Grove, IL, USA) in transmitted and reflected light modes. In the first assessment of these thin sections, the textures were documented along with a qualitative listing of minerals present in each sample. More in-depth qualitative observations were then made at this point, and the genetic sequencings of minerals along with their characteristics were noted. In the second extensive assessment of these samples using petrographic methods, the overall mineral modes were determined. This assumed that the mineral abundances in terms of thin section area are representative of the volume% abundances (true modes) of the mineral.

Figure 2. Example photographs of two samples from this study. (**a**) Bu-2 (Butte, Montana) hand sample showing sulfide vein stockwork characteristic of porphyry-type deposits and alteration haloes surrounding the veins. (**b**) Thin section billet cut from the sample in (a) and impregnated with epoxy. (**c**) Be-6 (Bethlehem, British Columbia) hand sample showing sulfide vein stockwork characteristic of porphyry-type deposits and alteration adjacent to the veins. (**d**) Thin section billet cut from the sample in (c) and impregnated with epoxy. (**e**) IP-12 (Ithaca Peak, Utah) hand sample showing a high-grade vein of Cu-Fe sulfides. (**f**) Y-3 (Yerington, Nevada) somewhat brecciated hand sample showing cross-cutting veinlet stockwork formed in multiple generations of fractures. American quarter for scale (~2.43 cm in diameter). Locations to be cut for thin-sectioning are outlined in black marker.

2.2.2. Portable X-ray Fluorescence

When the overall modes were determined for each sample, the bulk elemental compositions were then determined by pXRF analysis of the flat face of the corresponding billet the thin section was cut from. A Thermo Scientific Niton^{TM} XL3t-950 GOLDD+ (Thermo Fisher Scientific, Waltham, MA, USA) was the pXRF analyzer used in this study. All analyses were done in Mining Cu/Zn mode. A 10-minute warmup and system check were conducted prior to every session of analyses. Analysis quality was checked for instrument drift using two manufacturer-provided standards: CRM 180-649 NIST 2709a and blank 180-647 99.995% SiO_2 blank that were checked using the pXRF prior to the study and after the study. These standards are supplied as fine homogenized powders. No significant deviation from known values (>10%) was noted. Samples were analyzed with pXRF for a 120-s interval, with 30 s each for the main (Mo, Zr, U, Rb, Th, Pb, Se, As, Hg, Nb, Cu, Ni, Co, Fe, Mn, Sr, Y), low (Sc, Ca, K, V, Ti, Cr), high (Ba, Cs, W, Te, Sb, Sn, Cd, Ag, Pd, La, Ce, Pr, Nd), and light (Al, P, Si, Cl, S, Mg) characteristic X-ray ranges of elements using a series of filters. This allowed for fairly precise readings of elemental weight percent (wt%) for each sample. Each sample was analyzed three times in three separate locations on the sectioned surface of the billet in an attempt to address inhomogeneities within these billets. The average elemental composition was determined using all three readings per sample. In the case of three very different analyses, a fourth location was analyzed on the billet.

2.3. Data Calculation

Modes from petrographic microscopy observations were converted into estimates of bulk elemental composition for each sample. The modes, taken as volumetric percentages, were converted into mineral mass percentages using established densities. These mineral mass proportions were then converted into elemental proportions. Density and compositional values were sourced from webmineral.com. Elements not analyzable by pXRF (mostly elements lighter than Mg) were left out of calculations in the modal abundance conversion to allow for simpler comparison. Estimates of solid solution members were made for plagioclase, chlorite, and muscovite. In terms of atoms per formula unit (apfu) it is assumed that plagioclase has 0.5 Na, 0.5 Ca, 1 Al, and 3 Si, chlorite has 3.75 Mg and 1.25 Fe, and biotite has 2.5 Mg and 0.5 Fe.

The whole-rock compositions calculated from the modes were then compared with the average whole-rock compositions determined by pXRF for purposes of comparison and to audit the accuracy of each method against the other.

3. Results

3.1. Hand Sample Examination

At the hand sample scale, prior to the thin sectioning of each sample, porphyry-type copper deposit characteristics were evident (e.g., Figure 2; see [24] for a review). Characteristic veining associated with hydraulic fracturing was seen in many samples as mineral assemblages of silicates and sulfides with rare carbonates. Sulfides associated with the hypogene system are typically dominated by chalcopyrite and pyrite. Silicates in this zone displayed less alteration than that of more near-surface samples, maintaining the dominantly quartz monzonitic mineralogy seen in a majority of hypogene samples although chalky sericitization of feldspars is common in the host rocks. In hand samples, characteristic zoning of alteration could be seen with some ambiguity but overall a grade from hypogene to supergene alteration was clear (c.f., [7]). The oxidation and hydration of hypogene sulfide minerals was a clear indicator of supergene alteration.

3.2. Optical Petrography

In thin sections, the observations made of the hand samples were expressed in clearer detail, and the overall modes observed in hand samples are consistent with those observed at a microscopic level (Table 2, Figure 3). The main sulfides present in hypogene samples show characteristic veins/veinlets of chalcopyrite, bornite, and/or pyrite (e.g., Figure 3g,f), along with sulfide grains disseminated distally from veinlets (e.g., Figure 3a,h). These veinlets also comprise hydrothermal quartz along with sericite and chlorite, and rarely calcite (e.g., Figure 3g). Molybdenite was the only Mo-bearing sulfide seen in thin section (e.g., Figure 3b,c,h, Table 2). Samples affected by some supergene fluid alteration have distinct mineralogies. The sulfide assemblages of the hypogene zone, though starting rich in Cu sulfides, quickly grade into zones of pyrite showing a lack of Cu concentration distally (e.g., Figure 3f). In the supergene samples, chalcocite, covellite, and/or digenite are observed as rims on partially replaced pyrite and/or chalcopyrite (Figure 3a). Specular hematite (possibly with magnetite in some cases) occurs in some samples and also appears to be hydrothermal. This is distinct from the late hematite after sulfides seen in some partially oxidized samples (see below).

Table 2. Mineral modes (in volume%) of samples. Mineral abbreviations as in Whitney and Evans [25].

Sample:	Be-1	Be-4	Be-6	Bi-2	Bi-3	Bu-2	Bu-6	Bu-7	C-3	C-4	CC-1	F-1	F-2	F-3	IP-1	IP-2
Ccp	0.3	1	1.5	1	1	0.5	–	1.5	–	–	1	4	0.8	2.5	–	0.1
Cv	–	–	–	0.3	–	0.2	–	0.3	–	2	–	–	–	–	–	1
Cct	–	–	–	1.5	0.1	–	0.5	–	–	–	–	–	–	–	0.5	0.5
Bn	–	1	0.2	0.1	1	–	–	–	–	–	–	–	–	–	–	–
Dg	–	–	–	–	0.3	–	–	–	–	–	–	–	–	–	–	–
Mol	–	–	–	0.2	0.1	–	–	0.5	0.5		–	–	–	–	–	–
Py	1	–	–	2	–	9	20	45	4	7	–	–	2	2	3	4
Chl	8	4.5	7	1	–	–	–	–	–	–	–	–	0.4	0.3	–	–
Rt	0.3	–	0.3	0.5	0.2	1	–	–	–	–	0.3	0.5	1	0.4	1	–
Hem	2	0.1	–	–	–	3	–	–	–	0.5	–	0.5	0.5	–	–	–
Cal	–	2	–	–	–	–	–	–	–	–	–	–	–	3	–	–
Qz	57.4	64.9	60	48.7	61.3	56.3	74.5	48.7	60.5	50	67	58	56.3	59.8	64.5	67.4
Or	–	–	–	–	–	–	3						10	–	–	–
Ms	27	25.5	22	42.7	25	30	2	4	35	40.5	23.7	34	12	32	31	27
Bt	–	0.5	–	1	10	–	–	–	–	–	8	3	10	–	–	–
Pl	3	–	8	–	–	–	–	–	–	–	–	–	5	–	–	–
Ap	1	0.5	1	1	1	–	–	–	–	–	–	–	2	–	–	–
Total	100.0	100.0	100.0	100.0	100.0	100.0	100.0	100.0	100.0	100.0	100.0	100.0	100.0	100.0	100.0	100.0
Sample:	**IP-3**	**IP-10**	**IP-12**	**L-1**	**Mo-2**	**Mo-4**	**Mo-6**	**Mo-8**	**OH-1**	**S-1**	**S-2**	**S-4**	**SM-3**	**SM-4**	**Y-3**	
Ccp	0.1	0.05	22	6	1	–	2	1.5	1	1	2	0.5	3.5	4.5	2	
Cv	0.1	0.1	–	–	–	–	–	–	–	–	–	–	–	–	–	
Cct	–	0.2	–	–	–	–	–	–	–	–	–	–	–	–	–	
Bn	–	–	10	–	–	–	–	–	–	–	–	–	–	–	–	
Dg	–	–	–	–	–	–	–	–	–	–	–	–	–	–	–	
Mol	0.3	–	–	0.1	–	–	–	–	–	–	0.2	0.3	–	–	–	
Py	1	9	–	–	5	2.5	–	1.4	8	1.6	2	3	5	1	–	
Chl	–	–	5	–	13	–	1.5	–	11	11	2	0.5	–	–	4	
Rt	1	–	–	–	1	0.3	0.3	0.2	–	0.5	0.5	0.5	0.5	1.5	0.3	
Hem	1	–	0.5	–	–	–	–	–	0.5	–	–	0.3	–	1	1.5	
Cal	–	–	–	–	–	–	0.3	0.9	–	–	–	1	–	–	–	
Qz	43.3	85.65	44.5	85.4	45	94.9	54.4	53.4	26.5	35	62.5	50.9	43	36.3	55.7	
Or	16	–	1	–	–	–	–	7.1	–	8.9	4	3	–	–	5	
Ms	10	5	12	8	34	2.3	34.5	33	15	10	25	36	37	26	25	
Bt	24.5	–	1	–	–	–	2	2.5	3	–	–	–	4	25	–	
Pl	1	–	4	–	–	–	5	–	35	32	1.3	4	6.3	4	6	
Ap	1.7	–	–	0.5	1	–	–	–	–	–	0.5	–	0.7	0.7	0.5	
Total	100.0	100.0	100.0	100.0	100.0	100.0	100.0	100.0	100.0	100.0	100.0	100.0	100.0	100.0	100.0	

The host rocks vary depending on the particular deposit. Most are porphyritic hypabyssal/intrusive rocks that are quartz monzonitic/latitic, although granitic/rhyolitic and granodioritic/dacitic assemblages occur (c.f., [26]). Primary quartz is present in varying amounts but silicification is prevalent. Biotite, apatite, rare amphibole, and rare zircon are also present as primary minerals. In addition to silicification, potassic alteration, sericitization (mostly muscovite after plagioclase and orthoclase) and chloritization (of biotite) are prevalent (e.g., Figure 3e). Carbonation (of plagioclase and mafic minerals) occurs but is rare. In many samples, silicification and/or sericitization obliterated much of the primary phases and textures.

Supergene oxidation is evident in some samples by partial replacement of pyrite by masses of hematite ± goethite. These samples also showed some spots of decomposition of Cu-bearing sulfides as a greenish tinge or patina (i.e., chrysocolla-malachite) on hand samples, however, none of the areas thin sectioned contained these spots to any significance.

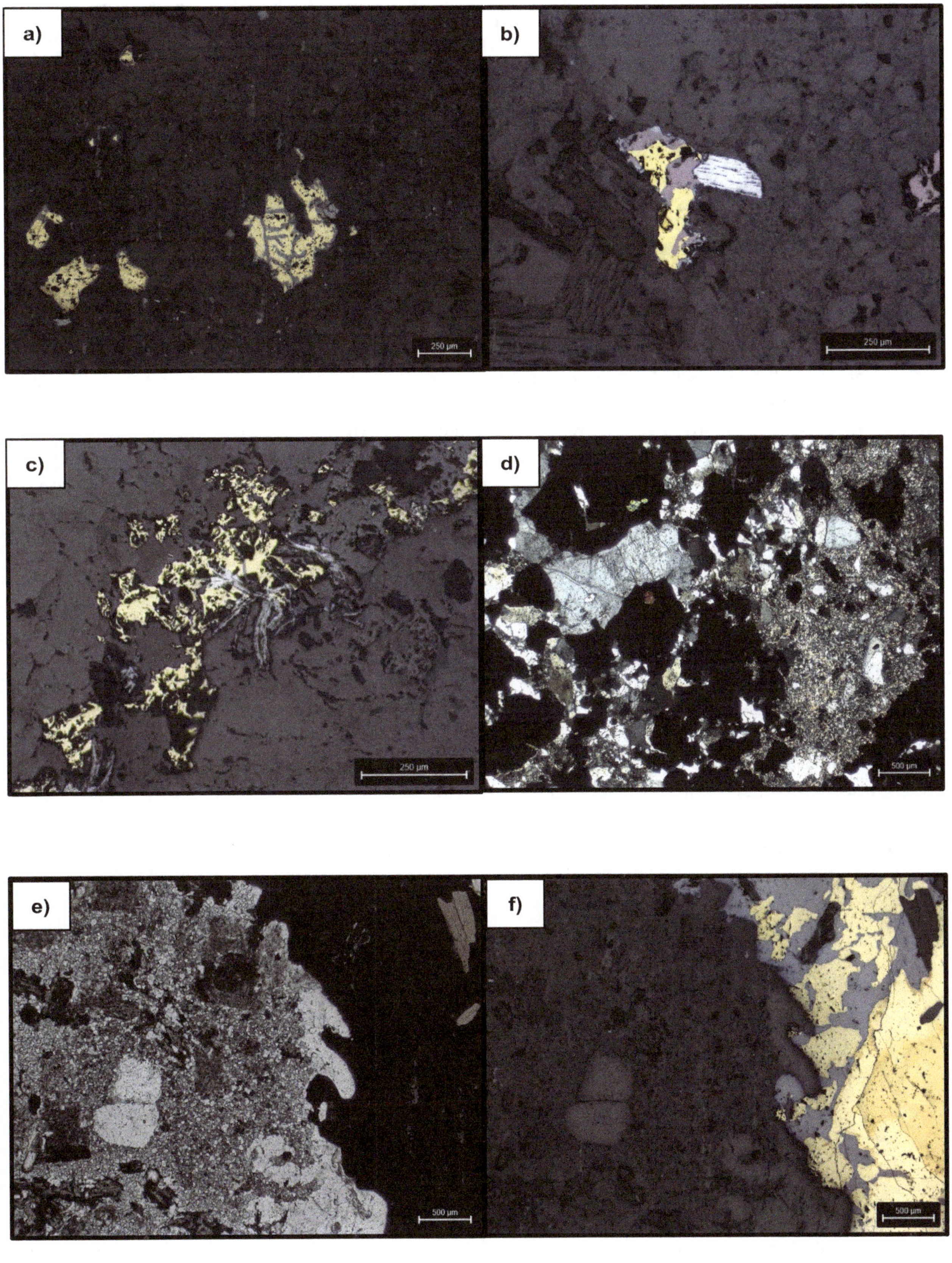

Figure 3. *Cont.*

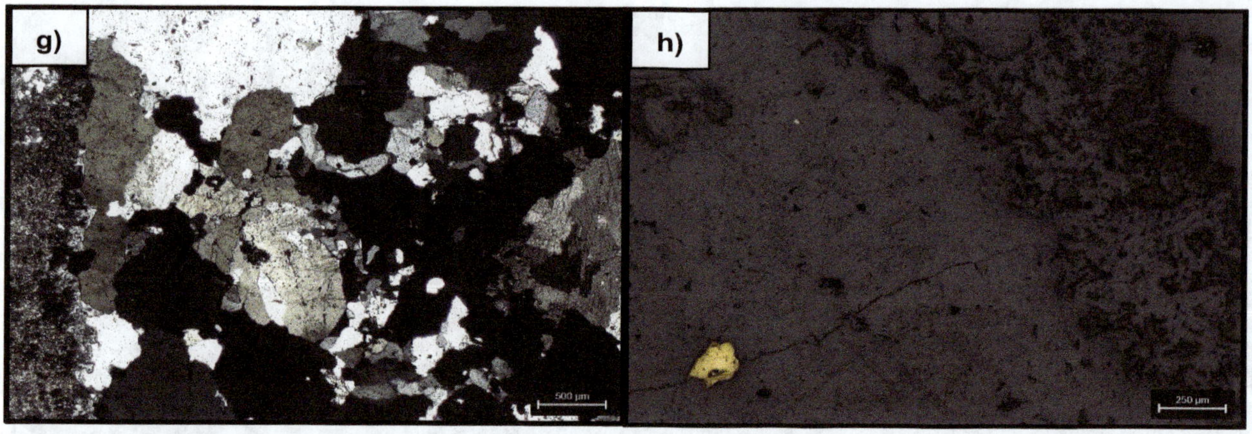

Figure 3. Example photomicrographs of two samples from this study. (**a**) Bi-2, chalcopyrite with alteration to chalcocite/covellite along crystal rims and fractures. Surrounding phases are quartz and sericite. RPPL (reflected plane-polarized light). (**b**) Bi-3, chalcopyrite, bornite, and molybdenite intergrown. Digenite alteration along the bornite and chalcopyrite crystal rims. Surrounding phases are biotite, quartz, and sericite. RPPL. (**c**) S-2, chalcopyrite and molybdenite in a quartz vein. RPPL (**d**) Bu-6, large pyrite crystals (some with possible zircon inclusions) in quartz + sericite + orthoclase. TXPL (transmitted cross-polarized light). (**e**) IP-12, vein of chalcopyrite + bornite + biotite in sericitized and chloritized porphyritic biotite quartz monzonite/latite. TPPL (transmitted plane-polarized light). (**f**) Same view as in (**e**), but in RPPL. (**g**) F-3, quartz + pyrite + chalcopyrite + calcite vein in sericitized monzonite host. (**h**) L-1, small grains of chalcopyrite (lower left) and molybdenite (upper right) in quartz and sericite. RPPL.

3.3. Estimation of Bulk Compositions from Modes

The estimated bulk rock compositions for these modes was calculated by converting the observed mineral abundances to weight proportions, and then elemental proportions based on idealized compositions of the minerals. Based on the observed minerals, only the elements Mg, Al, Si, P, S, K, Ca, Ti, Fe, Cu, and Mo were considered (Table 3).

Table 3. Bulk rock compositions (elemental wt%) for selected elements calculated from modes determined by optical petrography.

Sample	Be-1	Be-4	Be-6	Bi-2	Bi-3	Bu-2	Bu-6	Bu-7	C-3	C-4	CC-1	F-1	F-2	F-3	IP-1	IP-2
Mg	1.2	0.8	1.0	0.3	1.6	0.0	0.0	0.0	0.0	0.0	1.3	0.5	1.6	0.0	0.0	0.0
Al	6.3	5.8	6.1	8.7	5.8	5.7	0.6	0.6	7.2	7.9	5.5	7.2	4.6	6.6	6.4	5.5
Si	32.3	35.5	35.4	30.2	34.5	28.9	30.4	16.5	33.9	29.2	37.0	33.4	33.8	33.2	35.0	35.1
P	0.2	0.1	0.2	0.2	0.2	0.0	0.0	0.0	0.0	0.0	0.0	0.0	0.4	0.0	0.0	0.0
S	1.1	1.0	0.9	3.4	1.2	8.4	17.4	33.2	4.2	7.5	0.5	2.1	2.3	3.2	3.1	4.6
K	2.7	2.6	2.2	4.2	3.5	2.8	0.5	0.3	3.5	3.8	3.2	3.7	3.5	3.2	3.1	2.7
Ca	0.5	1.0	1.0	0.4	0.5	0.0	0.0	0.0	0.0	0.0	0.0	0.0	1.3	1.2	0.0	0.0
Ti	0.3	0.0	0.3	0.4	0.2	0.8	0.0	0.0	0.0	0.0	0.3	0.5	0.9	0.4	0.9	0.0
Fe	1.9	1.4	1.6	2.3	1.4	10.9	15.0	28.5	3.3	6.2	1.1	2.7	3.5	3.1	2.5	3.4
Cu	0.2	1.7	1.0	3.3	2.3	0.4	0.7	0.8	0.0	2.1	0.5	2.1	0.4	1.3	0.8	2.0
Mo	0.0	0.0	0.0	0.2	0.1	0.0	0.0	0.4	0.6	0.0	0.0	0.0	0.0	0.0	0.0	0.0
Balance	53.5	50.1	50.2	46.2	48.7	42.0	35.4	19.6	47.3	43.2	50.5	47.9	47.8	47.7	48.2	46.7
Sample	**IP-3**	**IP-10**	**IP-12**	**L-1**	**Mo-2**	**Mo-4**	**Mo-6**	**Mo-8**	**OH-1**	**S-1**	**S-2**	**S-4**	**SM-3**	**SM-4**	**Y-3**	
Mg	3.7	0.0	0.8	0.0	1.9	0.0	0.5	0.4	2.0	1.6	0.3	0.1	0.6	3.7	0.6	
Al	5.2	1.0	0.3	1.7	7.9	0.5	8.0	7.7	7.7	7.0	5.8	8.1	8.1	7.1	6.6	
Si	30.7	37.8	20.9	40.0	28.2	43.7	34.0	33.4	26.3	31.8	34.9	32.0	28.4	26.6	34.0	
P	0.3	0.0	0.0	0.1	0.2	0.0	0.0	0.0	0.0	0.0	0.1	0.0	0.1	0.1	0.1	
S	1.3	8.6	13.9	3.3	5.2	2.5	1.1	2.2	7.9	2.1	3.2	3.4	6.4	3.2	1.1	
K	5.4	0.5	1.2	0.8	3.3	0.2	3.7	4.5	1.7	2.2	3.0	3.9	3.9	4.8	3.2	
Ca	0.8	0.0	0.2	0.2	0.4	0.0	0.5	0.4	2.4	2.3	0.3	0.7	0.7	0.6	0.7	
Ti	0.9	0.0	0.0	0.0	0.9	0.3	0.3	0.2	0.0	0.5	0.5	0.5	0.4	1.3	0.3	
Fe	3.9	7.4	11.5	2.8	6.0	2.2	1.2	2.1	8.9	3.1	2.8	3.2	5.8	5.7	3.4	
Cu	0.2	0.5	19.7	3.2	0.5	0.0	1.1	0.8	0.5	0.5	1.0	0.3	1.7	2.2	1.1	
Mo	0.3	0.0	0.0	0.1	0.0	0.0	0.0	0.0	0.0	0.0	0.2	0.4	0.0	0.0	0.0	
Balance	47.2	44.3	31.5	47.8	45.5	50.6	49.7	48.5	42.7	48.8	47.8	47.6	43.7	44.8	49.1	

3.4. Portable X-ray Fluorescence Analysis of Bulk Compositions

Analysis by pXRF provided compositional information in terms of elements Mg and heavier (Table 4). Overall composition across the samples varies significantly. Significant variations in Mg, Si, P, S, K, Ca, Ti, Fe, Cu, and Mo content occur. Other elements analyzed for, but are low in concentration (<0.5 wt%) and do not show significant variation are: Cl, V, Cr, Mn, Co, Ni, Zn, As, Se, Rb, Sr, Y, Zr, Nb, Ag, Cd, Sn, Sb, Ba, La, Ce, Pr, Nd, W, Au, Pb, Th, and U. Concentrations for V, Co, Se, Y, Sn, Sb, Au, and U were essentially non-detectable (less than approximately 0.001 wt%). Uncertainty in analyses is dominated by the variance between spots compared to the actual instrumental error with the possible exception of Mg, the element with the highest uncertainty in terms of measurement.

In terms of the economic metals, Cu grades vary from 0 to 14.8 wt% (mean: 0.8 wt%) and Mo grades vary from 0 to 0.3 wt% (mean 0.02 wt%). For the other elements, silicon is the most abundant and ranges from 23.2 to 53.8 wt% (mean: 37.4 wt%). Al is the second most consistently abundant element (0.1 to 8 wt%, mean: 4.9 wt%) followed by K (0 to 8.5 wt%, mean: 3.5 wt%). Elements that are high abundance in some samples (those with high sulfide mineral content) are Fe (0.5 to 20.8 wt%, mean: 3.6 wt%) and S (0.1 to 40 wt%, mean: 5.4 wt%). Less significant variations are Mg (0 to 3.5 wt%, mean: 0.6 wt%), P (0 to 0.4 wt%, mean 0.1 wt%), Ca (0 to 3.2 wt%, mean 0.8 wt%), and Ti (0 to 1.2 wt%, mean 0.3 wt%). The sum of chalcophile element concentrations: Fe, Cu, and Mo, correlate with S content (Figure 4).

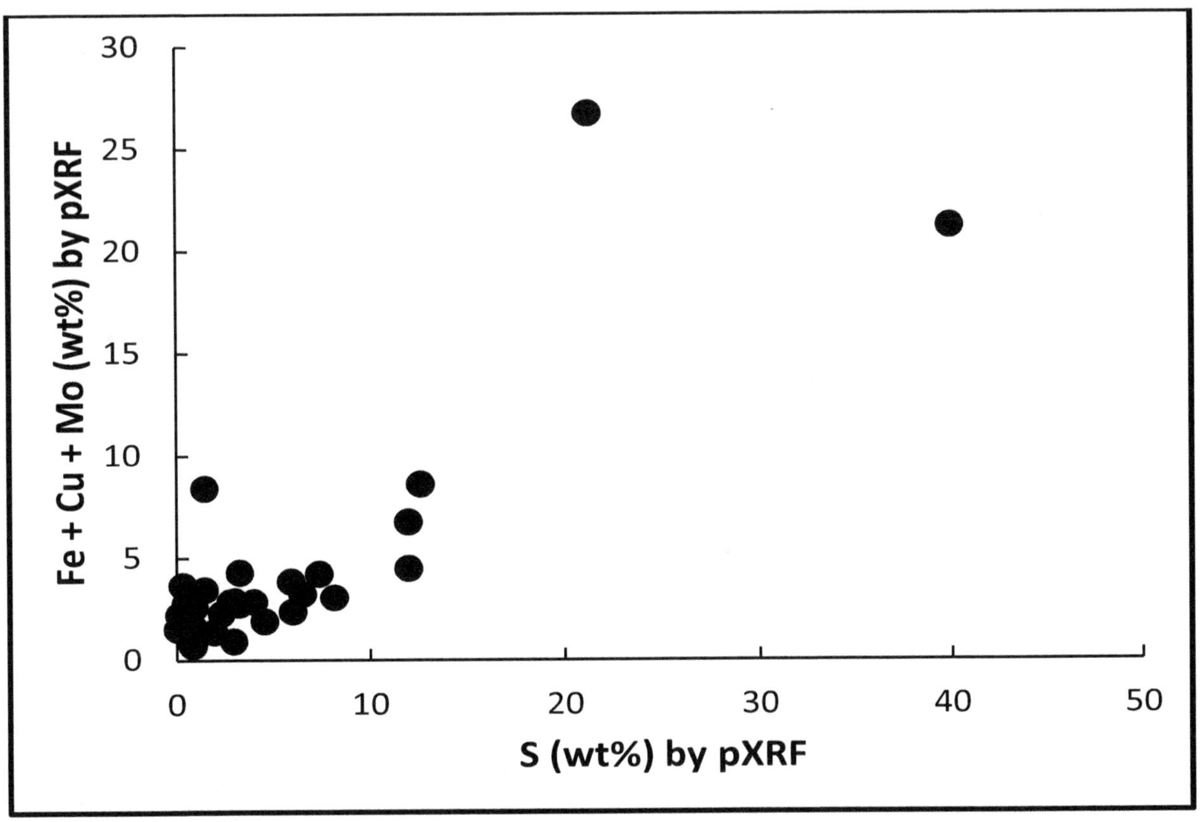

Figure 4. Plot of S content vs. sum of chalcophile elements determined by pXRF.

3.5. Comparison of Bulk Compositions from Both Methods

Calculated bulk rock composition (in terms of selected elements) based on estimated modes from optical petrography and the estimated bulk rock compositions based on pXRF analyses are compared on an element-by-element basis (Figure 5).

Table 4. Bulk rock compositions (elemental wt%) for selected elements determined by the average of point analyses for each thin section billet. Uncertainties are the averaged instrumental population standard deviation determined by the pXRF software: $(2\sigma)_{inst}$ and the sample standard deviation of the spot analyses across the billet: $(s)_{spot}$. Samples with more than three spots analyzed are Bu-7 (five spots), Bu-2 (four spots), and IP-3 (four spots).

Sample	Be-1	Be-4	Be-6	Bi-2	Bi-3	Bu-2	Bu-6	Bu-7	C-3	C-4	CC-1	F-1	F-2	F-3	IP-1	IP-2
Mg	0.558	0.422	0.921	0.610	0.776	0.332	0.000	0.000	0.000	0.328	0.559	0.141	0.851	0.151	0.000	0.000
$Mg(2\sigma)_{inst}$	0.197	0.201	0.203	0.284	0.237	0.668	0.409	1.405	0.318	0.356	0.209	0.326	0.213	0.391	0.327	0.437
$Mg(s)_{spot}$	0.306	0.081	0.470	0.107	0.088	0.423	0.000	0.000	0.000	0.294	0.247	0.244	0.488	0.262	0.000	0.000
Al	5.995	5.798	5.979	6.977	6.007	5.821	0.134	0.152	6.972	6.819	5.109	4.832	4.397	5.828	5.247	4.555
$Al(2\sigma)_{inst}$	0.134	0.136	0.130	0.173	0.147	0.185	0.076	0.197	0.153	0.163	0.131	0.121	0.119	0.156	0.127	0.123
$Al(s)_{spot}$	0.131	2.363	0.659	0.077	0.646	0.348	0.134	0.147	0.954	0.641	0.215	1.550	0.357	1.631	2.394	0.411
Si	36.175	41.352	32.875	33.531	37.622	34.213	49.225	23.243	39.127	36.880	40.978	38.249	40.287	40.416	41.148	36.948
$Si(2\sigma)_{inst}$	0.213	0.207	0.215	0.201	0.209	0.208	0.182	0.137	0.220	0.198	0.217	0.210	0.232	0.196	0.208	0.187
$Si(s)_{spot}$	1.925	7.025	1.674	1.190	0.584	7.114	2.716	20.339	2.561	2.991	1.034	2.019	3.793	6.274	5.434	5.187
P	0.085	0.110	0.201	0.372	0.150	0.071	0.015	0.000	0.000	0.020	0.000	0.000	0.085	0.000	0.080	0.000
$P(2\sigma)_{inst}$	0.020	0.023	0.019	0.026	0.023	0.035	0.045	0.070	0.033	0.033	0.039	0.029	0.026	0.041	0.024	0.038
$P(s)_{spot}$	0.047	0.018	0.045	0.100	0.079	0.057	0.027	0.000	0.000	0.034	0.000	0.000	0.075	0.000	0.012	0.000
S	0.136	0.581	0.472	2.778	0.564	12.577	8.156	39.920	1.933	6.521	0.056	0.903	0.885	3.250	4.570	6.029
$S(2\sigma)_{inst}$	0.008	0.014	0.012	0.036	0.015	0.100	0.064	0.451	0.028	0.060	0.007	0.017	0.018	0.037	0.045	0.052
$S(s)_{spot}$	0.155	0.625	0.127	0.470	0.187	12.607	0.753	23.920	0.798	2.363	0.006	0.425	0.596	1.270	3.526	5.011
K	2.768	3.145	1.272	8.525	7.306	3.381	0.030	0.028	4.740	3.737	5.275	4.364	1.939	6.701	3.242	3.886
$K(2\sigma)_{inst}$	0.041	0.039	0.034	0.083	0.069	0.063	0.010	0.036	0.055	0.051	0.056	0.044	0.039	0.064	0.042	0.044
$K(s)_{spot}$	0.484	1.162	0.140	0.160	0.781	0.502	0.005	0.043	0.493	0.246	0.651	3.827	1.012	1.825	1.416	0.881
Ca	1.410	1.282	1.086	0.137	0.231	0.021	0.009	0.014	0.054	0.080	0.460	0.599	1.185	0.341	0.000	0.028
$Ca(2\sigma)_{inst}$	0.034	0.029	0.030	0.026	0.023	0.027	0.004	0.019	0.016	0.017	0.024	0.022	0.032	0.025	0.026	0.014
$Ca(s)_{spot}$	0.529	0.276	0.077	0.029	0.088	0.024	0.000	0.019	0.006	0.010	0.135	0.247	0.168	0.210	0.000	0.007
Ti	0.262	0.113	0.448	0.414	0.228	0.336	0.000	0.000	0.429	0.201	0.087	0.070	0.594	0.042	0.068	0.034
$Ti(2\sigma)_{inst}$	0.088	0.059	0.105	0.115	0.083	0.179	0.062	0.163	0.094	0.092	0.083	0.080	0.109	0.102	0.101	0.082
$Ti(s)_{spot}$	0.155	0.017	0.086	0.026	0.146	0.224	0.000	0.000	0.027	0.035	0.083	0.121	0.331	0.073	0.118	0.058
Fe	2.187	1.012	2.561	1.187	0.760	8.580	2.783	20.806	1.311	3.047	1.207	0.726	2.522	2.053	1.887	1.657
$Fe(2\sigma)_{inst}$	0.031	0.024	0.033	0.027	0.024	0.066	0.029	0.422	0.026	0.035	0.026	0.023	0.033	0.029	0.029	0.027
$Fe(s)_{spot}$	1.083	0.158	0.304	0.210	0.169	5.749	0.373	15.386	0.483	0.597	0.116	0.324	1.217	0.811	1.088	1.582
Cu	0.027	0.442	0.212	1.658	0.445	0.019	0.273	0.414	0.064	0.186	0.309	0.257	0.063	0.664	0.006	0.674
$Cu(2\sigma)_{inst}$	0.002	0.006	0.005	0.016	0.007	0.002	0.005	0.014	0.003	0.005	0.005	0.005	0.003	0.008	0.002	0.007
$Cu(s)_{spot}$	0.034	0.632	0.032	0.513	0.116	0.010	0.069	0.262	0.019	0.045	0.061	0.115	0.040	0.439	0.006	1.114
Mo	0.000	0.000	0.000	0.005	0.000	0.000	0.000	0.002	0.000	0.000	0.000	0.000	0.000	0.002	0.005	0.043
$Mo(2\sigma)_{inst}$	0.002	0.002	0.002	0.001	0.002	0.002	0.002	0.002	0.002	0.002	0.002	0.002	0.002	0.002	0.002	0.001
$Mo(s)_{spot}$	0.000	0.000	0.000	0.002	0.000	0.000	0.000	0.003	0.000	0.000	0.000	0.000	0.000	0.003	0.008	0.042
balance	50.397	45.743	53.974	43.805	45.910	34.650	39.374	15.422	45.370	42.181	45.960	49.860	47.192	40.553	43.746	46.146

Table 4. *Cont.*

Sample	IP-3	IP-10	IP-12	L-1	Mo-2	Mo-4	Mo-6	Mo-8	OH-1	S-1	S-2	S-4	SM-3	SM-4	Y-3
Mg	3.466	0.000	0.000	0.061	0.269	0.000	0.377	0.303	1.051	1.292	0.202	0.595	0.497	2.504	0.646
$Mg(2\sigma)_{inst}$	0.349	0.460	1.146	0.201	0.261	0.243	0.273	0.307	0.357	0.317	0.440	0.261	0.287	0.401	0.208
$Mg(s)_{spot}$	2.952	0.000	0.000	0.106	0.233	0.000	0.327	0.270	0.493	0.421	0.350	0.166	0.038	0.410	0.315
Al	3.792	0.630	2.089	0.516	7.597	0.552	5.698	6.713	5.670	7.591	4.372	6.134	7.896	8.039	4.013
$Al(2\sigma)_{inst}$	0.132	0.075	0.153	0.058	0.165	0.066	0.148	0.167	0.167	0.187	0.150	0.156	0.182	0.212	0.114
$Al(s)_{spot}$	2.270	0.657	3.105	0.802	0.232	0.072	3.146	0.808	0.968	0.344	0.426	0.395	0.254	0.424	0.729
Si	33.855	42.872	27.146	46.957	39.581	53.778	39.729	36.751	29.638	32.075	36.292	36.938	34.163	28.977	38.218
$Si(2\sigma)_{inst}$	0.226	0.176	0.167	0.210	0.227	0.228	0.205	0.204	0.182	0.216	0.203	0.202	0.187	0.243	0.212
$Si(s)_{spot}$	10.773	6.181	4.108	4.859	0.925	2.258	7.880	3.732	4.311	1.404	1.954	2.825	0.809	1.918	0.839
P	0.417	0.045	0.047	0.085	0.137	0.000	0.030	0.056	0.014	0.108	0.239	0.071	0.167	0.153	0.057
$P(2\sigma)_{inst}$	0.025	0.038	0.043	0.035	0.024	0.042	0.031	0.024	0.051	0.024	0.027	0.024	0.026	0.023	0.028
$P(s)_{spot}$	0.436	0.042	0.081	0.023	0.044	0.000	0.052	0.016	0.024	0.004	0.161	0.028	0.044	0.067	0.059
S	1.424	11.951	21.206	0.825	2.322	2.969	0.866	3.018	11.947	3.284	5.937	4.012	7.376	1.453	0.323
$S(2\sigma)_{inst}$	0.023	0.080	0.150	0.014	0.031	0.033	0.018	0.037	0.091	0.041	0.057	0.041	0.064	0.027	0.011
$S(s)_{spot}$	1.185	8.437	6.641	0.642	0.295	3.238	0.515	1.362	8.311	0.953	1.410	3.240	0.765	0.552	0.197
K	5.778	0.301	0.087	0.361	4.530	0.296	5.385	5.082	1.676	3.363	3.247	5.932	4.304	5.508	3.039
$K(2\sigma)_{inst}$	0.075	0.015	0.026	0.010	0.057	0.014	0.057	0.058	0.043	0.056	0.050	0.065	0.056	0.093	0.043
$K(s)_{spot}$	2.875	0.358	0.077	0.573	0.073	0.042	2.497	1.210	0.058	2.083	0.498	1.774	0.105	0.334	0.905
Ca	0.558	0.007	0.032	0.011	0.257	0.030	2.443	2.069	2.309	3.203	2.203	0.888	1.359	0.498	0.729
$Ca(2\sigma)_{inst}$	0.033	0.009	0.012	0.004	0.022	0.005	0.044	0.043	0.054	0.062	0.046	0.034	0.039	0.040	0.026
$Ca(s)_{spot}$	0.624	0.012	0.014	0.001	0.082	0.003	0.892	1.472	0.387	0.719	0.473	0.402	0.696	0.085	0.215
Ti	0.698	0.000	0.000	0.000	0.358	0.126	0.263	0.241	0.137	0.404	0.465	0.168	0.307	1.228	0.232
$Ti(2\sigma)_{inst}$	0.148	0.066	0.150	0.031	0.102	0.045	0.084	0.091	0.131	0.135	0.110	0.106	0.102	0.235	0.088
$Ti(s)_{spot}$	0.334	0.000	0.000	0.000	0.032	0.016	0.171	0.030	0.120	0.018	0.325	0.185	0.036	0.027	0.118
Fe	3.156	4.299	11.918	0.498	2.275	0.856	1.167	2.134	6.721	4.254	2.922	2.792	3.619	8.000	3.543
$Fe(2\sigma)_{inst}$	0.039	0.037	0.089	0.018	0.034	0.023	0.026	0.031	0.055	0.045	0.036	0.035	0.037	0.066	0.037
$Fe(s)_{spot}$	2.385	3.306	1.197	0.247	0.097	0.986	0.576	0.828	2.205	0.253	1.337	1.392	0.222	0.907	1.221
Cu	0.022	0.122	14.803	0.199	0.000	0.061	0.409	0.757	0.024	0.033	0.930	0.032	0.607	0.400	0.116
$Cu(2\sigma)_{inst}$	0.002	0.003	0.160	0.004	0.003	0.003	0.006	0.009	0.002	0.002	0.010	0.002	0.008	0.007	0.004
$Cu(s)_{spot}$	0.006	0.139	3.822	0.166	0.000	0.055	0.235	0.545	0.008	0.017	0.640	0.038	0.097	0.218	0.085
Mo	0.273	0.065	0.000	0.009	0.000	0.000	0.007	0.008	0.009	0.000	0.000	0.034	0.002	0.001	0.000
$Mo(2\sigma)_{inst}$	0.003	0.001	0.002	0.002	0.002	0.002	0.001	0.001	0.001	0.002	0.002	0.001	0.001	0.002	0.002
$Mo(s)_{spot}$	0.372	0.069	0.000	0.016	0.000	0.000	0.004	0.006	0.006	0.000	0.000	0.005	0.002	0.002	0.000
balance	46.561	39.708	22.672	50.479	42.674	41.333	43.625	42.867	40.805	44.392	43.190	42.405	39.705	43.238	49.085

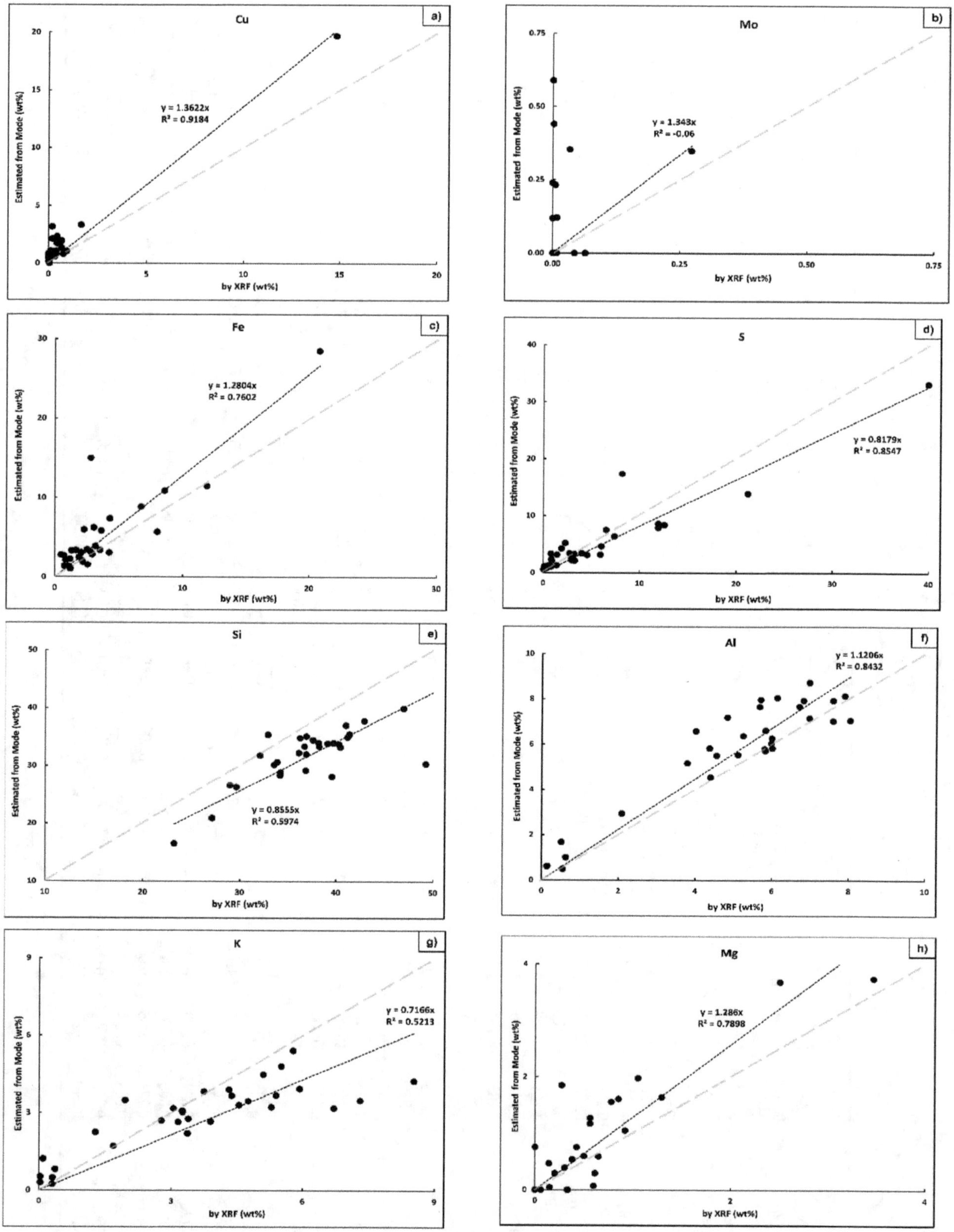

Figure 5. Plots comparing the composition of samples determined by pXRF with those values calculated from mineral modes: **(a)** Cu, **(b)** Mo, **(c)** Fe, **(d)** S, **(e)** Si, **(f)** Al, **(g)** K, and **(h)** Mg. The grey, coarsely dashed lines show 1:1 equivalency for both methods.

The degree of correlation between the two methods depends on the particular element. The lightest and the heaviest elements are systematically overestimated by optical petrography compared to pXRF: Cu, Mo, Fe, Al, Mg, and Ti (Figure 5a–c,f,h). In contrast, the mid-atomic number elements are underestimated by optical petrography compared to pXRF: S, Si, K, Ca, and P (Figure 5d,e,g). The element that shows the closest 1:1 correlation between the two methods is Al (Figure 5f), followed by Si (Figure 5e). The poorest correlations are shown for Cu and Mo (Figure 5a,b). The correlation for Cu is far worse when the single high-Cu sample (IP-12) is omitted from the fit (Figure 6). The best trendline fits are for Cu, S, and Al (Figure 5a,d,f). However, the fit is very poor in terms of Cu if sample IP-12 is omitted (Figure 6). The worst fit is shown by Mo ($R^2 = -0.06$, Figure 5b), followed by Ca ($R^2 = 0.443$), Ti ($R^2 = 0.4662$), and P ($R^2 = 0.4721$). The balances (remaining wt% not attributed to a single element) are slightly overestimated by mode compared to pXRF by a factor of 1.0623, and with an adequate fit ($R^2 = 0.7361$). Significant outliers from these trends are Bu-06: higher Fe and S, but lower Si by mode than by pXRF, Bi-02: lower K by mode than by pXRF, F-02, higher K by mode than by pXRF, L-01: higher Cu by mode than by pXRF, and Mo-02: higher Mg by mode than by pXRF.

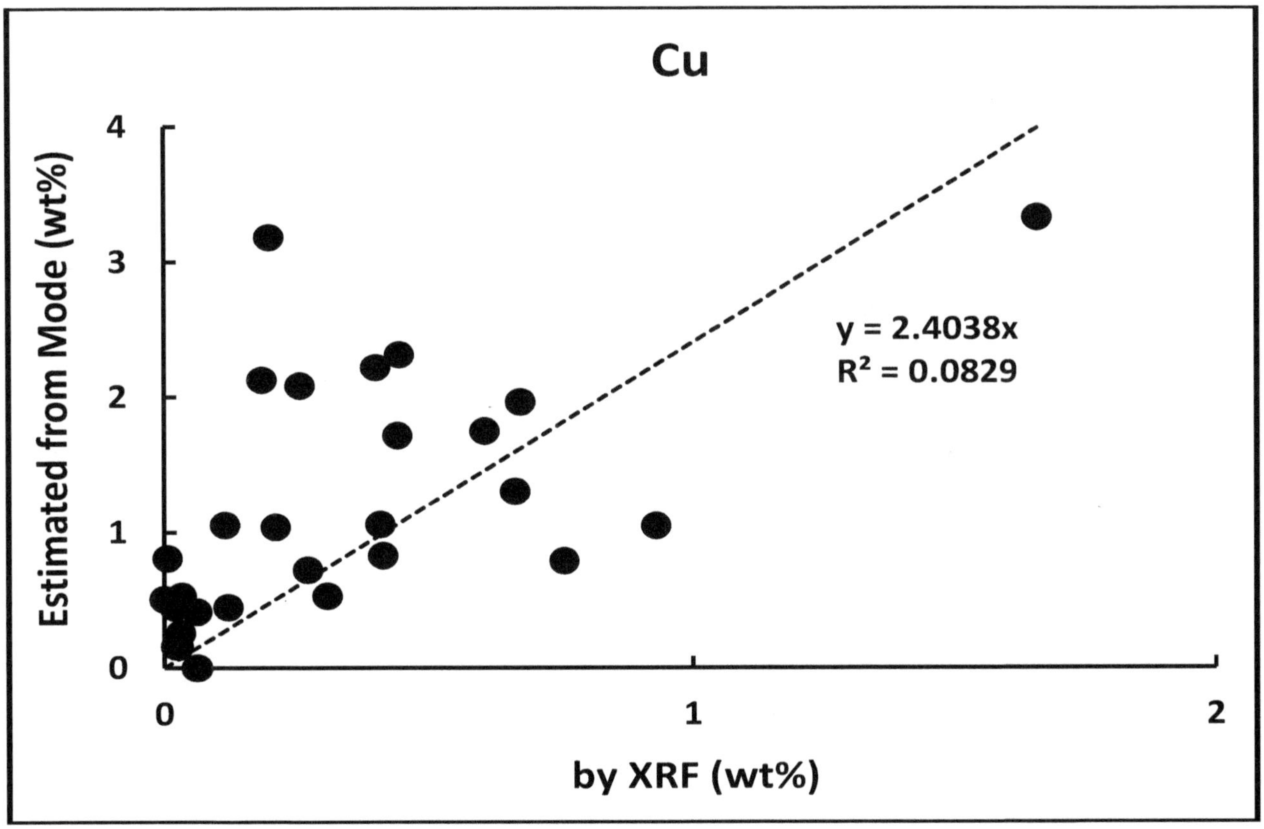

Figure 6. Plot comparing Cu content of samples determined by pXRF with those values calculated from mineral modes but omitting high-Cu sample IP-12.

4. Discussion

There are multiple factors affecting each method that may bias results. These affect the accuracy and precision of the mode-based or pXRF-based methods, and the correlation of the two.

4.1. *Causes of Poor Correlation*

In terms of optical petrography, melanocratic (i.e., mafic minerals, sulfides, and oxides) minerals are typically overestimated compared to leucocratic minerals (i.e., felsic minerals, carbonates, and phosphates; c.f., [27]). Sulfide minerals are exclusive hosts for Cu and Mo in the samples. Similarly, biotite-chlorite and rutile are the major hosts for Mg and Ti, respectively. Biotite, chlorite, hematite, and sulfides are also the hosts for Fe. These minerals are all dark-colored or opaque in TPPL and readily noted. Colorless or light-colored minerals in TPPL such as quartz, muscovite (as sericite), orthoclase, plagioclase, calcite, and apatite collectively host the bulk of Si, K, Ca, and P. The residency of Al is less clear as it is overestimated by optical petrography (Figure 5f) and is an important constituent (along with K and Si) of biotite-chlorite but is a more significant component of the feldpars and sericite. Human error as systematic visual overestimation of biotite and chlorite during optical petrography would give a positive bias to the Mg, Fe, and Al calculated by mode; particularly if it was at the expense of quartz content. The other element with unusual behavior is S in that it is underestimated by optical petrography. As the only S-bearing phases noted are sulfides, it would be expected that optical mineralogy would overestimate this element as with Fe, Cu, and Mo. These elements track with S (Figure 4), but it is possible another S-bearing phase was overlooked, such as barite, anhydrite, or gypsum. Barite is not likely as a negligible Ba content was noted by pXRF in all samples. Hypogene anhydrite occurs in porphyry-type deposits (e.g., [28]), and can be fairly common [5,29]. Another cause for S underestimation from the mode is misidentifying sulfide minerals that are S-rich (chalcopyrite and covellite) for S-poor phases (bornite, digenite, and chalcocite). However, the most abundant S-bearing mineral is pyrite, and a similar underestimation is not present with Fe; nor are those two groups of Cu-sulfides easily confused in RPPL. It may be that the pXRF systematically overestimates S content of the samples and further calibration should have been applied. Systematic overestimation of S (along with P, Cu, and Mg) by pXRF has been observed with field analysis of soils [30]. However, Roman Alday et al. [16] found that for samples from the Elatsite porphyry-type deposit, pXRF overestimates Fe and K, and underestimates Mg compared to more intensive laboratory methods. Matrix effects on pXRF vary with mineral assemblages and development of calibration standards is reliant on rock type [23,31]. The poorest correlation of the chalcophile elements is for Mo. Economic grades in porphyry-type deposits range from approximately 0.1 to 0.3 wt% Mo, although lower grades in the presence of another commodity such as Cu are common [32]. This is a lower grade than Cu due to the higher valuation of Mo, typically 2× to 3× the per-pound metal price. This grade range approaches the limit of detection for Mo by most pXRF units in a typical silicate–sulfide matrix. The result is that although molybdenite is fairly distinct and easy to identify in reflected light, it may be in trace concentrations that contribute grades too low to be detected by pXRF. Thus, mode-based estimates of grade will likely be higher than those detected by pXRF. The selection of the pXRF analytical spots may also play a role (see Section 4.4).

4.2. Outliers from Trends

Five samples, in particular, plotted as major outliers from the trends established by the rest of the data for one or more elements. These are most likely the result of errors in visual estimation of modes as discussed above. Bu-06 has 2.8 wt% Fe by pXRF and 15 wt% Fe by mode. This is accompanied by a similar overestimation of S by mode and underestimation of Si by mode. This is most likely due to the overestimation of pyrite at the expense of quartz. Bi-02 has 8.5 wt% K by pXRF and 4.2 wt% K by mode. This is likely from an underestimation of muscovite (as sericite). F-02 has 1.9 wt% K by pXRF and 3.5 wt% by mode. The overestimation of potassic phases is likely. Orthoclase is most probable as it has a higher proportion of K to Al than muscovite and there is not a significant Al outlier for this sample. There is no accompanying Fe and/or Mg outlier, so biotite-chlorite is not likely. L-01 has 0.2 wt% Cu by pXRF and 3.2 wt% Cu by mode. This is most likely due to an overestimate of the modal abundance of chalcopyrite. Mo-02 has 0.3 wt% Mg by pXRF and 1.9 wt% Mg by mode, most likely due to an overestimation of a mafic mineral, such as chlorite. It should be noted, that the element with the highest instrumental error is Mg (Table 4).

4.3. Assumptions of Mineral Composition

Other than errors in visually estimating the mode of a sample or in misidentifying phases, another source of error in calculating the bulk rock composition from the mode is the assumed composition of the mineral phases. Chlorite, biotite, and plagioclase all have assumed compositions in terms of solid solution components for the purposes of translating the modes into compositional data. These assumptions are not supported by any mineral analysis such as electron microprobe analysis. Minerals such as biotite can vary in composition between different deposits and within single deposits. Magmatic biotite of the host is typically more Fe-rich than the later hydrothermal biotite [6], largely depending on what alteration zone (i.e., potassic vs. phyllic) they are from (e.g., [33]). There was no distinction between these parageneses of the same minerals during the determination of the modes. As chlorite is largely a product of biotite alteration in these deposits, a similar degree of compositional variation is expected. Likewise, plagioclase in porphyry deposits varies in composition. For example, plagioclase from Sierrita ranges from Ab_{25} to Ab_{45} [34]. The simplified estimation of a molar 1:1 ratio of Na:Ca in plagioclase in this study is likely why bulk compositions based on modes overestimates Ca content. Muscovite and chlorite from the Highland Valley district (that includes the Lornex and Bethlehem deposits) range in composition 2.2–2.6 Al apfu and 0.5–3.5 Mg apfu, respectively [35].

4.4. Sampling Biases

A further source of disparity between the results of the two methods lies in sampling. Though the thin section is cut from the billet, there may be fairly distinct differences in mode between a 0.03 mm thick section and a ~10 mm thick billet. The thin section is essentially a 2D representation of a 3D rock, whereas the pXRF analysis is an average of three different spots on the billet, each 8 mm in diameter (the default aperture size on the Niton XL3t), with an effective sampling depth (d) expressed by Equation (1):

$$d = m/\rho \tag{1}$$

where d is the sampling depth from which 99% of the signal is sourced, m is the mass per unit area sampled and ρ is sample density [23]. For the samples, the least dense abundant mineral is plagioclase (ρ_{albite} = 2.61 g/cm^3) and the densest is pyrite (ρ_{pyrite} = 5.01 g/cm^3). Denser samples (i.e., sulfide-rich) will approach more surficial analyses, whereas sampling depth will be deeper for samples dominated by felsic minerals. If the billet has compositional variations in the vertical (depth) sense, then the spot analyses become pseudo-bulk analyses of a cylinder 8 mm in diameter a a depth that is a function of sample density. Similarly, in the horizontal sense, three spots at 50.27 mm^2 totals ~150.8 mm^2, whereas the thin section is up to 1196 mm^2 in total area. Accounting for 3 mm margins around the billet, this reduces to 800 mm^2 of sample area. Thus, this method in the best case, sampled only ~19% of the sample area and assumes it is representative of the other 21%. Sampling precision for finer-grained rocks by pXRF has been shown to be ≤5% relative standard deviation, but much higher for coarser-grained samples [36]. The thin section is a ~60 μm slice off of the top of the billet that is subsequently ground down to 30 μm. Sample heterogeneity is typically on a scale an order of magnitude larger. Differences in thin section and sample billet surface are negligible in comparison to the other discussed sources of uncertainty. Although most of the rocks in this sample suite are hypabyssal to volcanic, the presence of phenocrysts and veins/veinlets increases the number of measurements required to achieve the same precision that would be possible compared to an un-veined, aphyric sample. The large sample standard deviations across the spot analyses ($(s)_{spot}$) for many samples illustrates this (Table 4). Samples with a high degree of heterogeneity in terms of multiple elements are all the Ithaca Peak samples, Be-4, Bu-2, Bu-7, F-3. Mo-6, OH-1, and S-4.

4.5. Applicability of pXRF to Mineral Exploration

Compared to the classic standard of optical petrography, pXRF is faster, more accessible, portable, and more easily conducted with little training. It is not a replacement for laboratory methods as it cannot reliably reach the levels of precision and accuracy required to meet standards for mineral resource reporting in terms of NI 43-101 (National Instrument 43-101 is the Canadian code for the Standards of Disclosure for Mineral Projects. It sets standards for reporting mineral resource information such as grade, tonnage, etc., and what methods are suitable for measuring them to an adequate degree of certainty) or JORC (Joint Ore Reserves Committee Code is essentially the Australian equivalent to NI 43-101) guidelines, but with calibration, it can achieve a low but useful level of quantitative certainty (e.g., [16,20,21]). However, for quick decisions regarding unknown samples in the field or core shed, semi-quantitative analysis is satisfactory in order to make rough interpretations on whether to collect a sample for more detailed assay/examination. Similarly, pXRF can be used to guide petrographic examinations in terms of possible (and impossible) phases present as with this study.

Although this study attempts to simulate "field" use of pXRF, the analytical conditions for this study were more optimal and controlled than those typical of field or even core shed settings. Analyzed surfaces of the billets were fresh, clean, and flat. Samples from outcrop, float, or trenches may have soil, lichen, or other debris covering the rock. Thick weathering rinds may also be present. Surfaces are typically irregular. Although conditions for drill core analysis are more controlled, sample geometry is not flat, but curved. This is especially relevant for smaller diameter drill core such as AQ, BQ, or NQ. pXRF analysis of reverse circulation (RC) chips suffers from especially variable conditions in terms of sample geometry and open space between the chips. These sources of uncertainty can be mitigated by cleaning surfaces, preparing flat surfaces, and even pulping and packing them into homogenized pucks, but all at the cost of significant time.

4.6. Directions for Future Research

The certainty of pXRF results would be much improved from non-factory calibration specific to porphyry-type samples. On the petrographic side, determination of the variations in mineral composition, at least of biotite, chlorite, plagioclase, and muscovite is needed in order to provide more accuracy in determining a chemical composition from the mode. The use of a scanning electron microscope with energy-dispersive X-ray spectra would provide sufficient accuracy. More detailed petrographic work, along with cathodoluminescence would help distinguish between different parageneses of minerals and better characterize the range of compositions present (c.f., [29,37]).

Author Contributions: Conceptualization, A.D.V.; methodology, A.D.V.; formal analysis, C.A.G.; investigation, C.A.G.; resources, A.D.V.; data curation, C.A.G.; writing—original draft preparation, C.A.G.; writing—review and editing, A.D.V. and C.A.G.; supervision, A.D.V.; project administration, A.D.V.; funding acquisition, C.A.G. and A.D.V. All authors have read and agreed to the published version of the manuscript.

Acknowledgments: Samples were collected by Peter H. Kirwin and John Ray. Funding for the Leica microscope was provided by the Bloomsburg University College of Science and Technology Dean's office via Troy Prutzman and Robert Aronstam. I, Connor Gray, want to personally thank my undergraduate research advisor Adrian Van Rythoven for the opportunity to work on this project. Two anonymous reviewers and the editorial staff at *Minerals* provided comments that improved the quality of the manuscript.

References

1. Titley, S.R.; Hicks, C.L. *Geology of the Porphyry Copper Deposits, Southwestern North America*; University of Arizona Press: Tuscon, AZ, USA, 1966.
2. Lowell, J.D.; Guilbert, J.M. Lateral and vertical alteration-mineralization zoning in porphyry ore deposits. *Econ. Geol.* **1970**, *65*, 373–408. [CrossRef]
3. Sillitoe, R.H. A Plate Tectonic Model for the Origin of Porphyry Copper Deposits. *Econ. Geol.* **1972**, *67*, 184–197. [CrossRef]
4. Sillitoe, R.H.; Bonham, H.F. Volcanic landforms and ore deposits. *Econ. Geol.* **1984**, *79*, 1286–1298. [CrossRef]
5. Berger, B.R.; Ayuso, R.A.; Wynn, J.C.; Seal, R.R. *Preliminary Model of Porphyry Copper Deposits. Open-File Report 2008–1321*; USGS: Reston, VA, USA, 2008.

6. John, D.A.; Ayuso, R.A.; Barton, M.D.; Blakely, R.J.; Bodnar, R.J.; Dilles, J.H.; Gray, F.; Graybeal, F.T.; Mars, J.C.; McPhee, D.K.; et al. Porphyry Copper Deposit Model: Scientific Investigations Report 2010–5070–B. In *Mineral Deposit Models for Resource Assessment*; USGS: Reston, VA, USA, 2010; p. 169.
7. Sillitoe, R.H. Porphyry Copper Systems. *Econ. Geol.* **2010**, *105*, 3–41. [CrossRef]
8. Richards, J.P. Tectono-Magmatic Precursors for Porphyry Cu-(Mo-Au) Deposit Formation. *Econ. Geol.* **2003**, *98*, 1515–1533. [CrossRef]
9. Eliopoulos, D.G.; Economou-Eliopoulos, M. Platinum-group element and gold contents in the Skouries porphyry copper deposit, Chalkidiki Peninsula, northern Greece. *Econ. Geol.* **1991**, *86*, 740–749. [CrossRef]
10. Tarkian, M.; Stribrny, B. Platinum-group elements in porphyry copper deposits: A reconnaissance study. *Mineral. Petrol.* **1999**, *65*, 161–183. [CrossRef]
11. Babcock, R.C., Jr.; Ballantyne, G.H.; Phillips, H. Summary of the geology of the Bingham District, Utah. *Arizona Geol. Soc. Dig.* **1995**, *20*, 316–325.
12. Hou, Z.; Cook, N.J. Metallogenesis of the Tibetan collisional orogen: A review and introduction to the special issue. *Ore Geol. Rev.* **2009**, *36*, 2–24. [CrossRef]
13. Copper Alliance. *The Impacts of Copper Mining in Chile: Economic and Social Implications for the Country*; International Copper Association: Santiago, Chile, 2017.
14. Potts, P.J. Introduction, Analytical Instrumentation and Application Overview. In *Portable X-ray Fluorescence Spectrometry*; Potts, P.J., West, M., Eds.; Royal Society of Chemistry: Cambridge, UK, 2008; pp. 1–12, ISBN 978-0-85404-552-5.
15. Liangquan, G. Geochemical Prospecting. In *Portable X-ray Fluorescence Spectrometry*; Potts, P.J., West, M., Eds.; Royal Society of Chemistry: Cambridge, UK, 2008; pp. 141–173, ISBN 978-0-85404-552-5.
16. Roman Alday, M.C.; Kouzmanov, K.; Harlaux, M.; Stefanova, E. Comparative study of XRF and portable XRF analysis and application in hydrothermal alteration geochemistry: The Elatsite porphyry Cu-Au-PGE deposit, Bulgaria. In Proceedings of the 16th Swiss Geoscience Meeting, Bern, Switzerland, 30 November–1 December 2018; p. 2.
17. Lemière, B. A review of pXRF (field portable X-ray fluorescence) applications for applied geochemistry. *J. Geochemical Explor.* **2018**, *188*, 350–363. [CrossRef]
18. Kalnicky, D.J.; Singhvi, R. Field portable XRF analysis of environmental samples. *J. Hazard. Mater.* **2001**, *83*, 93–122. [CrossRef]
19. Uvarova, Y.; Cleverley, J.; Baensch, A.; Verrall, M. Coupled XRF and XRD analyses for rapid and low-cost characterization of geological materials in the mineral exploration and mining industry. *Explor. Newsl. Assoc. Appl. Geochemists* **2014**, *162*, 1–14.
20. Ahmed, A.; Crawford, A.J.; Leslie, C.; Phillips, J.; Wells, T.; Garay, A.; Hood, S.B.; Cooke, D.R. Assessing copper fertility of intrusive rocks using field portable X-ray fluorescence (pXRF) data. *Geochem. Explor. Environ. Anal.* **2019**, *20*, 81–97. [CrossRef]
21. Andrew, B.S.; Barker, S.L.L. Determination of carbonate vein chemistry using portable X-ray fluorescence and its application to mineral exploration. *Geochem. Explor. Environ. Anal.* **2018**, *18*, 85–93. [CrossRef]
22. Gallhofer, D.; Lottermoser, B.G. The Influence of Spectral Interferences on Critical Element Determination with Portable X-Ray Fluorescence (pXRF). *Minerals* **2018**, *8*, 320. [CrossRef]
23. Markowicz, A.A. Quantification and Correction Procedures. In *Portable X-ray Fluorescence Spectrometry*; Potts, P.J., West, M., Eds.; Royal Society of Chemistry: Cambridge, UK, 2008; pp. 13–38, ISBN 978-0-85404-552-5.
24. Taylor, R. *Ore Textures*; Springer: Berlin/Heidelberg, Germany, 2009; ISBN 978-3-642-01783-4.
25. Whitney, D.L.; Evans, B.W. Abbreviations for names of rock-forming minerals. *Am. Mineral.* **2010**, *95*, 185–187. [CrossRef]
26. Le Maitre, R.W.; Streckeisen, A.; Zanettin, B.; Le Bas, M.J.; Bonin, B.; Bateman, P.; Bellieni, G.; Dudek, A.; Efremova, S.; Keller, J.; et al. *Igneous Rocks: A Classification and Glossary of Terms: Recommendations of the International Union of Geological Sciences Subcommission on the Systematics of Igneous Rocks*, 2nd ed.; Le Maitre, R.W., Streckeisen, A., Zanettin, B., Le Bas, M.J., Bonin, B., Bateman, P., Eds.; Cambridge University Press: Cambridge, UK, 2002; ISBN 9780511535581.
27. Wright, T.L.; Peck, D.L. *Crystallization and Differnetiation of the Alae Magma, Alae Lava Lake, Hawaii. Professional Paper 935-C*; USGS: Reston, VA, USA, 1976.

28. Arnott, A.M.; Zentilli, M. Significance of calcium sulphates in the Chuquicamata porphyry copper system, Chile. In Proceedings of the Atlantic Geoscience Society Annual Meeting Abstracts - Atlantic Geology, Amherst, NS, Canada, 5–6 February 1999; Volume 35, p. 86.
29. Hutchinson, M.C.; Dilles, J.H. Evidence for Magmatic Anhydrite in Porphyry Copper Intrusions. *Econ. Geol.* **2019**, *114*, 143–152. [CrossRef]
30. Declercq, Y.; Delbecque, N.; De Grave, J.; De Smedt, P.; Finke, P.; Mouazen, A.M.; Nawar, S.; Vandenberghe, D.; Van Meirvenne, M.; Verdoodt, A. A Comprehensive Study of Three Different Portable XRF Scanners to Assess the Soil Geochemistry of an Extensive Sample Dataset. *Remote Sens.* **2019**, *11*, 2490. [CrossRef]
31. Pinto, A.H. Portable X-Ray Fluorescence Spectrometry: Principles and Applications for Analysis of Mineralogical and Environmental Materials. *Asp. Min. Miner. Sci.* **2018**, *1*, 42–47. [CrossRef]
32. Ludington, S.; Plumlee, G.S. *Climax-Type Porphyry Molybdenum Deposits. Open-File Report 2009-1215*; USGS: Reston, VA, USA, 2009.
33. Ayati, F.; Yavuz, F.; Noghreyan, M.; Haroni, H.A.; Yavuz, R. Chemical characteristics and composition of hydrothermal biotite from the Dalli porphyry copper prospect, Arak, central province of Iran. *Mineral. Petrol.* **2008**, *94*, 107–122. [CrossRef]
34. Preece, R.K., III. *Paragenesis, Geochemistry, and Temperatures of Formation of Alteration Assemblages of the Sierrita Depost, Pima County, Arizona*; University of Arizona: Tucson, AZ, USA, 1979.
35. Alva Jimenez, T.R. Variation in hydrothermal muscovite and chlorite composition in the Highland Valley Porphyry Cu-Mo District, British Columbia, Canada. Ph.D. Thesis, University of British Columbia, Vancouver, BC, Canada, 2011.
36. Potts, P.J.; Williams-Thorpe, O.; Webb, P.C. The Bulk Analysis of Silicate Rocks by Portable X-Ray Fluorescence: Effect of Sample Mineralogy in Relation to the Size of the Excited Volume. *Geostand. Geoanalytical Res.* **1997**, *21*, 29–41. [CrossRef]
37. Götze, J. Application of Cathodoluminescence Microscopy and Spectroscopy in Geosciences. *Microsc. Microanal.* **2012**, *18*, 1270–1284. [CrossRef] [PubMed]

8

Predicting the Logarithmic Distribution Factors for Coprecipitation into an Organic Salt: Selection of Rare Earths into a Mixed Oxalate

Harry Watts [1,2,*] and Yee-Kwong Leong [1]

1 Department of Chemical Engineering, University of Western Australia, Crawley, WA 6009, Australia; yeekwong.leong@uwa.edu.au

2 Watts & Fisher Pty Ltd., Crawley, WA 6009, Australia

* Correspondence: harry@wattsandfisher.com

Abstract: Thermodynamic modelling of a leaching system that involves concurrent precipitation depends on an understanding of how the metals distribute into the precipitate before an assessment of solubility can be made. It has been suggested in the past that a pair of rare earths (A and B) in solution will separate from each other by oxalate precipitation according to a logarithmic distribution coefficient (λ), determined by the kinetics of the precipitation. By contrast, the present study hypothesises that λ may be approximated from thermodynamic terms, including the solubility product (K_{Sp}) of each rare earth oxalate and the stability constant (β_1) for the mono-oxalato complex of each rare earth. The proposed model was used to calculate λ between pairs of rare earths. An experimental study was conducted to determine λ between selected pairs using homogenous precipitation through the hydrolysis of an oxalic acid ester, with fairly close agreement to the values under the proposed model. Though this model requires more thorough testing, as well as application to other organic salts, it may provide insight into distribution factors of a precipitate formed by a sequence of organic complexes.

Keywords: rare earths; distribution factor; selective precipitation; oxalates; organic complexes

1. Introduction

In rare earth extraction, a common technique for overcoming the poor solubility of rare earth minerals is to transfer the rare earth ions into a more stable solid. In sulfuric acid cracking, for example, the rare earths are transferred into solid sulphates. In caustic conversion, the rare earths are put into hydroxides. In more recent studies, minerals of rare earth phosphates have been dissolved in oxalic acid, while the rare earths concurrently precipitate as rare earth oxalates [1,2]. In each of these techniques, an effective transfer of the rare earths depends on each rare earth having a lower solubility or a higher stability in the destination compound.

The way that rare earths distribute into an oxalate precipitate has not been modelled, beyond describing the results of specific experimental conditions. This is despite oxalate coprecipitation being a common technique in a broad range of areas. Oxalate coprecipitation is a standard method for producing precursor powders that are calcined into mixed metal oxides. Examples in the literature abound, such as precursors for magnetic materials [3], piezoelectric oxides [4], superconducting materials [5,6] and alloys [7]. It is also used to separate rare earths and actinides from other ions in solution by precipitation. Breakdown of rare earth minerals such as monazite are often done in extreme conditions [8] that dissolve impurity ions along with the rare earths. The very low solubility of rare earth and actinide oxalates enables an isolation of rare earths from mixed solutions [9].

The most advanced modelling of oxalate co-precipitation appears to have come from experiments to separate rare earths from each other in controlled precipitation. Oxalic acid was generated homogeneously throughout the solution by hydrolysis of dimethyl oxalate, in order to minimize concentration gradients within the solution and allow separations closer to those predicted by solubility [10]. Feibush, Rowley and Gordon [11] found, however, that the separations were much lower than predicted, according to solubility for the rare earths tested in their study. There remained no method to predict the separation from fundamental terms, and the techniques for oxalate coprecipitation more generally appear to be based on trial and error.

Feibush, Rowley and Gordon [11] did find that the precipitation between pairs of rare earth oxalates follows a Doerner–Hoskins or logarithmic distribution coefficient. Such a distribution assumes that the precipitating solid does not re-order its whole self to be in equilibrium with the solution; rather, the precipitated surface is in equilibrium with the solution and once precipitated it forms an unchanging substrate for the next surface layer [12]. Doerner and Hoskins were the first to describe this running equilibrium mathematically [13]. It can be described between concentrations of rare earth ions $[A]$ and $[B]$ by Equation (1).

$$\frac{d[A]}{d[B]} = \lambda \frac{[A]}{[B]} \tag{1}$$

Here, $\frac{d[A]}{d[B]}$ is the ratio of the concentration of A to B lost from the solution in an infinitesimal increment, or, in other words, precipitated onto the surface in an infinitesimal increment. The letter λ signifies the logarithmic distribution coefficient. This expression can be integrated to give Equation (2).

$$ln\frac{[A]final}{[A]\ initial} = \lambda ln\frac{[B]\ final}{[B]\ initial} \tag{2}$$

Doerner and Hoskins [13] did not attempt to correlate the coefficient to fundamental terms, probably in the knowledge that different salts may be governed by different terms. Feibush, Rowley and Gordon [11] suggested that with rare earth oxalates, the coefficient should be the ratio between the square roots of the solubility products of the pure rare earth oxalates (square root because each unit of rare earth oxalate has two rare earth ions), as shown in Equation (3).

$$\lambda \cong \frac{\sqrt{K_{Sp,\ A}}}{\sqrt{K_{Sp,\ B}}} \tag{3}$$

This equation is based on $K_{Sp,\ A}$, as the solubility product for the equation shown in Equation (4).

$$A_2(C_2O_4)_3 \rightarrow 2A^{3+} + 3C_2O_4{}^{2-} \tag{4}$$

Their experiments, however, showed that the experimental values of the coefficients, while being logarithmic in nature, were far from what was predicted by solubility in the form of Equation (3). Equation (3) does not form a basis in the present study for further development of a model.

Some more recent work on the solubility of rare earth oxalates has determined that the saturated concentration of rare earth A in a solution of concentrated oxalic acid is determined by Equation (5) [14].

$$[A_{total}] = \frac{\sqrt{K_{Sp,\ A}}}{\gamma_3}\left(a_{C_2O_4{}^{2-}}\right)^{-3/2} + \frac{\sqrt{K_{Sp,\ A}}\,\beta_{1,A}}{\gamma_1}\left(a_{C_2O_4{}^{2-}}\right)^{-1/2} \tag{5}$$

In Equation (5), $\beta_{1,A}$ is the equilibrium constant for the reaction in Equation (6), with the subscript number denoting the number of oxalates in the complex formed. The expression for $\beta_{1,A}$ is shown in Equation (7), where a is the activity of the species denoted as subscript. The symbol γ_1 is the activity co-efficient for the rare earth mono-oxalato complex, and γ_3 is the activity co-efficient for the free rare earth ion, the subscript based on the magnitude of the charge.

$$A^{3+} + C_2O_4{}^{2-} \rightarrow AC_2O_4{}^{1+} \tag{6}$$

$$\beta_{1,A} = \frac{a_{AC_2O_4{}^{1+}}}{a_{A^{3+}}a_{C_2O_4{}^{2-}}} \quad (7)$$

It can be seen in Equation (5) that the total concentration of rare earth A is made up of two main species, the free or weakly-complexed rare earth ion in the first term, and the mono-oxalato complex in the second term. In one sense, Equation (5) represents the saturated concentrations of each of these species in the particular system. The challenge from a theoretical point of view is how a model of precipitation can be constructed if there are complexes present. For example, the precipitate may be formed from a sequence of complexation reactions, where the mono-oxalato complex is the precursor to the precipitate, rather than the free rare earth ion.

2. Materials and Methods

Rare earth oxides at a minimum of 99.9% purity were obtained from Treibacher Industries in Austria, along with certificates of assay. Dimethyl oxalate (99%) was obtained from Alfa Aesar, and diethyl oxalate (99%) from Sigma Aldrich. Analytical grade hydrochloric acid was used. Brand-new Duran 100 mL conical flasks were used for each experimental flask.

The method underwent considerable development. In the original method, each conical flask was initially weighed. To each conical flask was added at least two, but up to six, rare earth oxides. The amount of each rare earth oxide added was determined by a random number generator in Microsoft Excel, to be an amount between 0.00 and 0.10 g. As close to this number as possible was weighed and added to the flask. Hydrochloric acid was then added to dissolve the rare earth oxides into chlorides until a clear solution was obtained. The flask was then heated on a hotplate to evaporate the remaining hydrochloric acid and water, and leave a precipitate of rare earth chlorides. To the flask was added about 100 g of deionized water. The mass of dimethyl oxalate (DMO) required for a complete precipitation was calculated, and a fraction of this was chosen at random, weighed and added to the flask. The solution was swirled until all the dimethyl oxalate was dissolved. This also had the effect of dissolving any residual rare earth particles (either difficult crystals of chlorides or hydroxides). A stopper was placed in the opening of the flask. The flask was submerged in a water bath at 25 °C and left for a week. One week was chosen based on informal testing—after about four days, the solution could be decanted into another vessel without further precipitation. A sample of the solution was taken by pipette and diluted with hydrochloric acid (0.1 M), for analysis by ICP-AES. The density of the solution was checked to be 1 g/mL.

This method produced internally consistent results between yttrium, erbium and thulium. However, in the case of lanthanum or neodymium, the initial stages of the precipitation proceeded without the involvement of one or more of the elements. To minimize the impact of this initial precipitation, a higher initial concentration of rare earths was used so that the final concentration covered a greater range of concentration. Also, to minimize errors of measurement of the initial solution, a stock solution was prepared for use in every conical flask. To allow a slower, more gradual precipitation, diethyl oxalate (DEO) was chosen as the oxalic acid ester. This appears to have a slower rate of hydrolysis than dimethyl oxalate.

About 5 g of each of lanthanum, neodymium and yttrium oxides were dissolved together in the one flask in hydrochloric acid, and evaporated to rare earth chlorides. Deionized water and 1 mL of hydrochloric acid (10 M) was added to a total of 1000 g. The 1 mL of hydrochloric acid was to dissolve any rare earth hydroxides that may be present. The pH was not measured, since the acidity during the experiment increases as more of the oxalic acid ester hydrolyses, so that a constant pH could not be maintained without the use of a buffer that may complex the rare earth ions. To each 100 mL conical flask was added to about 100 g of rare earth solution and a weighed amount of diethyl oxalate solution. The diethyl oxalate was added by a glass dropper, as diethyl oxalate can be a solvent of plastic. A sample of the solution was taken by pipette and diluted with hydrochloric acid (0.1 M) for analysis by ICP-AES. The density of the solution was checked to be 1 g/mL.

It was decided to plot the results between two rare earths from every experimental vessel on the one graph. The graph was based on Equation (2), so that the logarithm of the quotient of the starting concentration and final concentration was plotted on an axis for one element, while the same was plotted on the other axis for the other element. In this way, the slope of the graph should give the logarithmic distribution coefficient. The reason why this graphical method was chosen rather than tabulating the coefficient for each experiment was because the graph allows checks of internal consistency. Firstly, the line of best fit should be passing through the origin. This is because any constants of integration would be cancelled out in the integration of Equation (1) to Equation (2). Secondly, the ratios between a pair should be expressible in terms of the others; for example, pairs in a triplet should multiply to one, as illustrated by Equation (8):

$$K_{\frac{A}{B}} = K_{\frac{C}{B}} \times K_{\frac{A}{C}} \tag{8}$$

This is because the logarithmic distribution coefficient should be consistent between a pair regardless of how many rare earths are in the system.

3. Results

Table 1 shows which rare earths and ester were in each experimental vessel or conical flask, as well as the solution masses used to calculate the initial concentration. The first five experiments used the original experimental method. In this procedure, the total solution mass was weighed after the rare earth chlorides and dimethyl oxalate were dissolved. The last four experiments used the stock solution. In this method, the assayed stock solution was weighed into the conical flask and then the addition of diethyl oxalate was weighed.

Table 1. Identity of solution components and masses for each experiment.

Experiment Number	Solution Components	Solution Mass, g	Ester	Ester Mass, g	Total Mass, g
1	La, Pr, Nd, Y, Er, Tm	100.00	DMO	-	100.00
2	Y, Er, Tm	100.16	DMO	-	100.16
3	La, Pr, Nd, Y, Er, Tm	100.01	DMO	-	100.01
4	Y, Er, Tm	100.04	DMO	-	100.04
5	Y, Er	100.84	DMO	-	100.84
6	La, Nd, Y	102.62	DEO	2.04	104.66
7	La, Nd, Y	102.24	DEO	4.20	106.45
8	La, Nd, Y	102.65	DEO	1.05	103.70
9	La, Nd, Y	101.50	DEO	3.15	104.65

Table 2 shows the calculated initial concentrations and the measured final concentrations in each experimental vessel. In Experiment 7, too much diethyl oxalate was added, and complete precipitation occurred, so that the rare earths were below the detection limit (BDL).

Figures 1–3 show the results of the experiments with diethyl oxalate and concentrations of lanthanum, neodymium and yttrium. In addition, Figure 4 shows the result of the experiments with dimethyl oxalate and yttrium and erbium, in which the intercepts were also small, so that the use of diethyl oxalate was not required. The graphical results involving thulium also had a small intercept but are not shown here, as no thermodynamic data for thulium could be found for correlating the slopes of the graphs.

It can be seen in Figures 1–4 that the results have an internal consistency, with the lines of best fit having an intercept close to the origin. Also, when the slopes of Figures 1–3 are multiplied together as a triplet, the product is very close to one, indicating another internal consistency. The data points also fit fairly closely to a straight line, although deviations will appear to be diminished somewhat by the logarithmic axes.

Table 2. Initial and final solution concentrations for each experiment.

Experiment Number		La (ppm)	Pr (ppm)	Nd (ppm)	Y (ppm)	Er (ppm)	Tm (ppm)
1	Initial	486	880	183	603	450	883
	Final	289	493	90	508	340	621
2	Initial	-	-	-	802	661	600
	Final	-	-	-	323	170	168
3	Initial	549	115	920	430	191	514
	Final	199	15	66	198	57	158
4	Initial	-	-	-	589	493	385
	Final	-	-	-	330	202	163
5	Initial	-	-	-	962	274	-
	Final	-	-	-	667	150	-
6	Initial	5603	-	4831	3681	-	-
	Final	1300	-	40	910	-	-
7	Initial	5488	-	4732	3606	-	-
	Final	*BDL*	-	*BDL*	*BDL*	-	-
8	Initial	5656	-	4877	3717	-	-
	Final	3885	-	1410	2600	-	-
9	Initial	5542	-	4779	3642	-	-
	Final	680	-	4	445	-	-

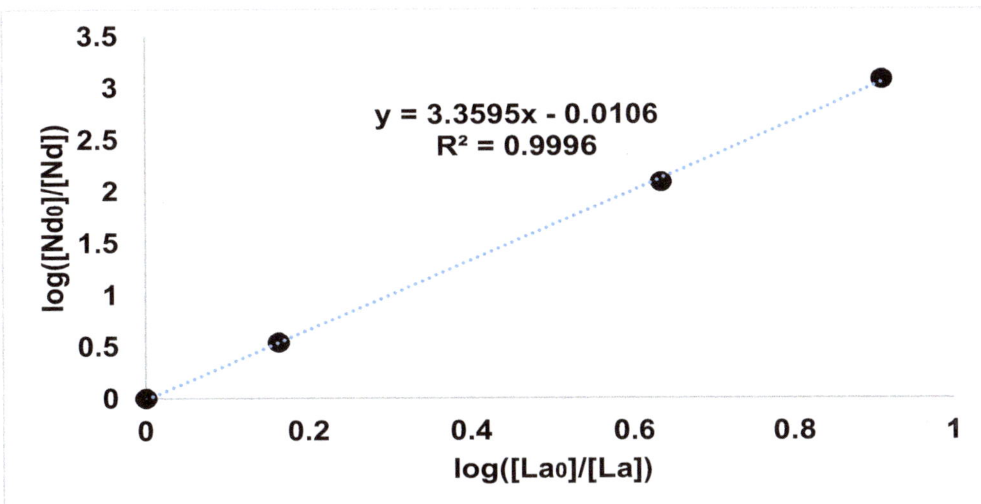

Figure 1. Enrichment of Nd against La: logarithm of the quotient of the starting concentration and final concentration for each rare earth.

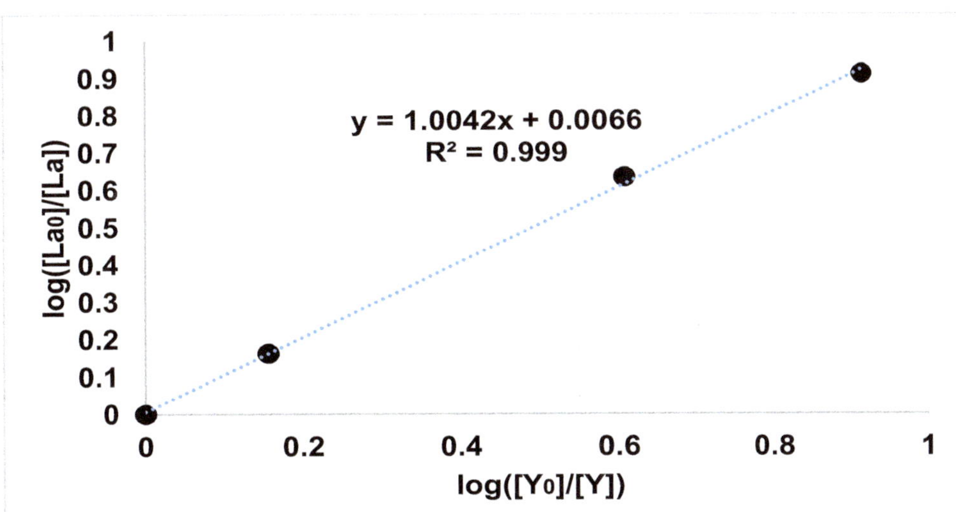

Figure 2. Enrichment of La against Y: logarithm of the quotient of the starting concentration and final concentration for each rare earth.

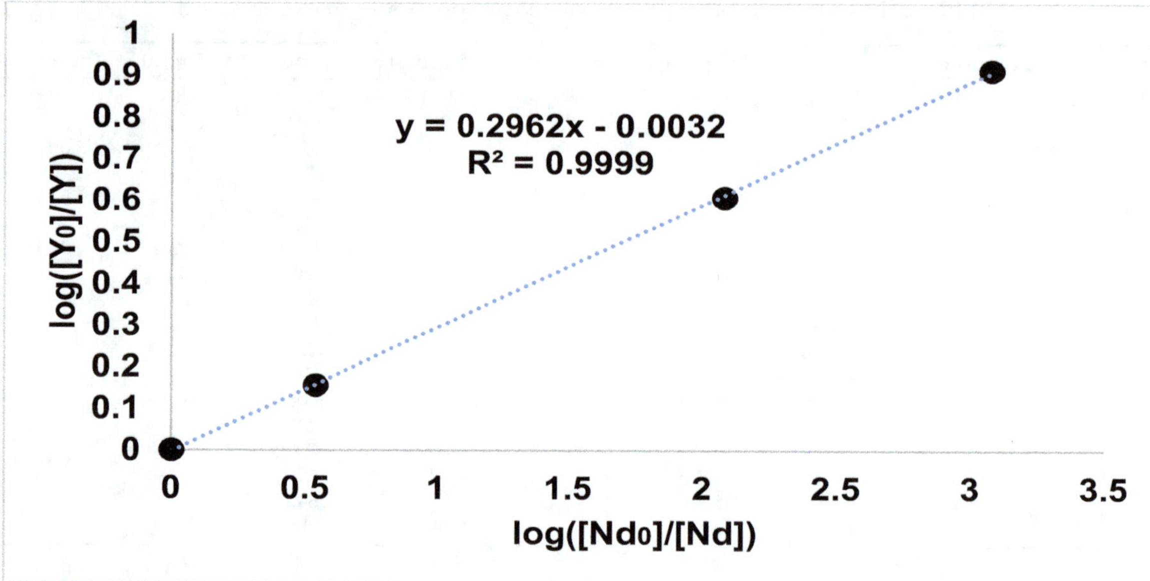

Figure 3. Enrichment of Y against Nd: logarithm of the quotient of the starting concentration and final concentration for each rare earth.

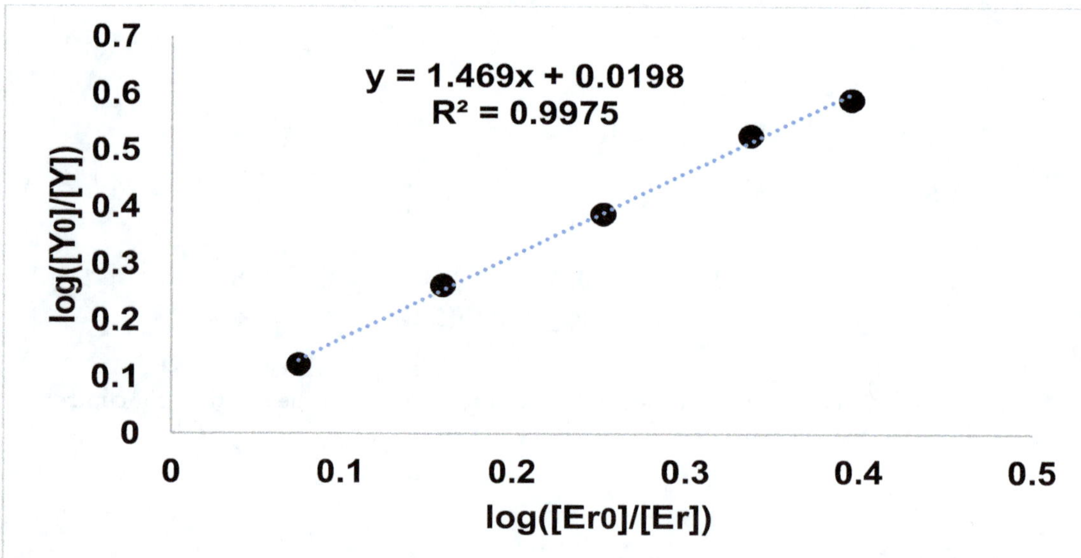

Figure 4. Enrichment of Y against Er: logarithm of the quotient of the starting concentration and final concentration for each rare earth.

It would have been preferable to have had a greater number of data points on each graph. However, each data point comes from a separate experimental vessel, and there was no exclusion of outliers, apart from Experiment 7, in which the precipitation was complete. In terms of repetition, in one sense each data point on the one graph is a repetition of the same separation, although repetition of the same specific conditions would also provide greater confidence. Overall, it seems that the slope values in Figures 1–4 can be described as the logarithmic distribution coefficient (λ), found experimentally.

4. Discussion

Given that Equation (3) did not describe the distribution coefficient λ [11], and that it is based on the solubility of the free or weakly-complexed rare earth ion, an alternative basis for describing λ was needed. It was hypothesized that the precipitation was governed by the energy difference between the solid rare earth oxalate and the mono-oxalato complex in Equation (9).

$$AC_2O_4{}^{1+} \leftrightarrow 0.5\, A_2(C_2O_4)_3 \tag{9}$$

Equation (10) was posed as a way to describe λ, because it represents the ratio of the equilibrium constants for Equation (9), which in turn is based on Equations (4) and (6). It also approximates the ratio of the saturated concentrations of the mono-oxalato complexes of each rare earth in question on the basis of Equation (5), assuming that the activity coefficient γ_1 is similar between rare earths. The term in Equation (5) for the free or weakly-complexed rare earth ion was neglected, as it seemed unlikely that solid rare earth oxalates would form directly from this simple species.

Equation (10) was used to calculate theoretical values in Table 3. The experimentally found values match these theoretical values fairly closely.

$$\lambda_{\frac{B}{A}} \cong \frac{\sqrt{K_{Sp,\,A}\,\beta_{1,A}}}{\sqrt{K_{Sp,\,B}\,\beta_{1,B}}} \tag{10}$$

Table 3. Comparison between theoretical and experimentally obtained values for the logarithmic distribution coefficient.

Pair	Theoretical	Experimental
Nd/La	3.25	3.36
La/Y	0.96	1.00
Y/Nd	0.32	0.30
Y/Er	1.47	1.47

The theoretical values for λ were calculated using Equation (10) on the results obtained by Chung, Kim, Lee and Yoo [14], shown in Table 4. The theoretical values for λ for the range of rare earths for which data is available are shown in Table 5.

Table 4. Solubility products [14], equilibrium constants [14] and their products $\sqrt{K_{Sp}}\,\beta_1$.

Element	Solubility Product, K_{Sp}	Equilibrium Constant, β_1	$\sqrt{K_{Sp}\cdot\beta_1}$
Y	5.1×10^{-30}	2.3×10^{7}	5.2×10^{-8}
La	6.0×10^{-30}	2.2×10^{7}	5.4×10^{-8}
Nd	1.3×10^{-31}	4.6×10^{7}	1.7×10^{-8}
Sm	4.5×10^{-32}	3.2×10^{7}	6.8×10^{-9}
Eu	4.2×10^{-32}	3.3×10^{7}	6.8×10^{-9}
Gd	4.25×10^{-32}	3.5×10^{7}	7.2×10^{-9}
Dy	2.0×10^{-31}	4.9×10^{7}	2.2×10^{-8}
Er	9.0×10^{-31}	8.0×10^{7}	7.6×10^{-8}

The reason why Equation (10) should approximate λ is a conundrum. Firstly, the most abundant form of each rare earth will be the mono-oxalato complex only in a concentrated solution of oxalic acid according to Equation (5). In the present case, a hydrolyzing ester will only produce a very dilute concentration of oxalic acid before precipitation occurs, meaning that the most abundant form of rare earth in solution will be the free or weakly-complexed ion according to Equation (5). This implies that the measured final concentrations are mainly made up of free or weakly-complexed rare earth ions, and that the model is still a relation between the concentration of these ions and the fractions on the surface of the precipitate. Yet, according to Equation (9), the free rare earth ion is not involved in that equilibrium directly.

Table 5. Calculated logarithmic distribution coefficients (precipitation of the element in the horizontal row for every one of the elements in the vertical column).

Element	La	Nd	Sm	Eu	Gd	Dy	Er
Y	0.96	3.1	7.7	7.7	7.2	2.4	0.68
La		3.25	7.9	8.0	7.5	2.5	0.71
Nd			2.4	2.5	2.3	0.76	0.22
Sm				1.0	0.94	0.31	0.09
Eu					0.94	0.31	0.09
Gd						0.33	0.10
Dy							0.29

Nor is there an obvious kinetic argument, as a higher stability of mono-oxalato complex will cause less precipitation into the solid oxalate compound under this particular model. It would perhaps be expected that a higher concentration of mono-oxalato complexes would lead to a higher rate of collisions leading to precipitation.

The notion of a stepwise formation of organic complexes into an organic salt also does not yield an adequate explanation. The concentration of the mono-oxalato complex of each rare earth would be determined by the stability constant in Equation (7), as well as the activity fraction of the free rare earth ion. Adding the chemical equations for the formation of the mono-oxalato complex (Equation (6)), and the formation of the precipitate from the mono-oxalato complex (Equation (9)) merely yields Equation (4), which represents the solubility product.

The system is behaving as though the ratio of the concentrations of the mono-oxalato complex of each rare earth is the same as that for the free rare earth ions. It may be conjectured that Equation (6) for the formation of the mono-oxalato complex is not attaining equilibrium before precipitation is occurring. It is possible that this may be connected to the very low concentration of oxalic acid and the very low solubility of the oxalate precipitate. Overall, it remains difficult to explain the model from kinetic and thermodynamic principles.

5. Conclusions

It was found that the logarithmic distribution co-efficient λ for precipitation between a pair of rare earths was fairly close to the ratio of the equilibrium constants for the step from mono-oxalato complex to the rare earth oxalate salt. This model relates the total concentration of rare earths (mostly free rare earth ions) to the fractions precipitating into the solid, so it is unclear why only the last step of the precipitation counts for λ. There appears to be no simple kinetic explanation either, with λ being derived from thermodynamic terms, and the curious result that a higher stability of mono-oxalato complex results in less precipitation.

Author Contributions: H.W. was responsible for designing and conducting the experiments and drafting. Y.-K.L. provided academic supervision and feedback. All authors have read and agreed to the published version of the manuscript.

Acknowledgments: Acknowledged is the Minerals Research Institute of Western Australia for its funding of this project. Also acknowledged is the Australian Government Research Training Program for its funding.

References

1. Lapidus, G.T.; Doyle, F.M. Selective thorium and uranium extraction from monazite: I. Single-stage oxalate leaching. *Hydrometallurgy* **2015**, *154*, 102–110. [CrossRef]
2. Lazo, D.E.; Dyer, L.G.; Alorro, R.D.; Browner, R. Treatment of monazite by organic acids 1: Solution coversion of rare earths. *Hydrometallurgy* **2017**, *174*, 202–209. [CrossRef]

3. Gadkari, A.B.; Shinde, T.J.; Vasambekar, P.N. Magnetic properties of rare earth ion (Sm^{3+}) added nanocrystalline Mg–Cd ferrites, prepared by oxalate co-precipitation method. *J. Magn. Magn. Mater.* **2010**, *322*, 3823–3827. [CrossRef]
4. Choy, J.H.; Han, Y.H.; Kim, S.J. Oxalate coprecipitation route to the piezoelectric $Pb(Zr,Ti)O_3$ oxide. *J. Mater. Chem.* **1997**, *7*, 1807–1813. [CrossRef]
5. Cui, L.J.; Zhang, P.X.; Li, J.S.; Yan, G.; Wang, D.Y.; Pan, X.F.; Liu, G.Q.; Hao, Q.B.; Liu, X.H.; Feng, Y. Preparation and Characterization of Large-Quantity Bi-2223 Precursor Powder by Oxalate Coprecipitation. *IEEE Trans. Appl. Supercond.* **2016**, *26*, 1–4. [CrossRef]
6. Shter, G.E.; Grader, G.S. YBCO Oxalate Coprecipitation in Alcoholic Solutions. *J. Am. Ceram. Soc.* **1994**, *77*, 1436–1440. [CrossRef]
7. Drozd, V.A.; Gabovich, A.M.; Kała, M.P.; Nedilko, S.A.; Gierłowski, P. Oxalate coprecipitation synthesis and transport properties of polycrystalline $Sr_{1-x}La_xPbO_{3-\delta}$ solid solutions. *J. Alloy. Compd.* **2004**, *367*, 246–250. [CrossRef]
8. Krishnamurthy, N.; Gupta, C.K. *Extractive Metallurgy of Rare Earths*, 2nd ed.; CRC Press: Boca Raton, FL, USA, 2015.
9. Rodríguez-Ruiz, I.; Teychené, S.; Vitry, Y.; Biscans, B.; Charton, S. Thermodynamic modeling of neodymium and cerium oxalates reactive precipitation in concentrated nitric acid media. *Chem. Eng. Sci.* **2018**, *183*, 20–25. [CrossRef]
10. Gordon, L.; Shaver, K.J. Fractionation of some rare earth pairs: By precipitation from homogeneous solution. *Anal. Chem.* **1953**, *25*, 784–787. [CrossRef]
11. Feibush, A.M.; Rowley, K.; Gordon, L. Coprecipitation in Some Binary Systems of Rare Earth Oxalates. *Anal. Chem.* **1958**, *30*, 1605–1609. [CrossRef]
12. Matsui, M. The Coprecipitation Behavior of Rare Earth Elements with Calcium Oxalate upon Precipitation from a Homogeneous System. *Bull. Chem. Soc. Jpn.* **1966**, *39*, 1114. [CrossRef]
13. Doerner, H.A.; Hoskins, W.M. Co-precipitation of radium and barium sulfates 1. *J. Am. Chem. Soc.* **1925**, *47*, 662–675. [CrossRef]
14. Chung, D.Y.; Kim, E.H.; Lee, E.H.; Yoo, J.H. Solubility of rare earth oxalate in oxalic and nitric acid media. *J. Ind. Eng. Chem.* **1998**, *4*, 277–284.

9

Platinum Group Elements in Arsenopyrites and Pyrites of the Natalkinskoe Gold Deposit Northeastern Russia

Raisa G. Kravtsova, Vladimir L. Tauson, Artem S. Makshakov *, Nikolay V. Bryansky and Nikolay V. Smagunov

A.P. Vinogradov Institute of Geochemistry, Siberian Branch of Russian Academy of Sciences, Favorskiy str 1A, 664033 Irkutsk, Russia; krg@igc.irk.ru (R.G.K.); vltauson@igc.irk.ru (V.L.T.); tridigron@yandex.ru (N.V.B.); nicksm@igc.irk.ru (N.V.S.)

* Correspondence: artem_m@mail.ru

Abstract: The peculiarities of the distribution and binding forms of platinum group elements (Pt, Pd, Ru, Rh, Os and Ir) in the arsenopyrites and pyrites of the Natalkinskoe gold ore deposit (Northeastern Russia) were examined using atomic absorption spectrometry with analytical data selections for single crystals (AAS-ADSSC), a "phase" chemical analysis (PCA) based on AAS of different size-fractions of minerals, scanning electron microscopy with energy dispersive X-ray spectrometry (SEM-EDX) and laser ablation inductively coupled plasma mass spectrometry (LA-ICP-MS). The arsenopyrites and pyrites of the Natalkinskoe gold deposit were found to concentrate not only Au but also platinum group elements (PGEs) such as Pt, Pd, Ru and Rh. The PCA showed that the highest contents (in ppm) were found in the monofractions of arsenopyrite—Pt up to 128, Pd up to 20, Ru up to 86 and Rh up to 21—and comparably lower in monofractions of pyrite—Pt to 29, Pd to 15, Ru to 58 and Rh to 5.9. The AAS-ADSSC method revealed two forms of uniformly distributed Pt, Pd and Ru corresponding to the chemically bound element in the structure of the mineral and in the superficial non-autonomous phase (NAP). The superficially bound form dominates over the structural form and presumably exists in a very thin surface layer of the crystal (~100–500 nm). The maximum contents of these PGE, chemically bound in the structure of arsenopyrite, reached values of (in ppm) 48, 5.9 and 48; and in pyrite structure, 68, 5.2 and 34 for Pt, Pd and Ru respectively. The contents of Pt, Pd and Ru related to NAP on the surface of the crystal were significantly higher and amounted (in ppm) for arsenopyrite to 714, 114 and 1083; and for pyrite 890, 62 and 690 for Pt, Pd and Ru, respectively. Preliminary results for the Rh form in arsenopyrite crystals suggest that the surface-related form (154–678 ppm) is more abundant than the structural form (17–45 ppm). Data from studying the surfaces of sulphide minerals by SEM-EDX and LA-ICP-MS confirmed the presence of Pt, Pd, Ru and Rh on the surface of arsenopyrite and pyrite crystals. These methods generated primary data on the content of Os and Ir in arsenopyrite and pyrite in the surface layer. The maximum content of Os and Ir found in arsenopyrites was up to 0.7 wt%. PGE-enriched fluids (up to ~3 ppm Pt) may exist in the gold ore deposit. It is assumed that there is a common mechanism of impurities uptake associated with the active role of the crystal surface and surface defects for gold-bearing arsenopyrites and pyrites. The surface enrichment is due to peculiarities in the crystal growth mechanism through the medium of NAP and the dualism of the element distribution coefficient in the system of mineral–hydrothermal solution, which is higher for NAP, compared to the volume of the crystal. Although mineral forms of Pt, Pd, Ru, Rh, Os and Ir have not been found at the Natalkinskoe gold deposit, their existence in the form of nano-scale particles is not excluded. This follows from the evolutionary model of surficial NAPs, assuming their partial transformation and aggregation with the formation of nano- and micro-sized autonomous phases of trace elements. The presence of PGE in the ores and the possibility of their extraction significantly increase the quality and value of the extracted raw gold materials at the Natalkinskoe deposit, and adds to the list of known platiniferous ore formations.

Keywords: Natalkinskoe gold deposit; arsenopyrite; pyrite; platinum group elements; distribution; binding forms

---

1. Introduction

Studying the platinum content of gold ore deposits and platinum group elements (PGEs) in general in the rocks of carbon-containing and black shale formations is currently a promising area of research, both in Russia and abroad. Many researchers see these deposits as a promising new mineral resource, although views on the presence and potential of extracting PGE from the ores of these deposits are somewhat contradictory [1–13]. The Natalkinskoe deposit (Northeastern Russia, Magadan region) is one of the largest gold deposits in Russia, and so is of particular interest. In addition to gold, there are quite high contents of Pt and Pd in the ores of the deposit. There are few data on other PGEs (Ru, Rh, Ir and Os). PGE contents and distribution studies are necessary to complete the most important tasks in the geochemistry of endogenous ore formation, such as the study of the material composition and the identification of genesis features of deposits, including giant and unique deposits. These studies are of great practical importance. The presence of Pt in ores and the possibility of its extraction would significantly complement the list of already known platiniferous ore formations and significantly increase the quality and value of the gold ore raw materials mined.

Work to detect platinum content at the Natalkinskoe deposit began in 1990 by a research group headed by V.I. Goncharov from the North–East Interdisciplinary Scientific Research Institute of the Far-East Branch of the Russian Academy of Sciences (Magadan, Russia) [1–3]. Much geochemical and mineralogical work has been performed over the years, but in terms of contents of PGE " ... the obtained analytical results remain ambiguous, and often contradictory" [3] (p. 132). The reliability of previously obtained high Pt content in ores (tens of ppm) and Pd (several ppm) [3,14] is still questionable. It was concluded that "Minerals-concentrators of platinum group elements are not found in the deposit" [3] (p. 136) nor in native mineral forms of PGE.

Subsequently, and until now, questions relating to this issue have been raised repeatedly, but there has been no clear response. Plyusnina and co-authors [14] (p. 840) concluded that " ... the industrial content of precious metals (Au and Pt) at the Natalkinskoe deposit is caused by metasomatic alteration and sulphidization of carbonaceous rocks". Unlike Au, the involvement of the carbonaceous substance in the accumulation of Pt has not been established by the authors of this work: the connection between this metal and the organic component of rocks and ores has not been found. It was assumed that sulphide minerals concentrate not only Au [3,15], but also Pt [14]. Later arsenopyrite was found to act as an Au, Pt and Pd concentrator at the Natalkinskoe deposit [11]. According to preliminary data obtained at that time, Pt contents in monofractions of this sulphide ranged from 23 to 62 ppm, and Pd from 2.3 to 9.5 ppm. Two main non-mineral forms of Pt and Pd, structural and surface-related, were identified. The nature of high contents of non-mineral Pt and Pd forms in arsenopyrite was found to be mainly superficial. Distinct mineral forms of Pt and Pd were not observed.

Considering all the above, the study of the binding forms and concentration levels of PGE in the most common sulphide minerals at the Natalkinskoe deposit, arsenopyrite and pyrite, seems to be an interesting and important task. This paper presents new data on the content of Pt and Pd in arsenopyrite and pyrite. The concentration levels of Ru, Rh, Ir and Os in these minerals are estimated for the first time. It is shown that the ambiguity of the results of previous studies may have been caused by the presence of hidden (so-called "invisible") forms of PGEs.

2. Research Objects

The Natalkinskoe gold ore deposit, where this research was conducted, is located in the territory of Northeastern Russia (see Figure 1, inset). It is part of the Omchak ore-placer cluster and one of the largest gold reserves in Russia. Structurally, the deposit is confined to the boundary part of the estimated granite pluton in the Tenka deep fault zone and is associated with the collision development stage of the Yana–Kolyma fold system. The primary objects of the study are arsenopyrites and pyrites selected from the most industrially significant vein, streaky-vein and veinlet-disseminated ores of the Natalkinskoe deposit, which belong to the productive gold-bearing quartz–pyrite–arsenopyrite (with native gold) mineral association.

The deposit falls under the category of orogenic gold–quartz ore formations, and it is characterized by a complex, long-term nature of development and, according to most researchers, by metamorphogenic-hydrothermal genesis. The ore hosted rocks are identified as late Palaeozoic (upper Permian), and ore mineralization is presumably Mesozoic (from late Jurassic to late Cretaceous) [3]. The ores of the Natalkinskoe gold deposit, unique in scale, despite the diversity, form a deposit with uniform internal structure consisting of zones of quartz, carbonate–quartz, sulphide–quartz veins and veinlets, surrounded by a wide halo of sulphidized rocks. They are generally characterized by the same elemental and mineral composition, differing only in the degree of their manifestation: the quantitative ratios. The host volcanogenic sedimentary strata include diamictites, argillites and siltstone shales with layers of sandstones and gravelites with high carbon content. The ores of the deposit, as already mentioned, demonstrate relatively high contents of Pt and Pd. Commercial ores emerged in the interaction of host rocks with low- and moderate-salinity water–bicarbonate fluids in the salinity range of 3–12 wt% eq. NaCl, at temperatures of 360–280 °C and pressures of approximately 2.4–1.1 kbar [16]. More details on the geology, mineralogy, geochemistry and conditions of formation of the Natalkinskoe giant gold deposit are available [1–4,11,14–28].

To study the distribution features and forms of occurrence of PGEs in arsenopyrites and pyrites, 35 large-volume ore samples were selected, each weighing up to 10 kg. The samples were taken from quarries and boreholes of different horizons of the deposit from those most rich in the number of sulphide mineral veins, veinlets and metasomatites classed as gold vein, streaky-vein and veinlet-disseminated ores (the areas named "Geological", "North–West", "Central" and "South–East"). The schematic geological map of the Natalkinskoe gold deposit with areas from which samples were taken is shown in Figure 1, ore types in Figure 2.

Non-metallic minerals such as quartz, carbonates, feldspar, sericite, chlorite and carbonaceous matter (micrographite) were found in the specimens characterizing these samples. Arsenopyrite and pyrite amounted to 95–99% of the ore minerals, with the former predominating. Ore minerals, along with arsenopyrite and pyrite, the amount of which in some areas reached ~7%, included galena, sphalerite, chalcopyrite, rutile and native gold (up to 1%). Native gold with a fineness from 750‰, to 900‰, less frequently electrum, mainly of large grain size, was located mainly in the vein quartz and intergrowths with sulphide minerals. Small quantities of pyrrhotite, sulphoarsenides of nickel and cobalt and ilmenite were identified. More than 70 minerals were found at the deposit in total [3,19,20].

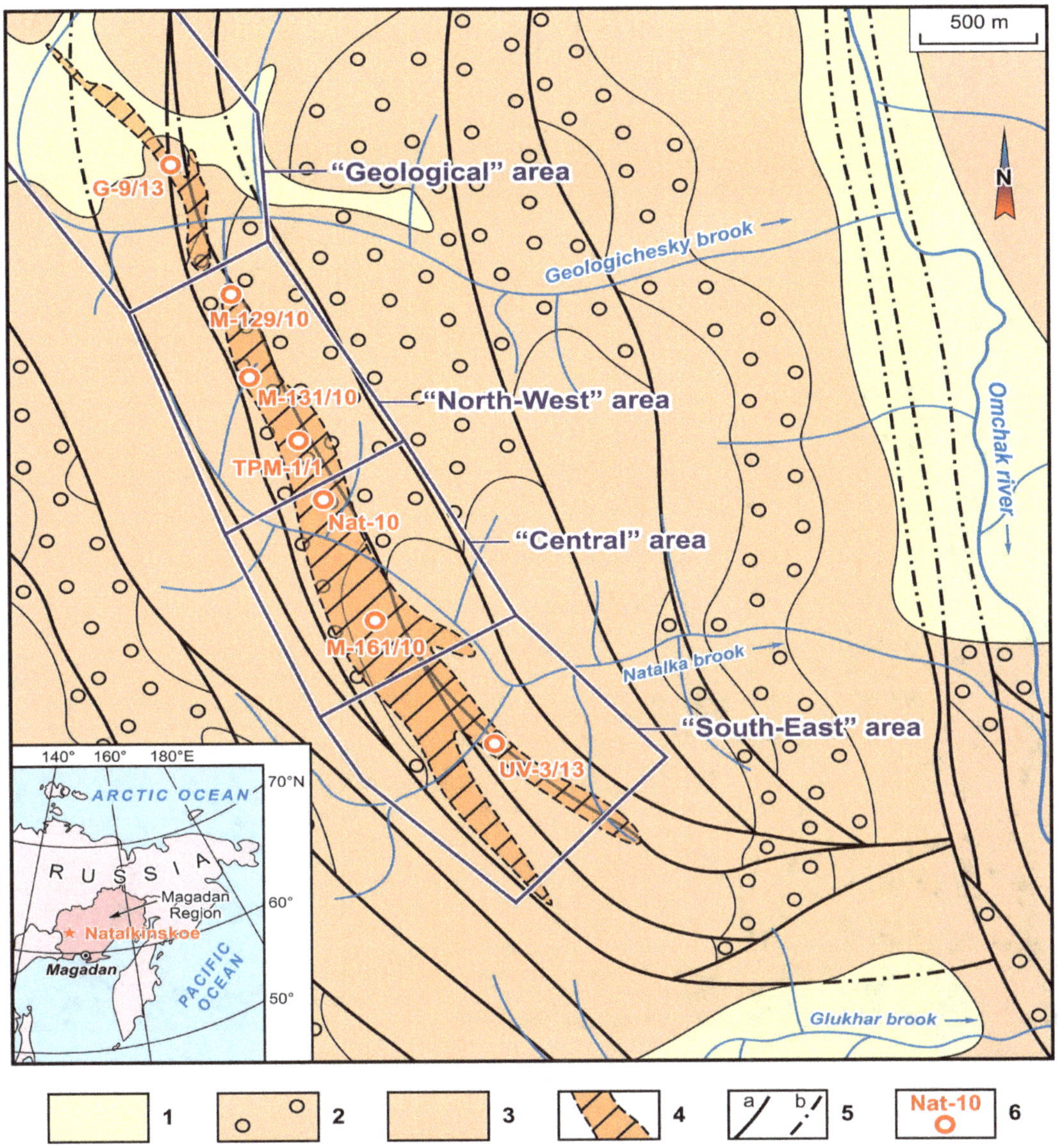

Figure 1. The schematic geological map of the Natalkinskoe gold deposit with areas from which ore samples were taken. Constructed by the authors of the present paper using the data [3,19,21,24,26]. Legend: 1—Quaternary sediments; 2 and 3—upper Permian ore hosted volcanogenic sedimentary rocks: 2—diamictites, 3—argillites and siltstone shales with layers of sandstones and gravelites with high carbon content; 4—ore deposit; 5—faults (a—reliable, b—estimated); 6—sampling points and their numbers. The inset map shows the location of the Natalkinskoe deposits in the territory of Northeastern Russia.

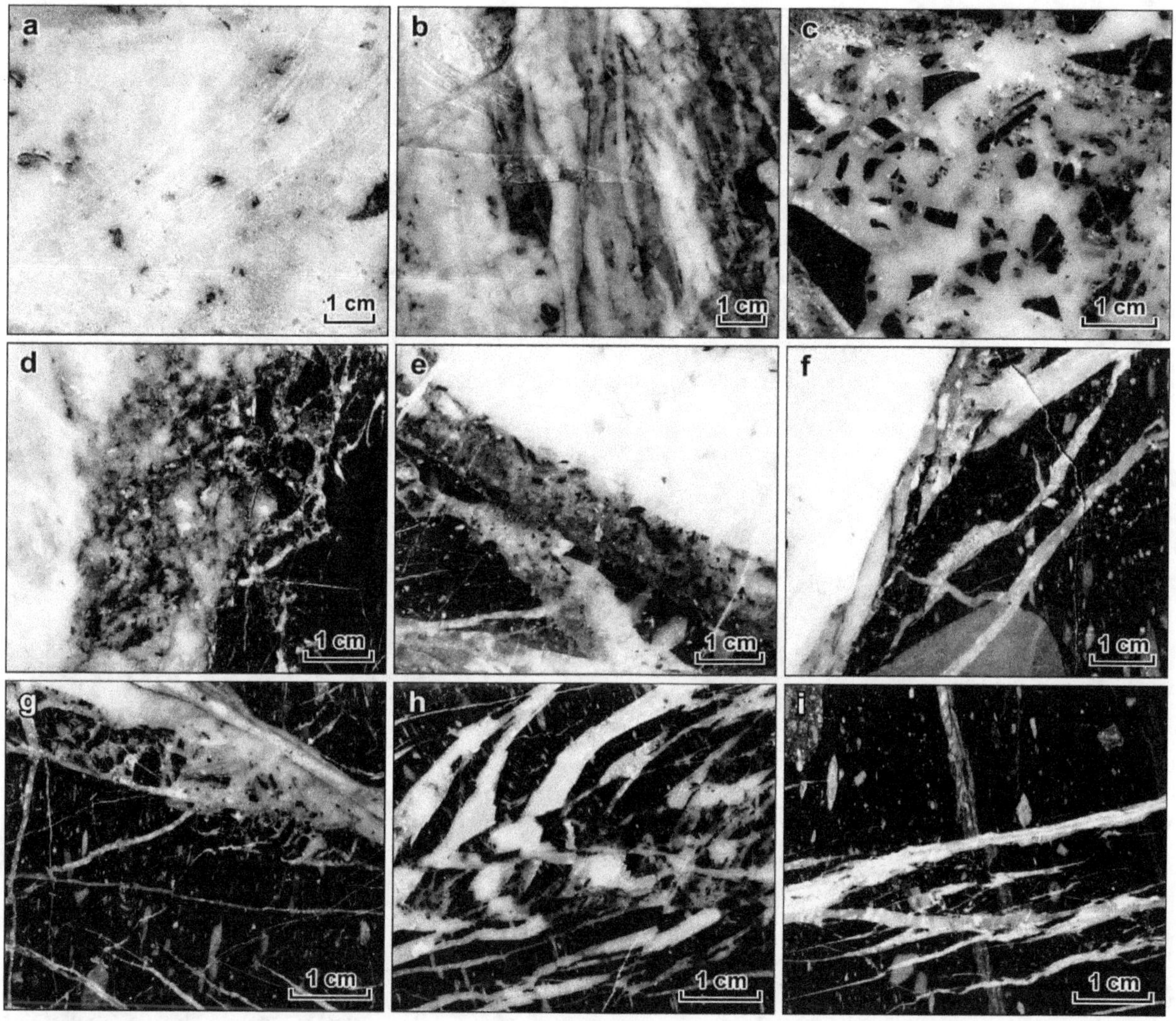

Figure 2. Types of ores in the Natalkinskoe deposit (a–c—vein, d–f—streaky-vein, g–i—veinlet-disseminated): (**a**) a massive quartz vein with a small number of fragments containing siltstones, nests and inclusions of arsenopyrite ("North–West" area, borehole DH-20/11n, interval 566–569 m); (**b**) quartz vein of laminated texture (bands are composed of completely silicificated siltstones) with the impregnations of small grains of arsenopyrite ("Central" area, quarry, elevation 810 m); (**c**) massive quartz–sulphide vein of breccia structure with fragments of host diamictites and enclaves of arsenopyrite and pyrite crystals ("North–West" area, quarry, elevation 900 m); (**d**) siltstones dissected by a net of quartz and carbonate–quartz veinlets with nests and individual crystals of arsenopyrite and pyrite, on the border with a massive quartz vein ("Central" area, mine, elevation 600 m); (**e**) silicificated diamictites dissected by quartz and carbonate–quartz veins with enclaves of arsenopyrite and pyrite crystals on the border with a massive quartz vein ("South–East" area, borehole DH50/12n, interval 452.0–454.9 m); (**f**) massive quartz vein on the border with diamictites, dissected by the carbonate–quartz veinlets, with fragments of sandy argillites of gravel size with enclaves of crystals and clusters of arsenopyrite and pyrite ("South–East" area, borehole DH50/12n, interval 460.8–463.8 m); (**g**) diamictites with fragments of argillites of gravel size dissected by a dense net of quartz and carbonate–quartz veins, with enclaves of arsenopyrite and pyrite crystals ("Central" area, borehole DH30/29, interval 273.5–276.5 m); (**h**) diamictites with enclaves of crystals, clusters and crosscuts of arsenopyrite and pyrite, from the zone of intense quartz and quartz–carbonate streaking ("South–East" area, borehole DH50/12n, interval 428.1–431.0 m); (**i**) diamictites dissected by transverse quartz and quartz–carbonate veins with fragments of argillites of gravel size and enclaves of arsenopyrite and pyrite crystals and clusters ("South–East" area, borehole DH50/12n, interval 528.4–531.4 m).

3. Research Methods

All ore samples were analyzed for Au and PGEs using atomic absorption spectrometry (AAS) and inductively coupled plasma mass spectrometry (ICP-MS) [29–31]. Twenty-two monofractions of arsenopyrite and 14 monofractions of pyrite were selected from the 7 ore samples richest in amount of sulphides, Au and PGE. They were all subjected to AAS analysis for Au, as were the ore samples [29,30]. All sulphide minerals were found to be gold-bearing. The level of Au content in arsenopyrite was 1.4–1383 ppm, in pyrite 0.8–158 ppm.

Further study of these gold-bearing arsenopyrites and pyrites for PGE was carried out using "phase" chemical analysis (PCA) based on AAS, coupled with analytical data selections for single crystals (ADSSC) [32,33], scanning electron microscopy with energy dispersive X-ray spectrometry (SEM-EDX) and laser ablation inductively coupled plasma mass spectrometry (LA-ICP-MS). These methods best characterize the peculiarities of minerals such as internal parts and surface layers, microinclusions, the so-called "invisible" impurity elements, their forms, distribution and concentration levels.

This set of methods was applied because PGEs in many deposits are represented by at least two forms—a solid solution in a carrier mineral and the submicroscopic inclusions of a PGE-rich material in an invisible form, undetectable by optical and scanning electron microscopy [34]. The results of the experiment and theoretical analysis suggest that not only Au, but also other noble metals (PGE and Ag) can form chemically bound forms of impurities in pyrite and other sulphides and chalcogenides [35]. The presence of these forms (at least one of them—a solid solution of PGE in sulphide or sulphoarsenide matrix) is identified even in low-temperature (~100 °C) platinum-bearing associations [36].

Current methods of "phase" chemical analysis for the determination of PGE in monofractions of arsenopyrite and pyrite, as well as the analysis of individual crystals of these minerals by AAS-ADSSC, allow a reliable determination of the contents of only three PGEs: Pt, Pd and Ru. The Rh analysis is under development, and the data are preliminary. All PGE—Pt, Pd, Ru, Rh, Ir and Os—can be analyzed using SEM-EDX and LA-ICP-MC methods. Restrictions are related only to their detection limits. Requirements for the selection of the test material are discussed below.

Monomineral samples of different sizes (0.07–0.14, 0.14–0.2, 0.2–0.25, 0.25–0.5, 0.5–1.0 and 1.0–2.0 mm) are taken for "phase" chemical analysis: individual and cluster crystals of arsenopyrite represented by pseudo-orthorhombic and monoclinic prisms; individual and cluster crystals of pyrite of cubic, pentagon-dodecahedral and cuboctahedral habits (Figure 3).

Figure 3. Monomineral fractions: (**a**) arsenopyrite, (**b**) pyrite.

In order to study PGEs in arsenopyrites and pyrites using the AAS-ADSSC method, euhedral individuals with well-defined morphology were selected from monomineral fractions of different sizes, from 0.25 to 2 mm. The crystals of arsenopyrite were predominantly shaped as pseudo-orthorhombic prisms, close to a slightly distorted parallelepiped, and less frequently as monoclinic prisms. Selected pyrite crystals generally had the shape of a cube or parallelepiped, although the samples were often complicated by pentagonal dodecahedron and octahedron facets (Figure 4). Shape is the characteristic required for calculation of the specific surface area of average crystal in size sample using the form coefficient for the true polyhedron. In our case, the coefficient was six for cubes and parallelepipeds. A total of 415 arsenopyrite crystals and 234 pyrite crystals were selected and studied.

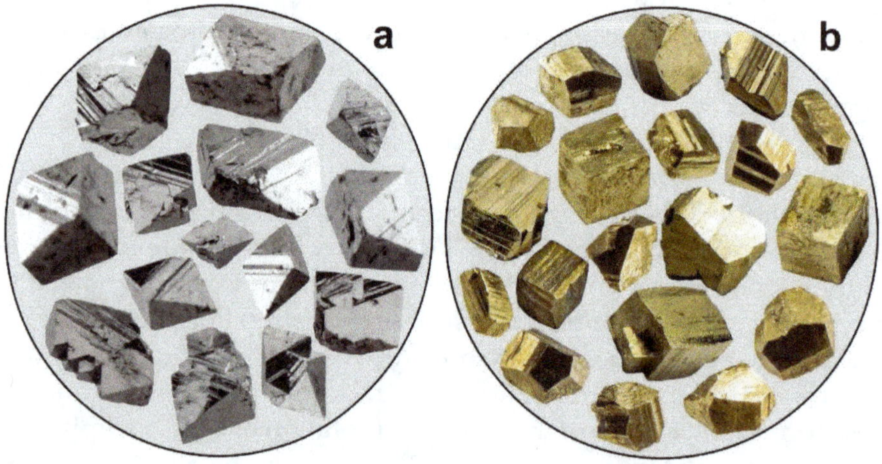

Figure 4. Crystals: (**a**) arsenopyrite, (**b**) pyrite.

Twenty arsenopyrite crystals and 14 pyrite crystals were selected and examined by SEM-EDX and LA-ICP-MS in order to study the PGE distribution in the surface layer. Euhedral pseudo-orthorhombic crystals of arsenopyrite and cubic crystals of pyrite 0.5–1 mm in size were used to study the surface of these minerals by SEM-EDX. Larger (1–2 mm) crystals of arsenopyrite and pyrite with at least one clearly manifested face belonging to certain simple form were selected for the LA-ICP-MS study (see Figure 4).

3.1. *"Phase" Chemical Analysis*

PCA was carried out with 10–20 mg samples with the view to determine the PGE in monofractions of arsenopyrite and pyrite. Monofractions were ground to powder and decomposed with Aqua Regia during heating. After decomposition, the material was treated with concentrated HCl and evaporation to dry state to remove nitric acid residues and convert the salts to chloride form. After cooling, the samples were diluted with 2M HCl to the condition required for the AAS method with electrothermal atomization on the M-503 device (Perkin-Elmer, Waltham, MA, USA) using graphite atomizer HGA-72 (Perkin-Elmer, Waltham, MA, USA) [29,30]. Au was determined in all the samples, along with Pt, Pd and Ru [29]. The accuracy of determination was ±10–12 rel. %.

3.2. *AAS-ADSSC Method*

The study of the so-called "invisible" uniformly distributed impurity component of PGE in arsenopyrites and pyrites was carried out using the AAS-ADSSC method. The method was developed by V.L. Tauson and co-authors for the study of structural and surface-bound forms of gold in ore minerals [32,33,37,38]. The method was further refined and successfully used to study structural and surface-bound forms of Pt, Pd, Ru and Rh [11,38,39]. The technique is designed to determine the content of the structural impurity of the element with an uncertainty of ±30 rel. % [40,41]. The determination of Pt, Pd, Ru and Rh in solutions obtained through the decomposition of individual crystals of arsenopyrite and pyrite was carried out following their preliminary extraction, concentration and separation from the matrix. Tristyrylphosphine, $(C_6H_5CH{-}CH)_3P$, was used as an extracting agent. Extraction was carried out using hydrochloric acid solutions (0.5 M HCl). The extracting agent concentration was

0.05 M (in toluene), and the phase contact time was 30 min. The ratio of aqueous to organic phases was 2:1. The extraction was carried out in static mode at room temperature and without labilising additives. The organic phase was used to measure element concentrations. Measurements were performed by AAS with electrochemical atomization on an M-503 (Perkin-Elmer, Waltham, MA, USA) device with graphite atomizer HGA-72 (Perkin-Elmer, Waltham, MA, USA) [29,30].

The data obtained were statistically processed according to the regulations of the distribution of different binding forms of the element [33,38]. We divided the dataset (>30–40 crystals with evenly distributed element) into the intervals of crystal masses (sizes), chosen to be as narrow as possible, although statistically representative, and determined the average crystal mass in every size fraction ($\overline{m}$), average size ($\overline{r}$), specific surface area ($\overline{S}_{sp}$) and element content ($\overline{C}$). The more size fractions and number of crystals in the final samples, the more reliable the results obtained. When constructing the dependences $\overline{C} = f(\overline{S}_{sp})$, we usually obtained a number of points best approximated with an exponent. The extrapolation of these curves to a zero-specific surface, that is, to a virtual infinite crystal, gave the bulk element content. In our model, this was equal to the structurally bound element content (C_{str}) because all other possible bulk forms were eliminated at the stage of initial dataset processing. The superficially bound element content ($\overline{C}_{sur}$) characterizes an average crystal among all size samples, that is, the surface-related excess content of the element, and can be calculated with the equation given in [38]. It is important to note that in such an approach, the amount of the material is normalized to the whole crystal, and this allows a comparison of the contribution of surface and bulk related forms of the element to its total content.

3.3. SEM-EDX Method

The study of the "invisible" impurity content of the element on the surface of arsenopyrite and pyrite crystals by SEM-EDX was carried out on a Quanta 200 scanning electron microscope (FEI Company, Hillsboro, OR, USA) with energy-dispersive accessory EDAX (FEI Company, Hillsboro, OR, USA) with nitrogen-free cooling for X-ray spectral microanalysis. The maximum spatial resolution of the device with a tungsten cathode and a standard detector of secondary and backscattered electrons was 3.5 nm. In elemental analysis, the resolution maximally amounted to 2 μm. The EDAX accessory with EdaxGenesis software enables not only semiquantitative but also quantitative analysis of elements in a wide range (from Be to U) with a resolution of 127 eV and a minimum detection limit of 0.5 wt% for the elements considered. Microphotography and crystal surface analyses were performed under high-vacuum conditions, without sputtering and at an accelerating voltage of 30 kV. Elemental analysis was performed on a flat area with a side of at least 2 μm. The contents were calculated using the three-correction ZAF method (atomic number, absorption and fluorescence). The ZAF method is based on measuring the X-ray intensity of the *i*-th element of the sample (J_i) and the standard of the known composition containing this element ($J_{(i)}$) under the same conditions, minus the intensity of the X-ray background radiation, followed by normalization by 100%. To study the chemical composition of the surface of arsenopyrite and pyrite crystals using the SEM-EDX method, the most "clean" areas, with virtually no inclusions or significant defects visible at maximum magnification were selected. It seems meaningless to use polished samples while examining natural surfaces. However, it is possible to minimize fluctuations of the probe current by selection of smooth-faced sections and their positioning in the SEM device. Sections with as little roughness as possible (2–3 nm, well comparable with ordinary polished surfaces) were selected using atomic force microscopy on a SMM-2000 scanning probe microscope (Proton-MIET, Moscow, Russia).

3.4. LA-ICP-MS Method

The study of the surface layer of arsenopyrite and pyrite crystals was continued using the LA-ICP-MS method. A quadrupole inductively coupled plasma mass spectrometer NexION 300D

(Perkin-Elmer, Waltham, MA, USA) was used in combination with a laser ablation system NWR-213 (New Wave Research, Fremont, CA, USA) with a laser wavelength of 213 nm. The working and carrier gases were Ar and He, at 99.999% purity. The laser burned a continuous groove (track) no longer than 1.4 mm on the surface of the crystal. The diameter of the laser beam was 100 μm, the frequency was 10 Hz and the energy on the surface of the sample was 0.4 J/cm^2. The laser movement speed was 70 μm/s, and the background measuring time was 20 s. The calibration graphs were constructed in compliance with the NISTSRM610, NISTSRM612, NISTSRM614, BHVO-2G, TB-1G and NKT-1G international standards. It was possible to verify the results for several elements (Au, Pd and Pt) using the in-house sulphide reference sample, which included highly homogeneous ferrous greenockite (α-CdS) crystals synthesized hydrothermally at 500 °C and 1 kbar pressure [42]. The instrument error determined for PGE did not exceed 10%. The contents of the elements in arsenopyrite and pyrite were calculated according to the depth of the track. The ablation depth was controlled by a Quanta 200 scanning electron microscope (FEI Company, Hillsboro, OR, USA). An analysis depth from 0.5 to 1 μm per laser pass was obtained for all samples. For one of the arsenopyrite crystals (sample M-129/10) selected for a detailed study, the laser parameters were adjusted in such a way as to achieve the burned layer of ~100 nm in depth in one pass. Depth was controlled using atomic force microscopy with a SMM-2000 scanning probe microscope (Proton-MIET, Moscow, Russia). In this case, the diameter of the laser beam was 5 μm, the frequency 20 Hz, and the energy on the surface of the sample was 10 J/cm^2 with similar parameters. The instrument error for Pt and Pd did not exceed 30%.

4. Results and Discussion

This section presents data for the study of distribution features and forms of occurrence of PGE (Pt, Pd, Ru, Rh, Ir and Os) in arsenopyrite and pyrite from the Natalkinskoe gold deposit. The data were obtained through local (SEM-EDX and LA-ICP-MS) and, so to speak, "semi-local" analyses (PCA and ADSSC). That is, the methods are not exactly local but were by no means bulk analysis, with a resolution at the level of different size fractions (PCA) or the size of individual crystals (ADSSC). The combination of local analysis and statistics for the compositions of individual monocrystals of different sizes is a unique feature of our approach.

4.1. Pt, Pd, Ru and Rh Content in Monomineral Fractions of Arsenopyrite and Pyrite According to PCA

All studied arsenopyrites and pyrites from the deposit were found to be concentrators not only of Au, but also of such elements as Pt, Pd and Ru. Data on their content are given in Table 1. The highest contents (in ppm) were observed in arsenopyrite: Pt up to 128, Pd up to 20 and Ru up to 86, with lower values identified in pyrite: Pt up to 29, Pd up to 15 and Ru up to 58. The first, although preliminary, results on the content in the monofractions of the studied sulphides of another platinum group element, Rh, have been obtained (Table 2). Maximum contents of this element, up to 21 ppm, as well as for Pt, Pd and Ru are found in arsenopyrite, and lower levels up to 5.9 ppm were found in pyrite. Not only gold-bearing arsenopyrites, but also pyrites were shown to act as concentrators of these elements. It can be argued that the most platinum-rich ores have a sulphide–quartz composition. The data on PGE content in bulk ore samples are in favor of this argument. Relatively high contents (in ppm) were observed for Pt (up to 0.260) and Pd (up to 0.029), and elevated contents for Os (up to 0.003), Ru (up to 0.004), Rh (up to 0.0012) and Ir (up to 0.0006).

Table 1. Platinum, palladium and ruthenium content in different-sized monofractions of arsenopyrite and pyrite selected from vein, streaky-vein and veinlet-disseminated ores of the Natalkinskoe deposit.

Fraction No.	Sample No.*	Characteristics of Samples for Selection of Monofractions	Fraction (mm)	Pt (ppm)	Pd (ppm)	Ru (ppm)
		Arsenopyrite				
1	G-9/13	Veinlet-disseminated ore type. Diamictites dissected by a dense net of quartz and carbonate–quartz veinlets with inclusions of arsenopyrite and pyrite crystals. Arsenopyrite crystals shaped as pseudo-orthorhombic and monoclinic prisms predominate.	0.5–1	60	12	44
2			0.25–0.5	70	20	55
3			0.2–0.25	128	5.4	25
4			0.14–0.2	64	14	53
5			0.07–0.14	63	17	86
6	TPM-1/1	Streaky-vein ore type. Veins and veinlets of quartz and carbonate–quartz in diamictites, with inclusions of crystals and nodules of arsenopyrite, less frequently—pyrite. The crystals of arsenopyrite are shaped as pseudo-orthorhombic and monoclinic prisms.	0.25–0.5	40	4.0	31
7			0.14–0.25	73	11	54
8			0.07–0.14	120	6.7	67
9	M-129/10	Vein ore type. Sulphidized quartz vein on contact with diamictites dissected by a dense net of quartz and carbonate–quartz veins. Sulphides are mainly represented by arsenopyrite dominated by crystals in the form of pseudo-orthorhombic and monoclinic prisms.	0.5–1	23	3.5	18
10			0.25–0.5	25	3.9	21
11			0.2–0.25	27	4.1	24
12	M-131/10	Streaky-vein ore type. Offset subparallel quartz and carbonate–quartz veins and veinlets in diamictite. Sulphides are mainly represented by arsenopyrite. The crystals are shaped as pseudo-orthorhombic and monoclinic prisms.	0.5–1	24	2.8	16
13			0.25–0.5	26	2.3	24
14			0.14–0.25	29	3.6	17
15	M-161/10	Vein ore type. Massive sulphide–quartz vein with a small number of fragments of diamictites, with nodes and inclusions of arsenopyrite, rarely pyrite. Arsenopyrite crystals shaped as monoclinic prisms dominate.	1–2	27	3.7	19
16			0.5–1	47	6.3	33
17			0.25–0.5	62	9.5	31
18			0.14–0.25	44	6.2	28
19	Nat-10	Veinlet-disseminated ore type. Diamictites interspersed with pyrite, rarely arsenopyrite, dissected by a dense net of quartz and carbonate veins from hair-thick to 5 mm. The crystals of arsenopyrite are shaped as pseudo-orthorhombic and monoclinic prisms.	0.2–0.25 **	20	3.4	20
				20	1.9	17
20	UV-3/13	Veinlet-disseminated ore type. Diamictites interspersed with arsenopyrite and pyrite, dissected by a dense net of quartz and carbonate veins. Arsenopyrite crystals shaped as pseudo-orthorhombic and monoclinic prisms predominate.	0.5–1	15	3.5	3.7
21			0.25–0.5 **	5.9	0.7	5.4
				8.0	0.4	4.0
22			0.2–0.25	10	0.7	4.2

Table 1. *Cont.*

Fraction No.	Sample No.*	Characteristics of Samples for Selection of Monofractions	Fraction (mm)	Pt (ppm)	Pd (ppm)	Ru (ppm)
		Pyrite				
23	TPM-1/1	Streaky-vein ore type. Veins and veinlets of quartz and carbonate–quartz in diamictites, with inclusions of crystals and nodules of arsenopyrite, less frequently—pyrite. Pyrite forms cubic and pentagon-dodecahedral crystals.	0.25–0.5	1.0	0.5	4.4
24			0.14–0.25	3.3	0.3	2.9
25			0.07–0.14	1.1	0.4	4.2
26	M-161/10	Vein ore type. Massive sulphide–quartz vein with a small number of fragments of diamictites, with nodules and inclusions of arsenopyrite, rarely pyrite. Cubic pyrite crystals predominate.	0.5–1	14	12	24
27			0.25–0.5	25	15	33
28			0.2–0.25	27	14	41
29	Nat-10	Veinlet-disseminated ore type. Diamictites interspersed with pyrite, rarely arsenopyrite, dissected by a dense net of quartz and carbonate veins from hair-thick to 5 mm. Pyrite forms cubic, pentagon-dodecahedral, rarely cuboctahedral crystals and their combined forms.	1–2 **	25	4.7	21
				10	1.0	28
30			0.5–1 **	15	1.0	57
				15	1.0	39
31			0.25–0.5 **	12	2.1	58
				10	5.8	48
32			0.2–0.25 **	10	2.8	22
				13	3.6	11
33			0.14–0.2	18	4.9	17
34	UV-3/13	Veinlet-disseminated ore type. Diamictites interspersed with arsenopyrite and pyrite, dissected by a dense net of quartz and carbonate veins. Pyrite forms cubic, pentagon-dodecahedral, rarely cuboctahedral crystals.	0.5–1 **	15	1.0	16
				8.0	0.5	2.8
35			0.25–0.5 **	11	1.0	28
				17	1.0	16
36			0.2–0.25	29	1.7	14

Note: The data for the "phase" chemical analysis are presented here and in Table 2. Here: elev.—elevation relative to sea level. * Ore sampling location: G-9/13—"Geological" area, borehole DH329n, interval 151.6–154.6 m (elev. 750 m); TPM-1/1—"North–West" area, quarry (elev. 860 m); M-129/10—"North–West" area, ore zone 33, surface (elev. 920 m); M-131/10—"North–West" area, ore zone 33, surface (elev. 900 m); M-161/10—"Central" area, mine (elev. 600 m, crosscut 11); Nat-10—"Central" area, quarry (elev. 790 m); UV-3/13—"South–East" area, borehole DH70/5n, interval 160.1–163.1 m (elev. 590 m). ** Measurements of Pt, Pd and Ru were carried out on two parallel samples.

Table 2. The content of rhodium in the differently sized monofractions of arsenopyrite and pyrite selected from vein and streaky-vein ores of the Natalkinskoe deposit.

Fraction No.	Sample No.	Monofraction (mm)	Rh (ppm)
		Arsenopyrite	
1	M-131/10	0.5–1	13
2		0.25–0.5	10
3		0.14–0.25	11
4	M-161/10	1–2	9.0
5		0.5–1	21
6		0.25–0.5	21
7		0.14–0.25	12
		Pyrite	
8	TPM-1/1	0.25–0.5	5.9
9		0.14–0.25	3.4
10		0.07–0.14	4.8

Note: Ore sampling location and brief characteristics are presented in Table 1.

4.2. Results of the Study of Pt, Pd, Ru and Rh in Arsenopyrites and Pyrites by AAS-ADSSC

Data on the so-called "invisible" uniformly distributed impurity of Pt, Pd and Ru and the ratio of their structural and surface-bound forms in arsenopyrites and pyrites of the Natalkinskoe deposit are presented in Figures 5–7 and Tables 3–5. In all cases, highly deterministic (R^2 = 0.82–1.0) dependences of the average element content in the size sample on the specific surface area of the average crystal in it were obtained. The highest R^2 values were set for arsenopyrite: Pt, 0.92–0.99; Pd, 0.93–1.0 and Ru, 0.97–1.0 and were slightly lower for pyrite: Pt, 0.82–0.99; Pd, 0.94–0.99 and Ru, 0.91–0.99.

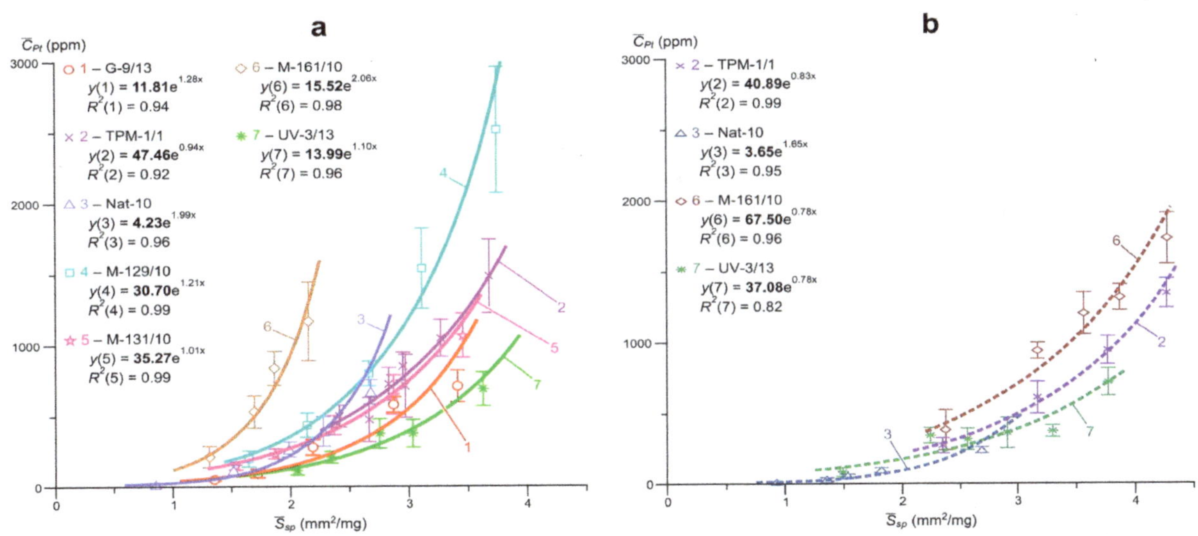

Figure 5. Dependences of the average content of evenly distributed platinum on the specific surface area of an average crystal in size fractions of arsenopyrite (**a**) and pyrite (**b**) from vein, streaky-vein and veinlet-disseminated ores of the Natalkinskoe deposit. The expressions for approximate curves are shown here and in Figures 6–8, in which the pre-exponential factor (in bold) is an estimate of the content of the structural form of the element. The contents of structurally and surface-bound forms of Pt, Pd, Ru and Rh in arsenopyrite and pyrite are given in Tables 3–6. Points 1–7 here and in Figures 6–8 correspond to the ore samples described in Table 1.

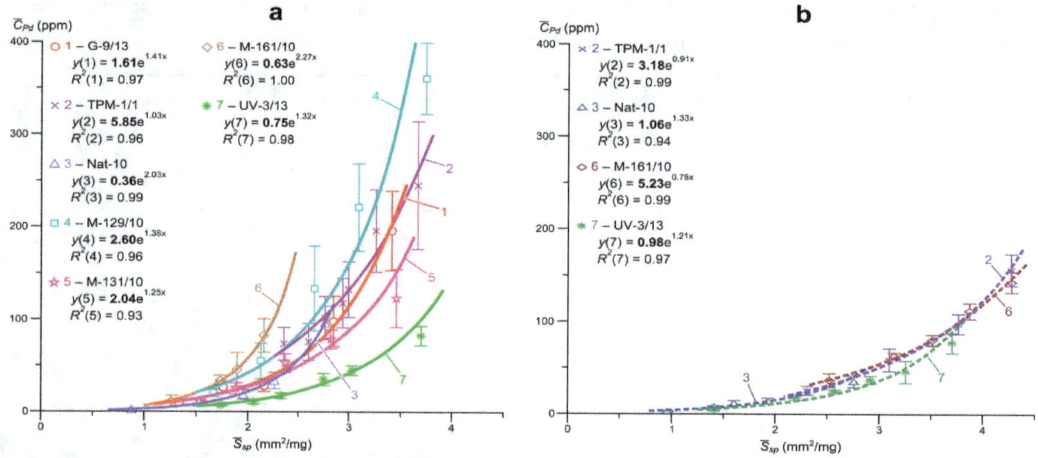

Figure 6. Dependences of the average content of evenly distributed palladium on the specific surface area of an average crystal in size fractions of arsenopyrite (**a**) and pyrite (**b**) from vein, streaky-vein and veinlet-disseminated ores of the Natalkinskoe deposit.

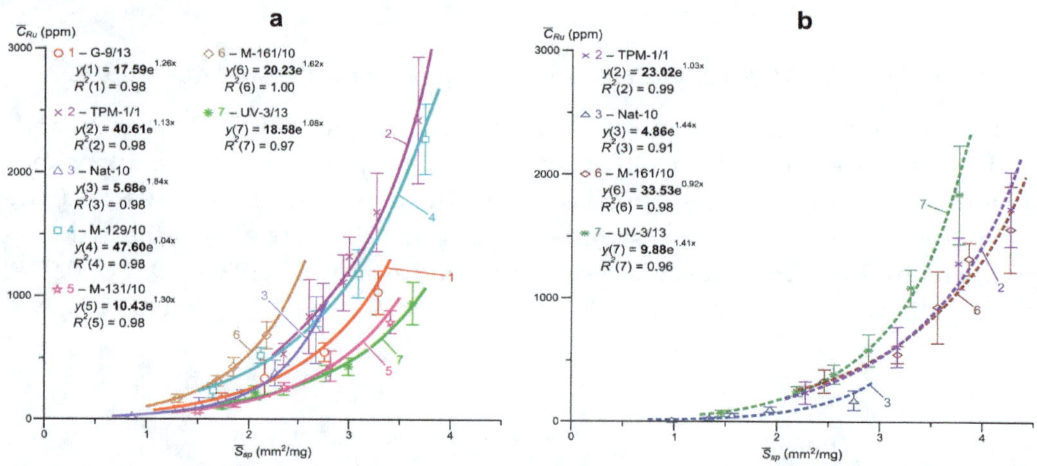

Figure 7. Dependences of the average content of evenly distributed ruthenium on the specific surface area of an average crystal in size fractions of arsenopyrite (**a**) and pyrite (**b**) from vein, streaky-vein and veinlet-disseminated ores of the Natalkinskoe deposit.

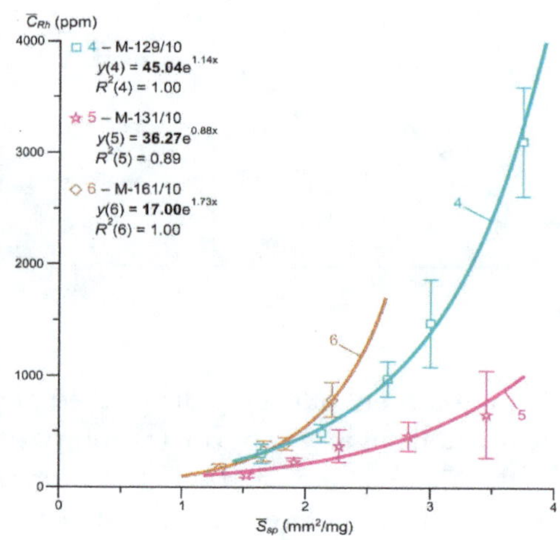

Figure 8. Dependences of the average content of evenly distributed rhodium on the specific surface area of an average crystal in size fractions of arsenopyrite from vein and streaky-vein ores of the Natalkinskoe deposit.

Table 3. Results of platinum analysis of single crystal size selections of arsenopyrite and pyrite from vein, streaky-vein and veinlet-disseminated ores of the Natalkinskoe deposit.

Sample No.	Number of Crystals (Initial–Final Sample)	Characteristics of the Final Sample						Contents (ppm)			
		Number of Crystals	Interval of Masses (mg)	$\bar{m}$ (mg)	$\bar{r}$ (mm)	$\bar{S}_{sp}$ (mm^2/mg)	$\bar{C}\pm\Delta$ (ppm)	$\bar{C}_{tot}$	$\bar{C}_{ev}$	C_{str}	$\bar{C}_{sur}$
					Arsenopyrite						
		8	0.12–0.18	0.15	0.292	3.411	710 ± 109				
		12	0.19–0.28	0.24	0.339	2.873	579 ± 56				
G-9/13	66–48	11	0.29–0.69	0.55	0.448	2.189	281 ± 54	184	145	12	140
		7	0.79–1.32	1.14	0.572	1.722	100 ± 30				
		10	1.34–3.64	2.34	0.727	1.355	55 ± 9				
		12	0.10–0.14	0.12	0.271	3.672	1488 ± 259				
		10	0.15–0.20	0.17	0.304	3.262	1052 ± 130				
		6	0.21–0.22	0.22	0.33	2.97	713 ± 221				
TPM-1/1	76–58	5	0.23–0.23	0.23	0.336	2.945	852 ± 98	844	728	48	714
		10	0.24–0.28	0.26	0.35	2.827	727 ± 115				
		6	0.29–0.33	0.31	0.371	2.664	477 ± 156				
		9	0.34–0.46	0.41	0.406	2.412	502 ± 77				
		9	0.10–0.12	0.11	0.262	3.745	2519 ± 443				
		7	0.13–0.24	0.19	0.314	3.114	1543 ± 282				
M-129/10	42–35	6	0.27–0.37	0.31	0.371	2.664	802 ± 87	669	536	31	581
		8	0.48–0.73	0.59	0.459	2.143	438 ± 86				
		5	1.01–1.45	1.3	0.597	1.645	202 ± 58				
		9	0.10–0.17	0.14	0.284	3.456	1067 ± 157				
		6	0.18–0.32	0.24	0.339	2.873	649 ± 136				
M-131/10	53–39	8	0.34–0.52	0.43	0.412	2.369	436 ± 67	320	288	35	244
		7	0.54–1.05	0.88	0.524	1.872	236 ± 35				
		9	1.07–2.15	1.58	0.637	1.541	152 ± 28				
		7	0.40–0.62	0.58	0.456	2.151	1168 ± 278				
M-161/10	40–31	8	0.66–1.01	0.9	0.529	1.866	838 ± 118	565	499	16	494
		9	1.03–1.39	1.2	0.582	1.694	532 ± 114				
		7	1.42–4.31	2.58	0.751	1.312	216 ± 75				

Table 3. *Cont.*

Sample No.	Number of Crystals (Initial–Final Sample)	Characteristics of the Final Sample						Contents (ppm)			
		Number of Crystals	Interval of Masses (mg)	$\bar{m}$ (mg)	$\bar{r}$ (mm)	$\bar{S}_{sp}$ (mm^2/mg)	$\bar{C}\pm\Delta$ (ppm)	$\bar{C}_{tot}$	$\bar{C}_{ev}$	C_{str}	$\bar{C}_{sur}$
Arsenopyrite											
		8	0.21–0.41	0.3	0.366	2.679	669 ± 85				
		8	0.42–0.56	0.49	0.431	2.275	397 ± 102				
Nat-10	60–39	9	0.58–0.98	0.74	0.495	1.987	272 ± 51	135	78	4.2	89
		8	1.17–2.31	1.65	0.646	1.518	125 ± 38				
		6	3.73–14.33	9.21	1.147	0.857	17 ± 14				
		11	0.11–0.24	0.18	0.33	3.63	688 ± 119				
		8	0.25–0.35	0.31	0.396	3.035	375 ± 97				
UV-3/13	78–50	8	0.36–0.51	0.41	0.434	2.756	375 ± 104	252	188	14	190
		9	0.52–0.75	0.67	0.512	2.348	210 ± 41				
		7	0.76–1.10	0.99	0.583	2.06	115 ± 31				
		7	1.19–2.16	1.7	0.698	1.72	90 ± 22				
Pyrite											
		8	0.10–0.12	0.11	0.28	4.276	1350 ± 100				
TPM-1/1	43–33	9	0.13–0.18	0.16	0.317	3.768	940 ± 100	574	534	41	462
		6	0.21–0.33	0.27	0.378	3.175	610 ± 110				
		10	0.41–1.65	0.67	0.512	2.348	272 ± 54				
		12	0.10–0.12	0.11	0.28	4.276	1730 ± 180				
		10	0.13–0.16	0.15	0.311	3.869	1320 ± 90				
M-161/10	58–45	8	0.17–0.21	0.19	0.336	3.565	1200 ± 150	990	940	68	890
		10	0.22–0.31	0.27	0.378	3.175	940 ± 57				
		5	0.34–1.13	0.65	0.507	2.373	380 ± 140				
		8	0.30–0.55	0.45	0.448	2.676	239 ± 21				
		8	0.60–1.96	1.46	0.663	1.806	97 ± 17				
Nat-10	54–34	7	2.24–2.71	2.51	0.795	1.511	56 ± 11	56	33	3.7	35
		6	2.84–4.14	3.65	0.9	1.332	31 ± 12				
		5	4.48–16.2	11.5	1.32	0.909	13 ± 5				

Table 3. *Cont.*

Sample No.	Number of Crystals (Initial–Final Sample)	Characteristics of the Final Sample						Contents (ppm)			
		Number of Crystals	Interval of Masses (mg)	$\bar{m}$ (mg)	$\bar{r}$ (mm)	$\bar{S}_{sp}$ (mm^2/mg)	$\bar{C}\pm\Delta$ (ppm)	$\bar{C}_{tot}$	$\bar{C}_{ev}$	C_{str}	$\bar{C}_{sur}$
Pyrite											
UV-3/13	79–52	10	0.10–0.18	0.16	0.317	3.768	716 ± 97	322	208	37	189
		10	0.20–0.29	0.24	0.363	3.294	372 ± 42				
		8	0.31–0.39	0.35	0.412	2.91	360 ± 107				
		9	0.41–0.58	0.52	0.47	2.549	316 ± 75				
		8	0.60–1.00	0.78	0.538	2.227	342 ± 54				
		7	1.05–3.60	2.69	0.813	1.474	84 ± 31				

Note: Here and in Tables 4–6: $\overline{m}$, the average mass of the crystal in size fraction; $\bar{r}$, the average size of the crystal in size fraction; $\bar{S}_{sp}$, the specific surface area of average crystal in size fraction; $\bar{C} \pm \Delta$, the average content of evenly distributed element ± one standard deviation; $\bar{C}_{tot}$, average total content, $\frac{\sum C_i m_i}{\sum m_i}$; $\bar{C}_{ev}$, average evenly distributed element content for all size fractions; C_{str}, structurally bound form content (C extrapolation to zero $\bar{S}_{sp}$); $\bar{C}_{sur}$, average content of surface-associated form.

Table 4. Results of palladium analysis of single crystal size selections of arsenopyrite and pyrite from vein, streaky-vein and veinlet-disseminated ores of the Natalkinskoe deposit.

Sample No.	Number of Crystals (Initial–Final Sample)	Characteristics of the Final Sample						Contents (ppm)			
		Number of Crystals	Interval of Masses (mg)	$\bar{m}$ (mg)	$\bar{r}$ (mm)	$\bar{S}_{sp}$ (mm^2/mg)	$\bar{C}\pm\Delta$ (ppm)	$\bar{C}_{tot}$	$\bar{C}_{ev}$	C_{str}	$\bar{C}_{sur}$
Arsenopyrite											
G-9/13	66–42	9	0.12–0.18	0.15	0.292	3.411	195 ± 43	40	27	1.6	26
		9	0.19–0.28	0.25	0.345	2.857	97 ± 27				
		7	0.29–0.69	0.57	0.453	2.16	27 ± 6				
		9	0.79–1.32	1.05	0.556	1.766	26 ± 6				
		8	1.34–3.64	2.42	0.735	1.339	9.5 ± 3.5				

Table 4. *Cont.*

Sample No.	Number of Crystals (Initial–Final Sample)	Characteristics of the Final Sample						Contents (ppm)			
		Number of Crystals	Interval of Masses (mg)	$\bar{m}$ (mg)	$\bar{r}$ (mm)	$\bar{S}_{sp}$ (mm^2/mg)	$\bar{C}\pm\Delta$ (ppm)	$\bar{C}_{tot}$	$\bar{C}_{ev}$	C_{str}	$\bar{C}_{sur}$
Arsenopyrite											
TPM-1/1	88–63	10	0.10–0.14	0.12	0.271	3.672	245 ± 69	142	114	5.9	114
		11	0.15–0.19	0.17	0.304	3.262	196 ± 45				
		8	0.20–0.22	0.21	0.324	3	132 ± 30				
		10	0.23–0.24	0.23	0.336	2.945	117 ± 27				
		9	0.25–0.29	0.27	0.353	2.769	101 ± 23				
		7	0.30–0.34	0.32	0.373	2.609	75 ± 20				
		8	0.35–0.65	0.44	0.416	2.36	73 ± 18				
M-129/10	41–31	9	0.10–0.12	0.11	0.262	3.745	361 ± 39	97	83	2.6	82
		6	0.13–0.24	0.2	0.321	3.091	221 ± 47				
		5	0.27–0.37	0.31	0.371	2.664	133 ± 46				
		6	0.48–0.73	0.6	0.461	2.125	54 ± 19				
		5	1.01–1.45	1.25	0.59	1.671	20 ± 4				
M-131/10	52–38	8	0.10–0.17	0.14	0.284	3.456	122 ± 31	32	30	2	27
		6	0.18–0.32	0.26	0.35	2.827	77 ± 10				
		8	0.34–0.52	0.43	0.412	2.369	52 ± 10				
		8	0.54–1.05	0.87	0.523	1.886	25 ± 6				
		8	1.07–2.15	1.58	0.637	1.541	10 ± 2				
M-161/10	40–24	8	0.40–0.62	0.58	0.456	2.151	83 ± 18	39	31	0.6	31
		5	0.66–0.98	0.85	0.518	1.894	46 ± 18				
		6	1.00–1.32	1.12	0.569	1.734	34 ± 6				
		5	1.38–4.31	2.87	0.777	1.262	11 ± 7				
Nat-10	60–39	9	0.21–0.41	0.27	0.353	2.769	96 ± 18	12	7.7	0.4	8.1
		8	0.42–0.56	0.51	0.438	2.257	33 ± 9				
		8	0.58–0.98	0.77	0.501	1.956	18 ± 5				
		7	1.17–2.31	1.64	0.646	1.527	9.8 ± 4.2				
		7	3.73–14.33	8.42	1.113	0.883	1.9 ± 0.8				

Table 4. *Cont.*

Sample No.	Number of Crystals (Initial–Final Sample)	Characteristics of the Final Sample						Contents (ppm)			
		Number of Crystals	Interval of Masses (mg)	$\bar{m}$ (mg)	$\bar{r}$ (mm)	$\bar{S}_{sp}$ (mm^2/mg)	$\bar{C}\pm\Delta$ (ppm)	$\bar{C}_{tot}$	$\bar{C}_{ev}$	C_{str}	$\bar{C}_{sur}$
Arsenopyrite											
UV-3/13	76–58	12	0.11–0.24	0.17	0.324	3.705	82 ± 11	21	18	0.8	25
		10	0.25–0.35	0.31	0.396	3.035	44 ± 5				
		9	0.35–0.51	0.41	0.434	2.756	35 ± 6				
		10	0.52–0.75	0.68	0.514	2.331	17 ± 3				
		8	0.76–1.10	0.99	0.583	2.06	10 ± 2				
		9	1.19–2.16	1.61	0.685	1.749	6.7 ± 1.1				
Pyrite											
TPM-1/1	43–33	10	0.10–0.12	0.11	0.28	4.276	156 ± 17	58	53	3.2	50
		8	0.13–0.18	0.16	0.317	3.768	97 ± 11				
		6	0.21–0.33	0.29	0.378	3.099	58 ± 12				
		9	0.41–1.65	0.7	0.519	2.309	25 ± 5				
M-161/10	54–44	10	0.10–0.12	0.11	0.28	4.276	142 ± 11	73	67	5.2	62
		9	0.13–0.17	0.15	0.311	3.869	111 ± 8				
		7	0.18–0.22	0.2	0.342	3.509	80 ± 6				
		9	0.24–0.31	0.28	0.383	3.143	62 ± 3				
		9	0.34–1.13	0.54	0.476	2.518	37 ± 7				
Nat-10	60–41	8	0.29–0.51	0.42	0.438	2.741	35 ± 6	8.7	6.3	1.1	4.9
		7	0.55–1.63	1.24	0.628	1.908	14 ± 3				
		8	1.72–2.57	2.17	0.757	1.584	11 ± 3				
		9	2.66–4.14	3.23	0.864	1.387	7.9 ± 0.8				
		9	4.48–16.2	10.66	1.287	0.932	2.7 ± 0.5				
UV-3/13	79–56	10	0.10–0.18	0.17	0.324	3.705	77 ± 11	22	15	1	18
		9	0.20–0.29	0.25	0.368	3.25	45 ± 12				
		11	0.31–0.39	0.35	0.412	2.91	37 ± 4				
		12	0.41–0.58	0.51	0.467	2.566	25 ± 2				
		9	0.60–1.00	0.83	0.55	2.187	18 ± 3				
		5	1.05–3.60	3.2	0.862	1.393	4.2 ± 1.2				

Table 5. Results of ruthenium analysis of single crystal size selections of arsenopyrite and pyrite from vein, streaky-vein and veinlet-disseminated ores of the Natalkinskoe deposit.

Sample No.	Number of Crystals (Initial–Final Sample)	Characteristics of the Final Sample						Contents (ppm)			
		Number of Crystals	Interval of Masses (mg)	$\bar{m}$ (mg)	$\bar{r}$ (mm)	$\bar{S}_{sp}$ (mm²/mg)	$\bar{C}\pm\Delta$ (ppm)	$\bar{C}_{tot}$	$\bar{C}_{ev}$	C_{str}	$\bar{C}_{sur}$
Arsenopyrite											
G-9/13	64–42	12	0.12–0.20	0.16	0.296	3.286	1034 ± 177	283	193	18	190
		9	0.22–0.33	0.27	0.353	2.769	547 ± 76				
		6	0.40–0.69	0.57	0.453	2.16	339 ± 149				
		7	0.79–1.34	1.1	0.565	1.741	160 ± 53				
		8	1.39–3.64	2.49	0.742	1.327	82 ± 26				
TPM-1/1	85–61	9	0.10–0.14	0.12	0.271	3.672	2424 ± 512	1318	1129	41	1083
		12	0.15–0.19	0.17	0.304	3.262	1678 ± 316				
		10	0.20–0.22	0.21	0.324	3	1321 ± 159				
		9	0.23–0.24	0.23	0.336	2.945	1108 ± 225				
		8	0.25–0.29	0.28	0.358	2.746	904 ± 197				
		6	0.30–0.34	0.32	0.373	2.609	832 ± 298				
		7	0.35–0.65	0.45	0.42	2.352	530 ± 86				
M-129/10	43–33	9	0.10–0.12	0.11	0.262	3.745	2269 ± 285	656	537	48	520
		6	0.13–0.24	0.2	0.321	3.091	1179 ± 196				
		5	0.27–0.37	0.3	0.366	2.679	812 ± 63				
		7	0.48–0.73	0.61	0.464	2.118	516 ± 63				
		6	1.01–1.45	1.27	0.592	1.656	231 ± 59				
M-131/10	40–29	6	0.10–0.17	0.15	0.292	3.411	789 ± 92	179	165	10	162
		5	0.18–0.32	0.26	0.35	2.827	432 ± 121				
		7	0.34–0.52	0.44	0.416	2.36	263 ± 33				
		6	0.54–1.05	0.9	0.529	1.866	129 ± 30				
		5	1.07–2.15	1.7	0.65	1.505	63 ± 15				
M-161/10	40–33	9	0.40–0.62	0.56	0.451	2.179	667 ± 107	331	297	20	290
		8	0.66–1.01	0.91	0.53	1.852	422 ± 75				
		9	1.03–1.39	1.2	0.582	1.694	309 ± 42				
		7	1.42–4.31	2.62	0.755	1.305	166 ± 33				

Table 5. *Cont.*

Sample No.	Number of Crystals (Initial–Final Sample)	Characteristics of the Final Sample						Contents (ppm)			
		Number of Crystals	Interval of Masses (mg)	$\bar{m}$ (mg)	$\bar{r}$ (mm)	$\bar{S}_{sp}$ (mm^2/mg)	$\bar{C}\pm\Delta$ (ppm)	$\bar{C}_{tot}$	$\bar{C}_{ev}$	C_{str}	$\bar{C}_{sur}$
Arsenopyrite											
		5	0.21–0.41	0.30	0.336	2.679	729 ± 262				
		8	0.42–0.56	0.50	0.434	2.260	375 ± 103				
Nat-10	59–35	8	0.58–0.98	0.80	0.508	1.936	187 ± 51	128	79	5.7	79
		8	1.17–2.31	1.65	0.646	1.518	124 ± 49				
		6	3.73–14.33	9.07	1.141	0.861	24 ± 10				
		10	0.11–0.24	0.18	0.33	3.63	941 ± 172				
		9	0.25–0.35	0.32	0.4	3	428 ± 69				
UV-3/13	78–58	10	0.36–0.51	0.4	0.431	2.787	372 ± 65	329	238	19	218
		10	0.52–0.75	0.66	0.509	2.355	245 ± 49				
		9	0.76–1.10	0.98	0.581	2.067	211 ± 56				
		10	1.19–2.16	1.63	0.688	1.742	103 ± 20				
Pyrite											
		8	0.10–0.12	0.11	0.28	4.276	1720 ± 300				
TPM-1/1	43–29	8	0.13–0.18	0.16	0.317	3.768	1280 ± 210	686	574	23	577
		6	0.21–0.33	0.27	0.378	3.175	605 ± 160				
		7	0.41–1.65	0.73	0.527	2.283	235 ± 90				
		10	0.10–0.12	0.11	0.28	4.276	1560 ± 350				
		10	0.13–0.16	0.15	0.311	3.869	1320 ± 130				
M-161/10	58–43	6	0.17–0.21	0.19	0.336	3.565	930 ± 290	850	700	34	690
		10	0.22–0.31	0.27	0.378	3.175	540 ± 70				
		7	0.34–1.13	0.58	0.488	2.464	330 ± 90				
		6	0.29–0.55	0.42	0.438	2.741	189 ± 86				
		10	0.60–1.72	1.23	0.627	1.918	109 ± 22				
Nat-10	58–40	7	1.81–2.66	2.32	0.774	1.549	58 ± 15	48	35	4.9	30
		9	2.71–4.48	3.58	0.895	1.343	33 ± 5				
		8	5.78–16.2	9.91	1.256	0.955	14 ± 4				

Table 5. *Cont.*

Sample No.	Number of Crystals (Initial–Final Sample)	Characteristics of the Final Sample						Contents (ppm)			
		Number of Crystals	Interval of Masses (mg)	$\bar{m}$ (mg)	$\bar{r}$ (mm)	$\bar{S}_{sp}$ (mm^2/mg)	$\bar{C}\pm\Delta$ (ppm)	$\bar{C}_{tot}$	$\bar{C}_{ev}$	C_{str}	$\bar{C}_{sur}$
Pyrite											
UV-3/13	79–55	10	0.10–0.18	0.16	0.317	3.768	1840 ± 400	454	272	9.9	330
		11	0.20–0.29	0.24	0.363	3.294	1085 ± 149				
		10	0.31–0.39	0.36	0.416	2.884	577 ± 129				
		9	0.41–0.58	0.52	0.47	2.549	384 ± 74				
		9	0.60–1.00	0.83	0.558	2.187	248 ± 39				
		6	1.05–3.60	2.9	0.834	1.439	67 ± 23				

Table 6. Results of rhodium analysis of size selections of arsenopyrite single crystals from vein and streaky-vein ores of the Natalkinskoe deposit.

Sample No.	Number of Crystals (Initial–Final Sample)	Characteristics of the Final Sample						Contents (ppm)			
		Number of Crystals	Interval of Masses (mg)	$\bar{m}$ (mg)	$\bar{r}$ (mm)	$\bar{S}_{sp}$ (mm^2/mg)	$\bar{C}\pm\Delta$ (ppm)	$\bar{C}_{tot}$	$\bar{C}_{ev}$	C_{str}	$\bar{C}_{sur}$
M-129/10	43–31	9	0.10–0.12	0.11	0.262	3.745	3108 ± 491	801	715	45	678
		5	0.13–0.24	0.21	0.324	3.000	1473 ± 394				
		6	0.27–0.37	0.31	0.371	2.664	973 ± 159				
		6	0.48–0.73	0.60	0.461	2.125	483 ± 78				
		5	1.01–1.45	1.29	0.595	1.647	296 ± 85				
M-131/10	38–28	5	0.10–0.17	0.14	0.284	3.456	653 ± 391	211	190	36	154
		4	0.18–0.31	0.26	0.350	2.827	453 ± 130				
		4	0.34–0.52	0.49	0.431	2.275	368 ± 146				
		7	0.54–1.04	0.83	0.514	1.910	224 ± 34				
		8	1.07–2.15	1.64	0.650	1.527	103 ± 19				
M-161/10	40–28	8	0.40–0.62	0.54	0.446	2.210	782 ± 155	360	309	17	299
		7	0.66–1.01	0.93	0.534	1.840	387 ± 55				
		7	1.03–1.39	1.26	0.592	1.669	320 ± 89				
		6	1.42–4.31	2.65	0.757	1.297	158 ± 46				

Extrapolation of these dependences to zero specific surface area ($\overline{S}_{sp}$, mm^2/mg, Tables 3–5), i.e., on a conditionally infinite crystal in the first case of arsenopyrite, and in the second pyrite, yields the following estimates of the contents of the structural form of the elements (C_{str}, ppm): for arsenopyrite Pt, 4.2–48; Pd, 0.4–5.9 and Ru, 5.7–48; for pyrite Pt, 3.7–68; Pd, 1.0–5.2 and Ru 4.9–34 (Figures 5–7). The average contents of surface-bound form of elements ($\overline{C}_{sur}$, ppm) are given in Tables 3–5 and for arsenopyrite amount to the following values: Pt, 89–714; Pd, 8.1–114 and Ru, 79–1083; for pyrite Pt, 35–890; Pd, 4.9–62 and Ru, 30–690.

By now, the range of PGE studied by the AAS-ADSSC method has expanded. The first results for the forms of occurrence of Rh, one more element of the platinum group in crystals of arsenopyrite, were obtained. These data are still limited and preliminary. The results of the study of the content of evenly distributed Rh and the ratio of its structural and surface-bound forms in arsenopyrites from vein and streaky-vein ores of the Natalkinskoe deposit are presented in Figure 8 and in Table 6. In all cases, as well as for Pt, Pd and Ru, highly deterministic (R^2 = 0.89–1.0) dependences of the average Rh content in the size fraction on the specific surface area of the average arsenopyrite crystal in it were obtained. Extrapolation of these dependences to a zero-specific surface, i.e., to a conditionally infinite crystal, yielded estimates of the contents of the structural form of this element of 17–45 ppm. The average grades of the surface-associated forms were 154–678 ppm for arsenopyrite (see Table 6).

The data presented (PCA and AAS-ADSSC) suggest that arsenopyrite also acts as a concentrator mineral for Rh. Structural and surface-bound forms in arsenopyrite (as well as in pyrite), as the main uniformly distributed forms, are characteristic not only of Pt, Pd and Ru, but also Rh. The data obtained confirms that arsenopyrites and pyrites are major mineral concentrators of PGE at the Natalkinskoe gold ore deposit. It can also be assumed that the size dependence of the content of evenly distributed Pt, Pd, Ru and Rh, and the confinement of a significant part to the surface of the crystals, is not a unique phenomenon for sulphide minerals of ore deposits. Structurally and superficially bound forms, as the main uniformly distributed forms, may be typical of both the remaining PGE (Os and Ir) and of other noble metals—Au and Ag [11,37–39,41–45].

New data for Pt and Pd forms in arsenopyrites, both structural and surface-bound, confirmed earlier conclusions [11]. The uniformly distributed Pt, Pd and Ru and, as we believe, Rh in arsenopyrites and pyrites are forms of elements which are chemically bound in the structure of these minerals and in nanoscale nonautonomous phases (NAPs) on the surface of arsenopyrite and pyrite crystals. The latter usually dominates over the former according to the content of these elements. Recent experimental and natural data [37,38,46] suggest that Pt, Pd, Ru and Rh, as well as Au, can be incorporated into NAP formed on the surface of arsenopyrite and pyrite crystals in a very thin surface layer (~100–500 nm). These elements are absorbed by NAP from the hydrothermal solution even more efficiently than gold, which is incompatible in pyrite.

There is currently high demand for studies of the surface layer of sulphide minerals as possible hosts for "invisible" forms of trace elements, the nature of their distribution and concentration levels. Nanoscale NAPs enriched with Pt, Pd, Ru and Rh form a part of the so-called impurity component on the surface of arsenopyrite and pyrite crystals. The importance of its investigation is obvious. It is assumed that there is a common mechanism of absorption of all noble metals for the studied sulphides due to the active role of the surface during the growth of the crystal from a supersaturated hydrothermal solution [38,39]. This mechanism is closely related to the conditions of ore formation. This is an important point that must be taken into account in practical terms, for example, in the concentration of ores that include sulphides, in our case arsenopyrite and pyrite.

The structurally bound form of PGE, primarily Pt, despite a lower content compared to the surface form, is also extremely important. As shown earlier, data on the structural form of Au, Ag and Pt make it possible to make comparative estimates of the contents of these elements in the ore-forming fluids that form gold deposits. It is the only form of the element occurrence that can be used as an indicator of element activity in hydrothermal solution [37,46]. For example, the experimental data on the distribution and segregation of trace elements in crystals of ore minerals grown in hydrothermal systems were used to assess the content of Au and Pt in the ore-forming fluid (C_{aq}) of the Natalkinskoe and Degdekan gold deposits. The content of the element in the structural form in natural pyrite, and the distribution coefficient (D_{str}) for the same form between pyrite and hydrothermal solution according to experimental data, were used for estimating element content in the ore-forming fluid [38]. In Table 7 these results are supplemented by the data from the present work on samples M-161/10, TPM-1/1 and Nat-10.

Table 7. Evaluation of gold and platinum grade in ore-forming fluids by pyrite composition using experimental distribution coefficients D_{str}.

Sample No.	Deposit	Number of Crystals (Initial–Final Sample)	Element	C_{str} (ppm)	$\overline{C}_{sur}$ (ppm)	C_{aq} (ppm)	Au/Pt_{aq}
M-163/10	Degdekan	95–59	Au	0.2	0.5	2.1	21
		92–71	Pt	2.0	7.1	0.1	
DG-10/14	Degdekan	67–44	Au	0.1	1.1	1.3	6.5
		34–22	Pt	5.1	102	0.2	
UV-3/13	Natalkinskoe	80–52	Au	0.3	1.4	2.9	1.6
		79–52	Pt	37	189	1.8	
M-161/10	Natalkinskoe	58–40	Au	0.2	3.6	1.8	0.6
		58–45	Pt	68	890	3.2	
TPM-1/1	Natalkinskoe	43–30	Au	0.3	3.4	3.2	1.6
		43–33	Pt	41	462	2.0	
Nat-10	Natalkinskoe	60–39	Au	1.8	2.7	18	90
		54–34	Pt	3.6	35	0.2	

Note: C_{str}, structurally bound element content; $\overline{C}_{sur}$, superficially bound element content; C_{aq}, element content in the ore-forming fluid.

The average value of D_{str} is 0.1 for Au and 21 for Pt [41,46]. The evaluation of Au and Pt content in the ore-forming fluid showed that Au content in the fluids of the Degdekan and Natalkinskoe gold deposits was mainly at the level of 1.3–3.2 ppm, and it reached a maximum value of 18 ppm for the Nat-10 sample. Pt grade ranged from 0.1 to 3.2 ppm and, on average, was clearly elevated at the Natalkinskoe deposit compared to Degdekan. The fluid of the Natalkinskoe deposit was rich in both elements compared to the fluid of the Degdekan deposit. Au and Pt contents in solution were also comparable for most samples from the Natalkinskoe deposit (with the exception of the anomalous Au in sample Nat-10), whereas Au dominated over Pt in the Degdekan deposit. This is another argument in favor of studying platinum-bearing gold deposits, of which the Natalkinskoe deposit is no exception. Information about the Pt content in the fluid is of particular value in view of the complex behavior of Pt and the heterogeneity of its distribution under the capture of gas–liquid inclusions, the composition of which is determined by LA-ICP-MS as a proxy of the ore fluid composition [47].

4.3. *Study of the Chemical Composition of the Surface of Crystals of Arsenopyrite and Pyrite Using the SEM-EDX Method*

Nearly isometric pseudo-orthorhombic crystals of arsenopyrite and cubic crystals of pyrite without visible inclusions of ore minerals and with a minimum number of non-metallic microinclusions (mainly quartz and carbonate) were selected for the study (Figure 9). The so-called "clean" areas virtually lacking visible inclusions and significant defects were then selected to study the composition of the "invisible" impurity component on the surface of these crystals (Figure 10). The size (in μm) of such areas was 10 × 10, 20 × 20, and it was at least 30 × 30. Sections with low roughness were selected using the scanning probe microscopy technique (see above).

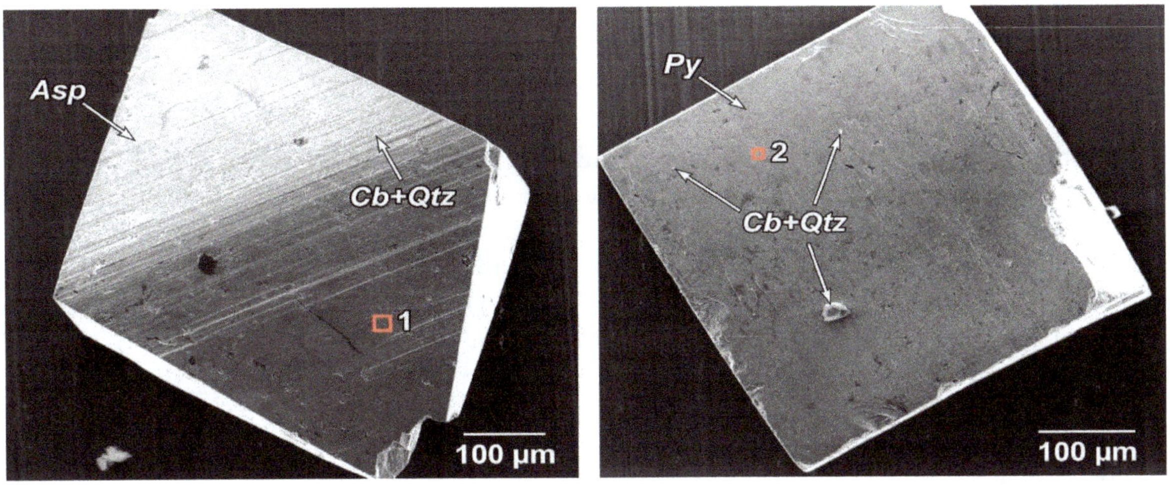

Figure 9. Crystals of arsenopyrite (Asp) and pyrite (Py) with microinclusions of carbonate (Cb) and quartz (Qtz). Surface areas of arsenopyrite (1) and pyrite (2) crystals, where measurements were carried out. Their enlarged fragments are shown in Figure 10.

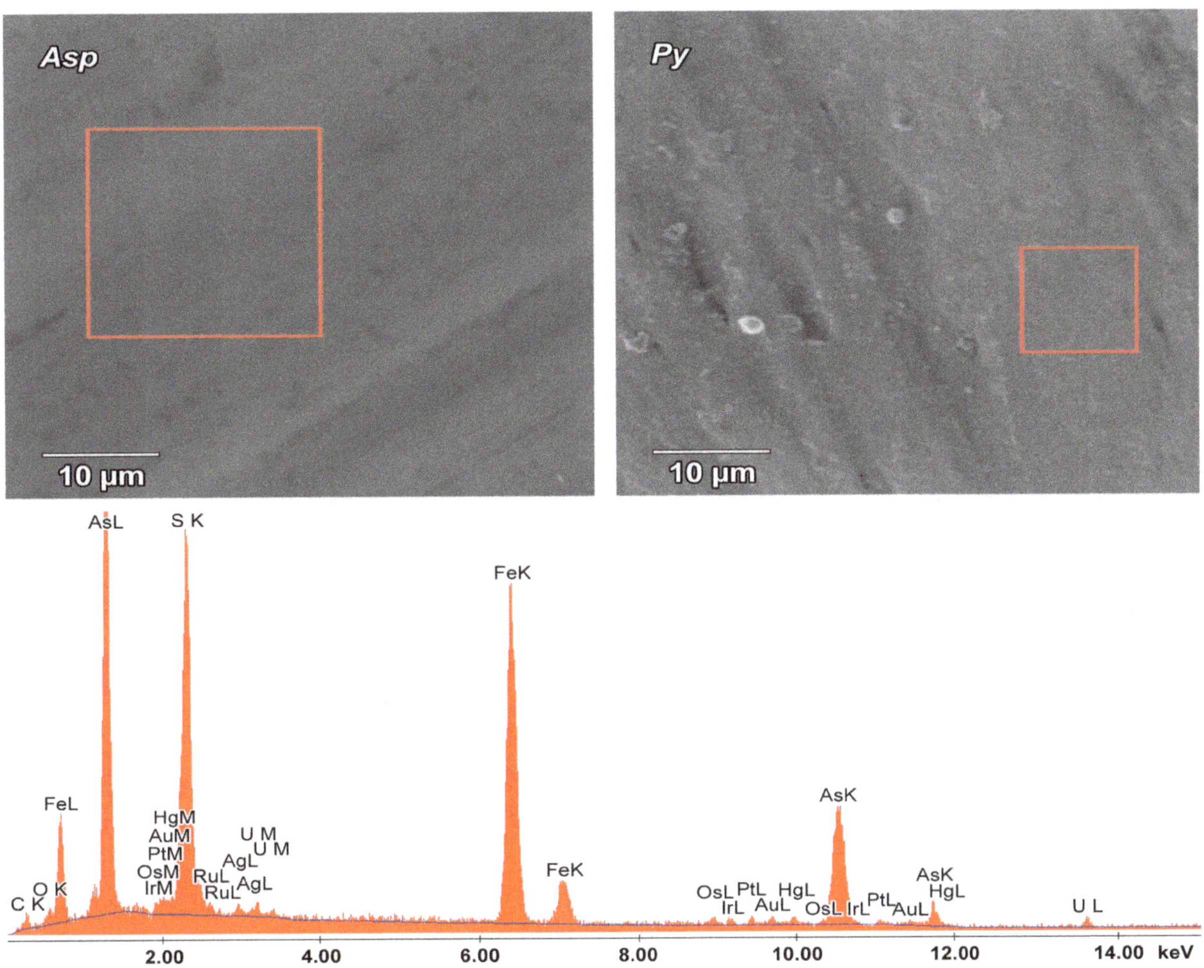

Figure 10. Enlarged fragments of arsenopyrite (Asp) and pyrite (Py) crystals with "clean" areas, where measurements were carried out (see Table 8, sample M-129/10, *N* 3, *n* 3 and Table 9, sample M-129/10, *N* 1, *n* 2). The bottom of the Figure 9 shows the X-ray energy spectra of arsenopyrite (see Table 8, *N* 3, *n* 3).

Faces with striation of the same crystallographic type were selected during scanning in order to eliminate the effect of capturing various impurities by faces of different crystallographic types during crystal growth: for arsenopyrite—prism {141}, for pyrite—cube {100}.

The surfaces of arsenopyrites and pyrites selected from the Natalkinskoe ore deposit (samples G-9/13, TPM-1/1, M-129/10, M-131/10, M-161/10 and Nat-10) clearly demonstrated the peculiarities of its elemental composition. On the "clean" surface of arsenopyrite, PGEs were thus represented by the following elements (in wt%): Pt (up to 2.1), Ru (1.4), Ir (up to 0.7) and Os (up to 0.7). Other ore elements were also determined (in wt%): U (up to 2.0), Hg (up to 1.1), Au (up to 1.0), rarely Ag (up to 0.7) and Cu (up to 0.6) (Table 8). The only PGE present on the "clean" surface of pyrite was Pt (up to 2.9 wt%), and other ore elements were represented by As (up to 2.2 wt%) and Cu (up to 0.7 wt%) (Table 9).

The SEM-EDX method confirmed the presence of Au and Pt on the surface of arsenopyrite crystals from the Natalkinskoe deposit, and high contents of them were detected earlier [11]. It is the first time such PGEs as Ru, Os, Ir and a number of ore elements—Ag, Hg, Cu and U—were detected on the surface of arsenopyrite. Pt and As are shown for the first time to be constantly present on the surface of pyrite crystals. Comparing the data from the AAS-ADSSC and SEM-EDX methods regarding the surface contents of Au, Pt and Ru, it is necessary to recall that the data on C_{sur} of AAS-ADSSC method are related to the entire average crystal, whereas the SEM-EDX method virtually defines local content of the element at the local area ("point") on the surface. It would be more correct to compare these results with the content of the element within NAP, which can be calculated under certain assumptions [46]. According to such calculations, Au in NAP on pyrite amounts to 0.3–0.5 wt% [38], which is comparable by an order of magnitude to the data in Tables 8 and 9 on Au and Pt. Increased platinum on pyrite (see Table 9) is qualitatively consistent with the fact that, according to AAS-ADSSC, C_{sur} Pt > C_{sur} Au (see Table 7).

Table 8. Chemical composition (in wt%) of arsenopyrite crystal surfaces with micron-size rare inclusions of non-metallic minerals. Samples from vein ores of the Natalkinskoe deposit.

Sample No.	*N*	*n*	Fe	As	S	C	O	Al	Si	Hg	Au	Ag	Cu	Ti	Pt	Os	Ir	Ru	U
M-129/10	1	1	**31.6**	**36.0**	**23.4**	2.8	2.0	1.2	*0.2*	0.7	0.7	*0.2*	-	-	*0.4*	-	-	-	0.8
		2	**32.4**	**35.6**	**20.2**	3.4	2.5	1.3	*0.2*	1.1	0.9	*0.3*	-	-	0.6	-	-	-	1.5
		3	**31.9**	**36.8**	**20.8**	3.7	1.5	0.6	*0.1*	1.0	0.9	*0.3*	-	-	0.8	-	-	-	1.6
	2	1	**30.0**	**36.3**	**20.4**	4.3	2.5	1.2	*0.1*	1.1	0.9	*0.1*	-	-	1.1	-	-	-	2.0
		2	**28.9**	**39.0**	**20.6**	4.6	2.5	0.5	*0.2*	0.9	0.8	*0.2*	-	-	*0.4*	-	-	-	1.4
	3	1	**32.6**	**41.8**	**16.1**	3.9	1.5	-	-	1.0	0.7	-	-	-	0.5	0.6	*0.4*	-	0.9
		2	**32.1**	**43.4**	**14.2**	2.3	1.4	-	-	0.9	1.0	*0.2*	-	-	0.9	0.6	0.7	1.0	1.3
		3	**32.4**	**43.1**	**15.4**	1.1	1.0	-	-	0.9	0.9	0.7	-	-	0.9	0.7	0.7	1.0	1.2
M-161/10	4	1	**34.0**	**40.2**	**15.1**	4.6	2.7	-	-	*0.4*	*0.3*	-	-	-	*0.4*	*0.2*	*0.3*	1.4	*0.4*
		2	**34.7**	**41.5**	**13.8**	4.8	1.8	-	-	*0.4*	*0.4*	-	-	-	*0.4*	*0.2*	*0.2*	1.2	0.6
	5	1	**30.8**	**41.2**	**17.1**	3.9	3.2	1.0	*0.1*	-	-	-	0.6	-	2.1	-	-	-	-
		2	**31.9**	**41.7**	**19.0**	1.9	2.1	1.4	*0.1*	-	-	*0.1*	0.5	*0.2*	1.1	-	-	-	-
	6	1	**33.3**	**42.4**	**19.0**	1.0	1.1	0.6	*0.2*	-	-	*0.3*	*0.2*	*0.2*	0.6	*0.2*	*0.3*	*0.3*	*0.3*
		2	**33.2**	**42.9**	**19.1**	0.6	0.8	0.5	*0.2*	-	-	*0.1*	*0.2*	*0.1*	1.2	*0.2*	*0.3*	*0.3*	*0.3*
		3	**33.1**	**43.2**	**19.1**	*0.4*	0.5	0.5	*0.1*	-	-	0.6	0.6	*0.1*	0.8	*0.2*	*0.3*	*0.2*	*0.3*

Note: Pd, Rh, F, Cl, Pb, Zn, Mo, W, Bi and REE were not detected. Analytical lines: FeK_α, AsK_α, SK_α, CK_α, OK_α, AlK_α, SiK_α, HgL_α, AuL_α, AgL_α, CuK_α, TiK_α, PtL_α, OsL_α, IrL_α, RuL_α and UL_α. Here and in Table 9: *N*—the serial number of the crystal, *n*—the serial number of the measurement area on the crystal surface; hyphen—element is not detected; italics—the elements content below the detection limit (<0.5 wt%), so their presence is problematic but without taking them into account the sum total does not reach 100%; the main composition is in bold.

Table 9. Chemical composition (in wt%) of pyrite crystal surfaces with micron-size rare inclusions of non-metallic minerals. Samples from vein ores of the Natalkinskoe deposit.

Sample No.	*N*	*n*	Fe	S	C	O	Al	Si	Ca	Na	As	Cu	Ti	Pt
M-129/10	1	1	**38.5**	**43.1**	5.3	7.7	1.2	*0.2*	-	-	1.3	0.7	*0.3*	1.7
		2	**36.9**	**39.2**	7.0	10.7	1.3	*0.3*	-	-	1.2	0.7	*0.4*	2.3
		3	**41.0**	**47.9**	2.9	2.6	1.4	*0.2*	-	-	1.7	0.5	*0.2*	1.6
	2	1	**39.8**	**44.1**	5.1	6.8	1.1	*0.2*	-	-	1.0	-	*0.2*	1.7
		2	**35.9**	**42.7**	5.6	10.2	0.8	1.1	-	-	1.0	0.6	*0.1*	2.0
		3	**38.8**	**47.0**	3.3	6.2	0.9	0.9	-	-	1.1	*0.3*	*0.1*	1.4
	3	1	**37.0**	**38.3**	4.3	13.2	0.9	0.7	*0.2*	*0.3*	1.8	0.6	-	2.7
		2	**39.5**	**48.4**	2.3	3.0	0.8	*0.1*	*0.1*	*0.2*	2.2	0.5	-	2.9
M-161/10	4	1	**37.4**	**37.5**	3.1	16.4	1.3	1.0	-	-	1.1	*0.4*	*0.1*	1.7
		2	**40.5**	**46.5**	3.6	5.6	1.4	*0.1*	-	*0.2*	1.4	*0.2*	-	0.5
	5	1	**41.7**	**47.2**	2.6	3.3	0.9	*0.1*	-	-	2.0	*0.3*	-	1.9
		2	**40.7**	**47.5**	3.0	4.0	0.9	*0.2*	-	-	2.0	*0.2*	-	1.5

Note: Au, Ag, Hg, Pd, Os, Ir, Ru, Rh, U, F, Cl, Pb, Zn, Mo, W, Bi and REE were not detected. Analytical lines: FeK_α, SK_α, CK_α, OK_α, AlK_α, SiK_α, CaK_α, NaK_α, AsK_α, CuK_α, TiK_α and PtL_α.

4.4. Study of the Chemical Composition of the Surface of Arsenopyrite and Pyrite Crystals by LA-ICP-MS

When studying the surface layers of arsenopyrite and pyrite crystals using the LA-ICP-MS method, attention was largely focused on Pt and Pd. Ten arsenopyrite crystals and six pyrite crystals were analyzed (Tables 10 and 11). For all crystals of arsenopyrite (samples M-129/10, M-131/10, M-161/10 and TPM-1/1) and pyrite (samples M-161/10 and TPM-1/1) a track depth of 0.5–0.6 μm per laser pass was obtained. Unfortunately, we do not know the NAP thickness of the studied crystals. If we focus on the previously obtained data [44,48], the thickness of NAP on pyrite is ≤0.5 μm, and for arsenopyrite it is, according to preliminary data, much less [11]. The first pass of the laser thus best represents the composition of NAP, although significant "contamination" by the material underlying NAP is likely. The parameters of the laser were adjusted for one of the arsenopyrite crystals selected for detailed study (see Table 10, sample M-129/10, crystal 2), in such a way that a layer of ~100 nm thick would be burned in one pass (see above). In this case, the maximum depth of the laser-dig groove was 600 nm. Figure 11 shows graphs of the detailed distribution of Pt and Pd contents in the surface layer of Crystal 2 from the M-129/10 sample.

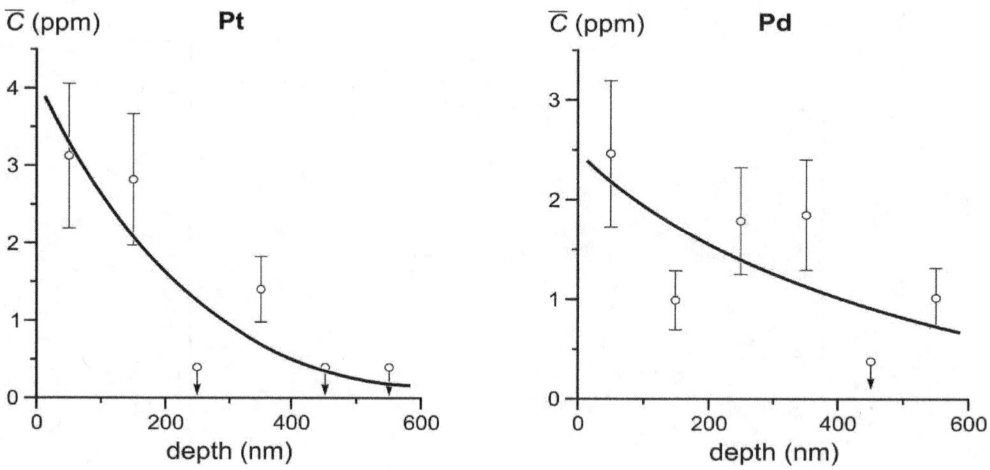

Figure 11. Distribution of the average content ($\overline{C}$) of platinum and palladium in the surface layer of arsenopyrite crystal (sample M-129/10, Crystal 2) according to the results of LA-ICP-MS analysis (the Natalkinskoe deposit).

Table 10. Results of LA-ICP-MS analysis of the surface of arsenopyrite crystals for Pt and Pd (the Natalkinskoe deposit).

Sample No.	Crystal No. (Size, mm)	Track No.	Contents of Elements (ppm)	
			Pt	Pd
M-129/10	1 (1.0)	1	2.9	1.8
		2	1.4	1.8
	2 (1.1)	1	**3.1**	**2.5**
			2.8	**1.0**
			-	**1.8**
			1.4	**1.8**
			-	-
			-	**1.0**
M-131/10	1 (1.3)	1	0.4	0.7
		2	0.4	0.5
	2 (1.3)	1	-	0.1
		2	-	0.2
	3 (1.0)	1	<0.4	0.1
		2	-	0.1
M-161/10	1 (1.0)	1	0.4	0.4
		2	23	3.8
		3	0.4	0.2
		4	<0.4	0.1
		5	6.2	1.4
	2 (1.2)	1	<0.4	<0.1
		2	-	0.6
TPM-1/1	1 (1.3)	1	0.4	0.8
	2 (1.5)	1	0.4	1.2
	3 (1.5)	1	<0.4	0.9
		2	0.4	4.8

Note: Here and in Tables 11–13:hyphen—element not detected; sign "<"—below detection limit. Bold—results of a detailed study of Pt and Pd contents to the depth of 600 nm with a step of 100 nm (see Figure 11).

Table 11. Results of LA-ICP-MS analysis of the surface of pyrite crystals for Pt and Pd (the Natalkinskoe deposit).

Sample No.	Crystal No. (Size, mm)	Track No.	Contents of Elements (ppm)	
			Pt	Pd
M-161/10	1 (1.0)	1	1.6	5.9
		2	0.4	0.1
		3	-	0.2
		4	-	0.1
	2 (1.2)	1	<0.4	0.9
		2	<0.4	0.2
		3	<0.4	0.4
	3 (1.5)	1	-	0.2
		2	-	0.4
TPM-1/1	1 (1.8)	1	0.4	3.8
		2	-	0.1
	2 (1.7)	1	-	0.7
		2	-	0.1
	3 (1.9)	1	0.4	0.1
		2	-	0.7

A generalization of the results of previous studies [11] and the new data suggest that the increased content of Pt and Pd is confined to the surface layer of not only arsenopyrite crystals, but also pyrite. It can also be assumed that the tendency for a Pt and Pd concentration in the surficial layer at <500 nm thick is common for both minerals (see Figure 11 and Table 11). Several exceptions can be noted in Table 10, however, which may be due to inhomogeneity in the impurity distribution because of the evolution of NAP during crystal growth [39].

Determination of the contents of a wider range of elements belonging to the group of PGEs—Pt, Pd, Rh and Ir—was carried out in the example of studying the surfaces of arsenopyrite and pyrite crystals taken from the UV-3/13 sample (Figures 12 and 13). The surface layer of these crystals was studied to a depth of 6 μm (depth per pass is 1 μm). Ru and Os contents were not determined due to their absence in the standards we used. According to LA-ICP-MS analysis, the surface layers of arsenopyrite and pyrite, along with Pt and Pd, concentrated PGEs such as Rh and Ir. The contents of elements (ppm) in the surface layers of arsenopyrite were Pt up to 5.2, Pd up to 8.4, Rh up to 8.9 and Ir up to 0.2, and in the surface layers of pyrite were Pt up to 13, Pd up to 896, Rh up to 6.7 and Ir up to 4.1. The maximum contents of Pt, Pd and Ir were established in pyrite, and Rh was established in arsenopyrite. Like Pt and Pd, all the highest contents of Rh and Ir were confined to the first removed layer (1 μm thick). There was a distinct tendency for the content of all these elements to decrease with depth (Tables 12 and 13).

High Pd content and Pd/Pt ratios on the pyrite surface (see Table 13) contradict the AAS-ADSSC and SEM-EDX data presented above, according to which the Pt content was more than an order of magnitude higher than Pd. The reason for this contradiction requires special research and may be related to the existence of an unidentified surface form of Pd (for example, Pd nanoparticles), which is inhomogeneously distributed over crystals. The SEM-EDX method was also local (locality is even higher than for LA-ICP-MS), and therefore cannot serve to confirm this hypothesis. In the AAS-ADSSC method, crystals with such an excess of Pd over the average content simply will not be included in the final sample.

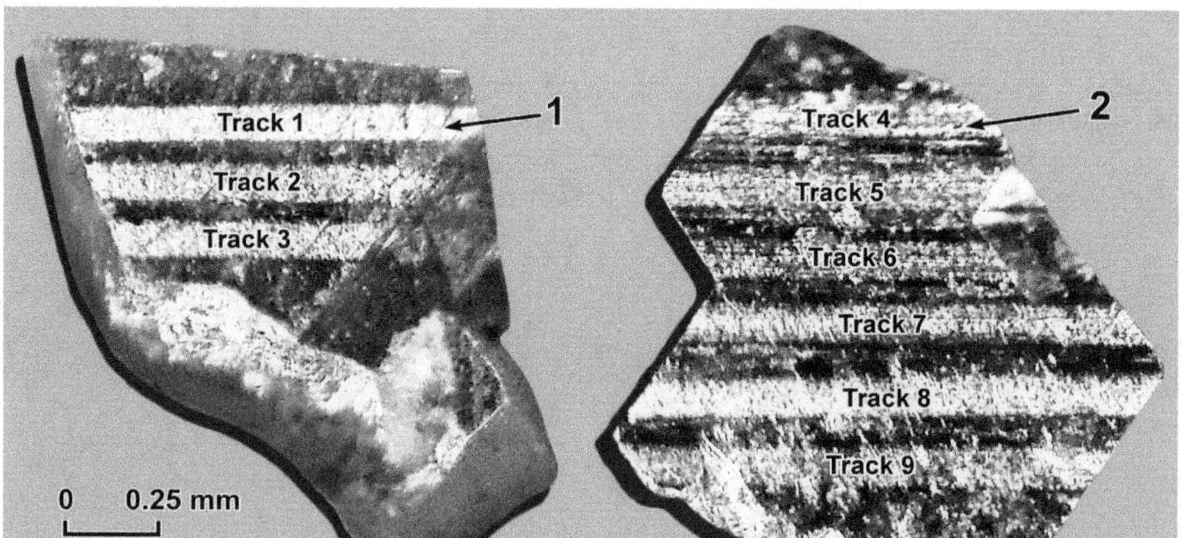

Figure 12. Arsenopyrite Crystals 1 and 2 (sample UV-3/13) from ores of the Natalkinskoe deposit. Tracks 1–9, numbers of laser passes over the crystal surface. The results of LA-ICP-MS analysis are presented in Table 12.

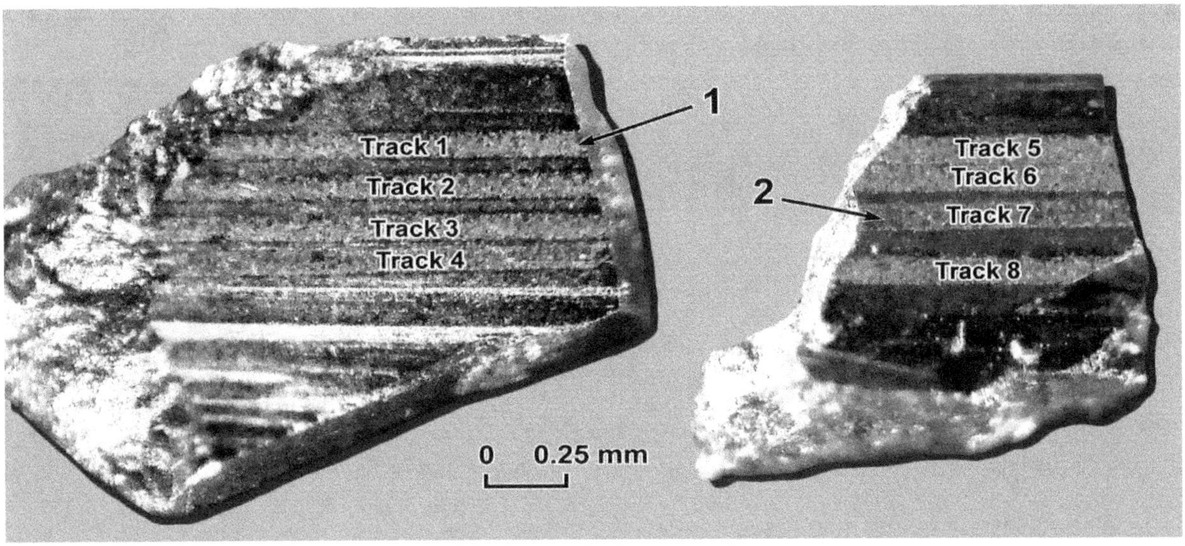

Figure 13. Pyrite Crystals 1 and 2 (sample UV-3/13) from ores of the Natalkinskoe deposit. Tracks 1–8, numbers of laser passes over the crystal surface. The results of LA-ICP-MS analysis are presented in Table 13.

Table 12. Results of LA-ICP-MS analysis (in ppm) of the surface of arsenopyrite crystals (1 and 2) for Pt, Pd, Rh and Ir (sample UV-3/13) (the Natalkinskoe deposit).

n	Pt	Pd	Rh	Ir	Pt	Pd	Rh	Ir	Pt	Pd	Rh	Ir
						Crystal 1						
	Track 1				**Track 2**				**Track 3**			
1	0.7	1.6	7.6	0.2	<0.7	5.2	1.6	-	<0.7	7.9	2.3	0.1
2	-	<0.2	2.5	<0.1	0.9	2.2	1.0	-	-	3.4	0.5	-
3	-	0.3	0.2	-	-	7.6	0.3	-	-	3.5	1.6	<0.1
4	-	0.4	<0.2	-	-	3.2	0.5	<0.1	-	2.9	0.4	-
5	-	0.2	0.3	-	-	2.0	0.4	<0.1	-	1.4	0.3	-
6	-	<0.2	0.2	-	-	2.0	0.6	-	-	1.5	-	-
						Crystal 2						
	Track 4				**Track 5**				**Track 6**			
1	<0.7	0.8	1.2	0.1	5.2	2.9	0.6	-	0.7	5.8	7.5	-
2	-	0.2	0.5	0.1	-	1.2	0.2	-	-	4.3	0.7	-
3	-	0.2	0.2	-	-	<0.2	0.2	-	-	1.7	0.5	-
4	-	0.3	0.5	-	-	1.4	0.7	-	-	1.5	0.2	-
5	-	<0.2	<0.2	-	-	0.4	0.4	-	-	1.3	2.1	-
6	-	-	<0.2	-	-	0.9	<0.2	-	-	0.7	0.2	-
	Track 7				**Track 8**				**Track 9**			
1	1.3	8.2	1.9	-	<0.7	2.8	6.4	<0.1	1.5	8.4	8.9	-
2	-	0.6	1.7	-	-	1.5	0.5	-	-	3.7	3.6	-
3	<0.7	<0.2	0.8	-	<0.7	1.8	0.8	-	-	2.5	3.7	-
4	-	0.4	0.9	-	-	<0.2	0.3	-	-	3.7	1.1	-
5	-	<0.2	0.4	-	-	0.2	0.9	-	-	3.6	1.5	-
6	-	0.4	<0.2	-	-	<0.2	-	-	-	2.7	0.7	<0.1

Note: Here and in Table 13: *n*—numbers of laser passes along the track (depth per pass is 1.0 μm).

Table 13. Results of LA-ICP-MS analysis (in ppm) of the surface of pyrite crystals (1 and 2) for Pt, Pd, Rh and Ir (sample UV-3/13) (the Natalkinskoe deposit).

n	Pt	Pd	Rh	Ir	Pt	Pd	Rh	Ir	Pt	Pd	Rh	Ir	Pt	Pd	Rh	Ir
	Crystal 1															
	Track 1				**Track 2**				**Track 3**				**Track 4**			
1	2.5	118	0.9	0.5	3.6	135	1.6	0.4	1.8	134	0.3	1.1	2.2	70	5.1	0.4
2	1.0	54	-	0.3	1.1	108	<0.2	0.4	1.1	89	0.3	0.5	2.2	109	1.7	0.2
3	0.5	26	0.7	0.1	<0.7	54	0.3	0.6	<0.7	41	-	0.2	1.1	47	0.4	-
4	0.3	10	-	-	0.9	56	<0.2	<0.2	-	18	<0.2	<0.2	<0.7	63	0.7	-
5	-	7.7	1.8	-	0.7	28	<0.2	-	-	14	-	-	<0.7	31	0.5	0.2
6	-	2.6	-	-	0.9	39	<0.2	-	-	17	<0.2	-	-	34	0.2	-
	Crystal 2															
	Track 5				**Track 6**				**Track 7**				**Track 8**			
1	11	453	2.4	1.8	13	855	2.9	3.1	13	896	6.7	4.1	7.7	693	1.2	3.0
2	5.3	820	0.5	3.6	9.1	845	1.1	4.1	8.4	624	2.4	1.9	5.3	715	0.4	3.0
3	2.7	503	1.2	0.7	5.9	521	0.9	1.8	2.8	378	0.5	1.5	8.3	461	0.4	2.8
4	1.1	101	0.4	0.5	4.7	365	0.2	1.3	3.8	210	1.6	0.8	0.9	301	<0.2	0.7
5	0.8	50	0.2	0.4	3.1	206	1.0	0.6	2.5	130	0.2	0.5	2.6	135	0.6	0.5
6	0.5	35	<0.2	0.5	1.3	89	0.3	0.7	0.9	112	0.2	0.5	2.2	143	<0.2	1.1

The maximum PGE contents at the Natalkinskoe deposit are currently known to be associated with sulphide minerals. High levels of PGE in the sulphide gravity concentrate of this deposit ore were established by sample decomposition using fluoro oxidants [49] and amounted to (in ppm) Pt, 12.9–92.8; Pd, 0.62–2.97; Ru, 0.11–3.27; Rh, 0.95–1.54 and Ir, 0.1–0.3. The speculation about the possible role of sulphide minerals as Pt concentrators at the Natalkinskoe gold deposit was previously made by L.P. Plyusnina and co-authors [14]; however, microprobe analysis did not reveal high content of noble metals in conventional preparations of arsenopyrite and pyrite [3]. Arsenopyrite concentrates of not only Au, but also Pt and Pd, were later demonstrated at the Natalkinskoe deposit [11]. The new data confirmed and supplemented the earlier conclusion. The highest contents of PGE (in ppm) observed in arsenopyrite were Pt up to 128, Pd up to 20, Ru up to 86 and Rh up to 21, and they were less high in pyrite: Pt up to 29, Pd up to 15, Ru up to 58 and Rh up to 5.9. AAS-ADSSC found that the high content of PGE largely was due to the existence of non-autonomous phases, causing their surface concentration. The fact that the nature of high contents of PGE in arsenopyrite and pyrite is mainly superficial was confirmed by the data obtained by SEM-EDX and (to some extent) LA-ICP-MS. If the data from the first two methods are comparable (taking into account the above considerations), the latter gives different results, which can be explained by the following reasons. First, there is some dispersion of the material on the surface after the first and each subsequent laser pass, which increases the error of determination. Second, the analysis of surface areas of several crystals does not have the necessary statistical reliability for determining the average contents, as is the case in the analytical technology of AAS-ADSSC. It is quite possible to meet with an "empty" crystal or, conversely, a crystal with the surface strongly enriched with impurity. This is all the more likely because the heterogeneity of the distribution of PGE is well known. For example, in pyrite from the intercumulus sulphide phases of the main sulphide zone of Great Dyke (Zimbabwe), the ion-microprobe analysis revealed a strong heterogeneity in the Pt distribution—from 0.4 to 244 ppm in different grains of the same sample [50]. Thirdly, significant errors can occur due to the lack of external calibration on such surface structures, for example, deposited on the crystal films of sulphide material with a certain content of admixture of the studied element. In the current state of the study of the surface composition of sulphides by LA-ICP-MS, although the history of such studies seems to date back 12 years [51], it is possible to speak only about the general tendency consisting of the enrichment of PGE of a surface of pyrite and arsenopyrite crystals; however, this trend is fully confirmed by other methods.

The reason for this phenomenon involves the features of the crystal growth mechanism and the dualism of the element distribution coefficient in the mineral-hydrothermal solution system, which is an order of magnitude higher for NAP compared to the crystal volume [38,39]. Non-autonomous phases are extremely important, but they are still rarely considered components of synthetic materials and minerals, the geochemical role of which lies in their ability to accumulate impurity elements in ultra-high contents [52]. Such phases occur on the surface of a growing crystal due to the chemical modification and structural reconstruction of its surface layer, which is in local equilibrium with a supersaturated solution. NAP contain defects, chemical bonds and valence states of elements unusual for crystal volume, for example AsV, AsIII, AsII and FeIII on pyrite [41], which provides additional opportunities for the incorporation of minor and trace elements, especially incompatible, into their structures. Since the stability field of arsenopyrite is adjacent to the pyrite field, according to the principle of continuity of phase formation on mineral surfaces and the correspondence of chemical forms of elements on them [53], the surface NAP on arsenopyrite can have a defective pyrite structure. Since platarsite $Pt(As,S)_2$ has the structure of pyrite *Pa*3, we should therefore expect an increase in the solubility of Pt in the surficial phase compared to the volume of the crystal.

Gold and PGE nanoparticles are increasingly, especially recently, found in ores of different genesis deposits [7,8,54–56]. Although no PGE mineral forms were identified at the Natalkinskoe deposit, the need to continue research in this direction is obvious. This may be demonstrated by the Degdekan deposit, another gold project in the northeast of Russia, which is in the territory of Pre Kolyma. Its ores, classed as black shale formation, yielded the first findings of a number of platinum group minerals of submicron dimensions: native osmium, routheniridosmine, osmirid, ruthenosmirid, laurite, iridarsenite, ruthenium arsenide and arsenide of osmium and iridium [8]. The studies by [5,6] established the existence of different mineral forms of platinum metals, including native platinum, phases of Pt–Cu–Fe system, sperrylite, cooperite, and minerals of the Pd–Bi system in gold ores of the Sukhoi Log deposit (Lena ore district, Irkutsk region, Russia). Their size mainly ranges within 1–10 µm. Later, native Pt nanoparticles, with a dominant size of 0.5–20 nm, were found in the concentrates of insoluble carbonaceous matter of shales containing ore mineralization [7]. It follows that there is a high probability of nanoscale mineral forms of PGE existing at the Natalkinskoe deposit, according to the model of surface NAP evolution in sulphide minerals, suggesting their partial transformation and aggregation with the formation of nano- and microinclusions of separate (autonomous) phases of trace elements [39]. The possibility of finding nano-sized particles of PGE in sulphides is confirmed by experimental data. For example, nanometer-size PtS_2 inclusions in synthetic Pt-containing pyrite were detected with high-resolution transmission electron microscopy [57].

5. Conclusions

Arsenopyrites and pyrites of the Natalkinskoe gold deposit selected from vein, streaky-vein and veinlet-disseminated ores belonging to the most productive gold-rich hydrothermal mineral formations concentrated not only Au, but also platinum group elements. The highest contents (in ppm) were detected in monofractions of arsenopyrite: Pt up to 128, Pd up to 20, Ru up to 86 and Rh up to 21, and less high in monofractions of pyrite: Pt up to 29, Pd up to 15, Ru up to 58 and Rh up to 5.9.

The original AAS-ADSSC analytical technology in arsenopyrites and pyrites from the deposit established that there were two forms of uniformly distributed Pt, Pd, Ru and Rh—structural and surface-bound chemically bound in the structure of these minerals and in non-autonomous phases located on the surface of arsenopyrite and pyrite crystals. The surficially bound form dominates and probably exists in a very thin surface layer of the crystal (~100–500 nm). This phenomenon arises due to the features of the crystal growth mechanism and the dualism of the element distribution coefficient in the mineral-hydrothermal solution system, which is an order of magnitude higher for NAP compared to the crystal volume. It is assumed that for gold-bearing arsenopyrites and pyrites there is a common mechanism of impurity absorption associated with the active role of the crystal surface and surface defects. Structural forms of Pt, Pd, Ru and Rh, despite a lower content of the associated element compared to the surface, are also extremely important. They are a reliable indicator of element activity in ore-forming fluids forming gold deposits and a criterion of their potential ore content.

Restoration of the ore-forming fluid composition using the content of the structural admixture of Au and Pt in pyrite and experimentally established coefficients of their distribution in the pyrite–hydrothermal solution system showed that ore-forming fluid that deposited pyrite locally contained comparable concentrations of gold and platinum. Unlike gold, which mostly occurred in visible native form, the majority of PGEs are in so-called hidden or invisible forms. Previously, they were not recognized due to the inability to establish the nature of the carrier without the use of methods to determine the surficial and structural forms of the element.

The data resulting from studying the surface of sulphide minerals from the Natalkinskoe deposit using SEM-EDX and LA-ICP-MS methods confirmed the surface presence on arsenopyrite and pyrite crystals of Pt, Pd, Ru and Rh and such PGE as Os and Ir, as well as other ore trace elements including Ag, Hg, Pb, Cu and U. The maximum contents of invisible forms of PGE, as in the case of gold, are confined to the surface of crystals. An early assumption that surface-bound forms, as the main uniformly distributed form of elements in ore minerals, can be characteristic not only for Au, Pt and Pd was confirmed [11]. They are typical of other PGE (Ru, Rh, Os and Ir) and, in general, for ore trace elements.

The contradiction of LA-ICP-MS data with other methods with respect to the distribution of Pt and Pd in pyrite was revealed. This is assumed to be due to the presence of an unknown surface form of Pd, inhomogeneously distributed over crystals (Pd nanoparticles or any of its compounds). Although no proper mineral forms of PGE (Pt, Pd, Ru, Rh, Ir and Os) were established at the Natalkinskoe gold deposit, the probability of nano- or micro-sized mineral forms of PGE at this deposit is very high. This conclusion is based on the model of the evolution of surface NAP, assuming their partial transformation and aggregation with the formation of nano- and microinclusions of separate (autonomous) phases of trace elements.

The results obtained through the study of PGEs in sulphide minerals are of interest both theoretically and practically. Data on the nature of distribution, concentration level and forms of Pt, Pd, Ru, Rh, Ir and Os in arsenopyrites and pyrites significantly supplement existing ideas concerning peculiarities in the composition of gold mineralization, and they serve as a criterion for assessing the potential ore content of fluids forming gold deposits. The association of maximum contents of invisible forms of PGEs with the surface of sulphide mineral crystals can be used to develop methods of PGE recovery, primarily Pt, without destruction of the mineral structure, through transformation and dissolution of the surface layer only. This is an important point that must be taken into account in the concentration of ores, which includes sulphides, and in our case arsenopyrite and pyrite. The presence of PGE in ores and the possibility of their extraction significantly enhances the value of the gold ore mined, thereby increasing the prospects of the deposit under study.

Author Contributions: Conceptualization, R.G.K.; methodology, R.G.K. and V.L.T.; validation, R.G.K.; formal analysis, V.L.T., A.S.M., N.V.B. and N.V.S.; investigation, R.G.K. and V.L.T.; resources, R.G.K. and A.S.M.; writing—original draft preparation, R.G.K. and V.L.T.; writing—review and editing, R.G.K., V.L.T., A.S.M. and N.V.B.; visualization, R.G.K., A.S.M. and N.V.S. All authors have read and agreed to the published version of the manuscript.

Acknowledgments: The authors thank T.M. Pastushkova, I.Yu. Voronova, V.N. Vlasova, G.A. Shcherbakova, S.V. Lipko and K.Yu. Arsent'ev for their assistance in analytical studies. We would like to express our special gratitude to all geological services of the JSC "RiM" (Magadan, Russia), its chief geologist R.N. Ovsov and leading geologist E.M. Nikitenko, for comprehensive assistance in conducting fieldwork in the period from 2008 to 2016.

References

1. Goncharov, V.I.; Voroshin, S.V.; Sidorov, V.A. Platinum mineralization of gold ore deposits in black shale sediments of the North–East of Russia: Challenges and prospects. In *Platinum of Russia. Possibilities of Development of the Raw Material Base of Platinum Metals (Third meeting of the Scientific-methodical Council of the program "Platinum of Russia")*; Orlov, V.P., Ed.; Geoinformmark: Moscow, Russia, 1995; Volume 2, Part 2; pp. 156–161. (In Russian)
2. Goncharov, V.I.; Sidorov, V.A.; Pristavko, V.A. Platinum metal mineralization of the Natakinskoye gold ore deposit: Research findings. *Kolyma* **2000**, *2*, 49–53. (In Russian)
3. Goncharov, V.I.; Voroshin, S.V.; Sidorov, V.A. *Natalkinskoe Gold Lode Deposit*; NEISRI FEB RAS: Magadan, Russia, 2002; p. 250. (In Russian)
4. Voroshin, S.V.; Sidorov, V.A.; Tyukova, E.E. Geology, geochemistry and prospects of platinum mineralization at the Natakinskoye gold ore deposit (North–East of Russia). In *Platinum of Russia. Possibilities of Development of the Raw Material Base of Platinum Metals (Third meeting of the Scientific-Methodical Council of the Program "Platinum of Russia")*; Orlov, V.P., Ed.; Geoinformmark: Moscow, Russia, 1995; Volume 2, Part 2; pp. 161–176. (In Russian)
5. Distler, V.V.; Mitrofanov, G.L.; Nemerov, V.K.; Kovalenker, V.A.; Mokhov, A.V.; Semeikina, L.K.; Yudovskaya, M.A. Forms of occurrence of platinum group metals and their genesis in the Sukhoi Log gold ore deposit (Russia). *Geologiya Rudnykh Mestorozhdenij* **1996**, *38*, 467–484. (In Russian)
6. Distler, V.V.; Yudovskaya, M.A.; Razvozzhaeva, E.A.; Mokhov, A.V.; Trubkin, N.V.; Mitrofanov, G.L.; Nemerov, V.K. New data on PGE mineralization in gold ores of the Sukhoi Log deposit, Lensk gold-bearing district, Russia. *Dokl. Earth Sci.* **2003**, *393*, 1265–1267.
7. Nemerov, V.K.; Razvozzhaeva, E.A.; Spiridonov, A.M.; Sukhov, B.G.; Trofimov, B.A. Nanodispersed state of metals and their migration in carbonaceous natural media. *Dokl. Earth Sci.* **2009**, *425*, 334–337. [CrossRef]
8. Goryachev, N.A.; Sotskaya, O.T.; Goryacheva, E.M.; Mikhalitsyna, T.I.; Man'shin, A.P. The first discovery of platinum group minerals in black shale gold ores of the Degdekan deposit, Northeast Russia. *Dokl. Earth Sci.* **2011**, *439*, 902–905. [CrossRef]
9. Khanchuk, A.I.; Plyusnina, L.P.; Nikitenko, E.M.; Kuzmina, T.V.; Barinov, N.N. The noble metal distribution in the black shales of the Degdekan gold deposit in northeast Russia. *Russ. J. Pac. Geol.* **2011**, *5*, 89–96. [CrossRef]
10. Liu, J.; Chen, Y.; Liao, Z.; Zhan, Y.; Guan, Y. Progress in platinum group element (PGE) in black shale series. *Appl. Mech. Mater.* **2013**, *353–356*, 1183–1186. [CrossRef]
11. Kravtsova, R.G.; Tauson, V.L.; Nikitenko, E.M. Modes of Au, Pt, and Pd occurrence in arsenopyrite from the Natalkinskoe deposit, NE Russia. *Geochem. Int.* **2015**, *53*, 964–972. [CrossRef]
12. Pasava, J.; Ackerman, L.; Halodova, P.; Pour, O.; Durisova, J.; Zaccarini, F.; Aiglsperger, T.; Vymazalova, A. Concentrations of platinum-group elements (PGE), Re and Au in arsenian pyrite and millerite from Mo–Ni–PGE–Au black shales (Zunyi region, Guizhou Province, China): Results from LA-ICPMS study. *Eur. J. Mineral.* **2017**, *29*, 623–633. [CrossRef]
13. Vasil'eva, I.E.; Shabanova, E.V.; Goryacheva, E.M.; Sotskaya, O.T.; Labusov, V.A.; Neklyudov, O.A.; Dzyuba, A.A. Determination of precious metals in geological samples from four gold ore deposits of the north–east of Russia. *J. Anal. Chem.* **2018**, *73*, 539–550. [CrossRef]
14. Plyusnina, L.P.; Khanchuk, A.I.; Goncharov, V.I.; Sidorov, V.A.; Goryachev, N.A.; Kuz'mina, T.V.; Likhoidov, G.G. Gold, platinum, and palladium in ores of the Natalkinskoe deposit, upper Kolyma region. *Dokl. Earth Sci.* **2003**, *391A*, 836–840.
15. Volkov, A.V.; Genkin, A.D.; Goncharov, V.I. The forms of the presence of gold in the ores of the Natalka and May deposits (Northeast Russia). *Tikhookeanskaya Geol.* **2006**, *25*, 18–29. (In Russian)
16. Goryachev, N.A.; Vikent'eva, O.V.; Bortnikov, N.S.; Prokof'ev, V.Y.; Alpatov, V.A.; Golub, V.V. The world-class Natalka gold deposit, northeast Russia: REE patterns, fluid inclusions, stable oxygen isotopes, and formation conditions of ore. *Geol. Ore Depos.* **2008**, *50*, 362–390. [CrossRef]
17. Eremin, R.A.; Osipov, A.P. On genesis of the Natalkinskoe gold ore deposit. *Kolyma* **1974**, *6*, 41–43. (In Russian)
18. Kalinin, A.I.; Kanishchev, V.K.; Orlov, A.G.; Gashtol'd, V.V. Structure of the Natalkinskoe ore field. *Kolyma* **1992**, *10–11*, 10–14. (In Russian)

19. Voroshin, S.V.; Shakhtyrov, V.G.; Tyukova, E.E.; Gashtold, V.V. Geology and genesis of the Natakinskoe gold ore deposit. *Kolyma* **2000**, *2*, 22–23. (In Russian)
20. Goryachev, N.A.; Sidorov, V.A.; Litvinenko, I.S.; Mikhalitsyna, T.I. Mineral composition and petrogeochemical peculiarities of ore zones of the Natalkinskoe deposit deep horizons. *Kolyma* **2000**, *2*, 38–49. (In Russian)
21. Mezhov, S.V. Geological structure of the Natalkinskoe gold ore deposit. *Kolymskie Vesti* **2000**, *9*, 8–17. (In Russian)
22. Pristavko, V.A.; Sidorov, V.A.; Mikhalitsyna, T.I.; Burova, A.S.; Krasnaya, E.N. Geological-geochemical model the Natalkinskoe gold ore deposit. *Kolymskie Vesti* **2000**, *9*, 18–24. (In Russian)
23. Grigorov, S.A. Genesis and formation dynamics of the Natalkinskoe gold ore deposit according to the system analysis of the geochemical field. *Rudy Met.* **2006**, *3*, 44–48. (In Russian)
24. Struzhkov, S.F.; Natalenko, M.N.; Chekvaidze, V.B.; Isaakovich, I.Z.; Golubev, S.Y.; Danil'chenko, V.A.; Obuskhov, A.V.; Zaitsev, M.A.; Kryazhev, S.G. Multi-factor model of the Natalkinskoe gold ore deposit. *Rudy Met.* **2006**, *3*, 34–44. (In Russian)
25. Volkov, A.V.; Murashov, K.Y.; Sidorov, A.A. Geochemical peculiarities of ores from the largest Natalka gold deposit in Northeastern Russia. *Dokl. Earth Sci.* **2016**, *466*, 161–164. [CrossRef]
26. Grigorov, S.A.; Vorozhbenko, V.D.; Kushnarev, P.I.; Markevich, V.Y.; Tokarev, V.N.; Chichev, V.I.; Yagubov, N.P.; Mikhailov, B.K. Geology and key signatures of the Natalkinskoe gold deposit. *Otechestvennaya Geol.* **2007**, *3*, 43–50. (In Russian)
27. Sharafutdinov, V.M.; Khasanov, I.M.; Mikhalitsyna, T.I. Petrophysical zoning of the Natalka ore field. *Russ. J. Pac. Geol.* **2008**, *2*, 441–453. [CrossRef]
28. Sotskaya, O.; Goryachev, N.; Goryacheva, E.; Nikitenko, E. Micromineralogy of "Black Shale" Disseminated-Sulphide Gold Ore Deposits of the Ayan-Yuryakh Anticlinorium (North–East of Russia). *J. Earth Sci. Eng.* **2012**, *2*, 744–753.
29. Bunakova, N.Y.; Zakharenko, V.M. *Gold, Platinum, Palladium, Rhodium, Iridium and Ruthenium Detection in Rocks of Various Composition by Atomic Absorption with Electrothermal Atomization Following Concentration*; NSAM No 430-Kh Methods; VIMS: Moscow, Russia, 2005; p. 24. (In Russian)
30. Lebedeva, M.I.; Rogozhin, A.A. *Gold Determination in Rocks, Ores and Their Processed Products by Extraction–Atomic–Absorption Method with Organic Sulfides*; NSAM No 237-S Methods; VIMS: Moscow, Russia, 2016; p. 18. (In Russian)
31. Men'shikov, V.I.; Vlasova, V.N.; Lozhkin, V.I.; Sokol'nikova, I.V. Determination of platinum-group elements in rocks by ICP-MS with external calibration after cation exchange separation of matrix elements by KU-2-8 resin. *Anal. Kontrol* **2016**, *20*, 190–201. (In Russian)
32. Tauson, V.L.; Bessarabova, O.I.; Kravtsova, R.G.; Pastushkova, T.M.; Smagunov, N.V. Separation of forms of gold occurrence in pyrites by studying statistic samples of analytical data. *Russ. Geol. Geophys.* **2002**, *43*, 57–67.
33. Tauson, V.L.; Lustenberg, E.K. Quantitative determination of modes of gold occurrence in minerals by the statistical analysis of analytical data samplings. *Geochem. Int.* **2008**, *46*, 423–428. [CrossRef]
34. Ripley, E.M.; Chryssoulis, S.L. Ion microprobe analysis of platinum-group elements in sulfide and arsenide minerals from the Babbitt Cu–Ni deposit, Duluth Complex, Minnesota. *Econ. Geol.* **1994**, *89*, 201–210. [CrossRef]
35. Trigub, A.L.; Tagirov, B.R.; Kvashnina, K.O.; Chareev, D.A.; Nickolsky, M.S.; Shiryaev, A.A.; Baranova, N.N.; Kovalchuk, E.V.; Mokhov, A.V. X-ray spectroscopy study of the chemical state of "invisible" Au in synthetic minerals in the Fe–As–S system. *Am. Mineral.* **2017**, *102*, 1057–1065. [CrossRef]
36. Parviainen, A.; Gervilla, F.; Melgarejo, J.-C.; Johanson, B. Low-temperature, platinum-group elements-bearing Ni arsenide assemblages from the Atrevida mine (Catalonian Coastal Ranges, NE Spain). *Neues Jahrb. Mineral. Abh.* **2008**, *185*, 33–49. [CrossRef]
37. Tauson, V.L.; Kravtsova, R.G.; Smagunov, N.V.; Spiridonov, A.M.; Grebenshchikova, V.I.; Budyak, A.E. Structurally and superficially bound gold in pyrite from deposits of different genetic types. *Russ. Geol. Geophys.* **2014**, *55*, 273–289. [CrossRef]
38. Tauson, V.L.; Lipko, S.V.; Smagunov, N.V.; Kravtsova, R.G. Trace element partitioning dualism under mineral-fluid interaction: Origin and geochemical significance. *Minerals* **2018**, *8*, 282. [CrossRef]

39. Tauson, V.L.; Lipko, S.V.; Smagunov, N.V.; Kravtsova, R.G.; Arsent'ev, K.Y. Distribution and segregation of trace elements during the growth of ore mineral crystals in hydrothermal systems: Geochemical and mineralogical implications. *Russ. Geol. Geophys.* **2018**, *59*, 1718–1732. [CrossRef]
40. Tauson, V.L.; Babkin, D.N.; Akimov, V.V.; Lipko, S.V.; Smagunov, N.V.; Parkhomenko, I.Y. Trace elements as indicators of the physicochemical conditions of mineral formation in hydrothermal sulfide systems. *Russ. Geol. Geophys.* **2013**, *54*, 526–543. [CrossRef]
41. Tauson, V.L.; Lipko, S.V.; Arsent'ev, K.Y.; Mikhlin, Y.L.; Babkin, D.N.; Smagunov, N.V.; Pastushkova, T.M.; Voronova, I.Y.; Belozerova, O.Y. Dualistic distribution coefficients of trace elements in the system mineral–hydrothermal solution. IV. Platinum and silver in pyrite. *Geochem. Int.* **2017**, *55*, 753–774. [CrossRef]
42. Tauson, V.; Lipko, S.; Kravtsova, R.; Smagunov, N.; Belozerova, O.; Voronova, I. Distribution of "invisible" noble metals between pyrite and arsenopyrite exemplified by minerals coexisting in orogenic Au deposits of North–Eastern Russia. *Minerals* **2019**, *9*, 660. [CrossRef]
43. Tauson, V.L.; Kravtsova, R.G. Evaluation of the gold admixture in structure of pyrite of epithermal gold-silver ore deposits (the North–East Russia). *Zap. Ross. Mineral. Obshch.* **2002**, *131*, 1–11. (In Russian)
44. Tauson, V.L.; Kravtsova, R.G. Chemical typomorphism of mineral surfaces: Surface composition specifics (by the example of gold-bearing pyrite from epithermal deposit). *Russ. Geol. Geophys.* **2004**, *45*, 204–209.
45. Tauson, V.L.; Kravtsova, R.G.; Grebenshchikova, V.I. Chemical typomorphism of the surface of pyrite crystals of gold ore deposits. *Dokl. Earth Sci.* **2004**, *399*, 1291–1295.
46. Tauson, V.L.; Babkin, D.N.; Pastushkova, T.M.; Krasnoshchekova, T.S.; Lustenberg, E.E.; Belozerova, O.Y. Dualistic distribution coefficients of elements in the system mineral-hydrothermal solution. I. Gold accumulation in pyrite. *Geochem. Int.* **2011**, *49*, 568–577. [CrossRef]
47. Zhitova, L.M.; Kinnaird, J.A.; Gora, M.P.; Shevko, E.P. Magmatogene fluids of metal-bearing reefs in the Bushveld Complex, South Africa: Based on research data on fluid inclusions in quartz. *Geol. Ore Depos.* **2016**, *58*, 58–81. [CrossRef]
48. Tauson, V.L.; Kravtsova, R.G.; Grebenshchikova, V.I.; Lustenberg, E.E.; Lipko, S.V. Surface typochemistry of hydrothermal pyrite: Electron spectroscopic and scanning probe microscopic data. II. Natural pyrite. *Geochem. Int.* **2009**, *47*, 231–243. [CrossRef]
49. Mitkin, V.N.; Galizky, A.A.; Korda, T.M. Some observations on the determination of gold and the platinum-group elements in black shales. *Geostandard Newslett.* **2000**, *24*, 227–240. [CrossRef]
50. Oberthür, T.; Cabri, L.J.; Weiser, T.W.; McMahon, G.; Müller, P. Pt, Pd and other trace elements in sulfides of the main sulfide zone, Great Dyke, Zimbabwe: A reconnaissance study. *Can. Mineral.* **1997**, *35*, 597–609.
51. Öhlander, B.; Müller, B.; Axelsson, M.; Alakangas, L. An attempt to use LA-ICP-SMS to quantify enrichment of trace elements on pyrite surfaces in oxidizing mine tailings. *J. Geochem. Explor.* **2007**, *92*, 1–12. [CrossRef]
52. Kovalenko, A.N.; Tugova, E.A. Thermodynamics and kinetics of non-autonomous phase formation in nanostructured materials with variable functional properties. *Nanosyst. Phys. Chem. Math.* **2018**, *9*, 641–662. [CrossRef]
53. Tauson, V.L. The principle of continuity of phase formation at mineral surfaces. *Dokl. Earth Sci.* **2009**, *425*, 471–475. [CrossRef]
54. Palenik, C.S.; Utsunomiya, S.; Reich, M.; Kesler, S.E.; Wang, L.; Ewing, R.C. "Invisible" gold revealed: Direct imaging of gold nanoparticules in a Carlin-type deposit. *Am. Mineral.* **2004**, *89*, 1359–1366. [CrossRef]
55. Gonzalez-Jimenez, J.M.; Reich, M. An overview of the platinum-group element nanoparticles in mantle-hosted chromite deposits. *Ore Geol. Rev.* **2017**, *81*, 1236–1248. [CrossRef]
56. Gonzalez-Jimenez, J.M.; Deditius, A.; Gervilla, F.; Reich, M.; Suvorova, A.; Roberts, M.P.; Roque, J.; Proenza, J.A. Nanoscale partitioning of Ru, Ir, and Pt in base-metal sulfides from the Caridad chromite deposit, Cuba. *Am. Mineral.* **2018**, *103*, 1208–1220. [CrossRef]
57. Filimonova, O.N.; Nickolsky, M.S.; Trigub, A.L.; Chareev, D.A.; Kvashnina, K.O.; Kovalchuk, E.V.; Vikentyev, I.V.; Tagirov, B.R. The state of platinum in pyrite studied by X-ray absorption spectroscopy of synthetic crystals. *Econ. Geol.* **2019**, *114*, 1649–1663. [CrossRef]

10

Leaching Kinetics of Rare Earth Elements from Fire Clay Seam Coal

Xinbo Yang * and Rick Q. Honaker

Department of Mining Engineering, University of Kentucky, Lexington, KY 40506, USA; rick.honaker@uky.edu
* Correspondence: xinbo.yang@uky.edu

Abstract: Recovery of rare earth elements (REEs) from coal samples collected from the Fire Clay coal seam using diluted mineral acid solutions was investigated. The initial processing step was coal recovery using conventional froth flotation which concentrated the REEs in tailing material resulting in an upgrade to values around 700 ppm on a dry whole mass basis. Leaching experiments were performed on the flotation tailings material using a 1.2 M sulfuric acid solution adjusted to a temperature of 75 °C to study the extractability of REEs from coal material. The effect of particle size, leaching time, leaching temperature, and solid concentration on REE leaching recovery were evaluated. The kinetic data obtained from leaching over a range of temperatures suggested that the leaching process follows the shrinking core model with possibly a mixed control mechanism that may be a result of several heterogenous materials leaching simultaneously. Leaching recovery increased rapidly at the beginning of the reaction then slowed as the system reached equilibrium. The apparent activation energy determined from test data obtained over a range of temperatures using 1 M sulfuric acid was 36 kJ/mol for the first 20 min of reaction time and 27 kJ/mol for the leaching period between 20 and 120 min. The leaching of light REEs during the initial stage was determined to be driven by a chemical reaction, followed by the formation of a product layer, which required lower activation energy in the later stage of leaching. In regards to the heavy REEs, the major mechanism for leaching is desorption and the product layer formation does not affect the heavy REEs significantly.

Keywords: rare earth elements; coal; leaching; kinetics; apparent activation energy

---

1. Introduction

Rare earth elements (REEs) exist in over 200 different mineral types. A portion of the REEs are associated with carbonates and oxides that are easy to extract while others are bonded with silicates and phosphates that are difficult to penetrate [1]. Commercially, REEs are produced from monazite, xenotime, bastnaesite, and clay adsorbed REEs, which is a unique source commonly associated with a commercial production operation in southern China [2,3]. Many studies have focused on developing REE extraction processes from secondary resources including mine waste streams, industrial wastes or byproducts, electronic waste recycling and magnet recycling industries, coal ash, slags and waste water, etc. [3–8]. The limitation of recovering REE from secondary resources is low grade, which limits the effort and cost that can be applied toward the extraction process. However, these resources take advantage of eliminating the mining cost as well as other associated costs such as the chemical cost already expended for treating the waste. Coal-based materials represent a potential secondary source for REEs which may be extracted and concentrated by the use of physical and/or chemical processes [9–11].

Several studies have investigated the REE mode of occurrences and distributions in coals of varying ranks, associated mineral matter, acid mine drainage and sludge, and coal combustion byproducts. Arbuzov et al. (2018) studied peat material collected from western Siberia and concluded that: (1) a maximum of 25% of the REEs had a mineral association, such as monazite, xenotime, and zircon, etc.; (2) 10%–30% of REEs was in humic substances, and (3) 40%–80% was in a water-soluble form [12]. Studies conducted on the ash product of pre-combustion coal material using a muffle furnace showed that 70% of the light REEs and 50% of the heavy REEs in bituminous coals are predominantly associated with phosphate minerals, and 50%–60% of the REEs in low rank coals are associated with clays [13]. The REEs in low rank coals (i.e., lignite and subbituminous) are primarily associated with the organic matter whereas the association with higher rank coals are with the associated mineral matter including grains of rare earth phosphate minerals that are micro-dispersed within the organic matrix [12,14,15].

The Fire Clay coal seam material appears to be a very suitable source for REEs extraction due to its high concentration of REEs on a dry ash basis as compared to other coal sources. The origin of the REEs in the Fire Clay coal has been associated with volcanic deposition during the coalification period [16]. Mineralogy studies performed on the bituminous coal source using scanning electron microscopy with energy dispersive spectroscopy (SEM-EDS) detected micro-dispersed rare earth phosphate mineral particles, which are generally difficult to dissolve in strong acid solutions [9]. The finely dispersed inorganic material contained in Fire Clay coal has relatively high concentrations of REEs Liberation of the rare earth minerals by size reduction provides an opportunity for recovering clean coal from middling streams which is generally discarded or partially blended into coal products to meet contract specifications. This study focused on the leaching characteristics of the REEs associated with various fractions of the Fire Clay coal and associated mineral matter.

2. Materials and Method

2.1. Materials

A representative sample (~200 kg) of a targeted fraction of the Fire Clay seam coal was collected from a coal preparation plant located in eastern Kentucky. The Fire Clay coal source is a high-volatile (30%–40% dry basis), bituminous coal well documented as being enriched with REEs and other critical elements. The sampled process stream was the product of a secondary dense medium vessel, which targeted the recovery of the 1.4 × 1.6 specific gravity (SG) fraction, commonly referred to as middlings, in the 150 × 9 mm particle size fraction of the plant feed.

Analysis of a representative sample of the bulk revealed that the middlings material contained 242 ppm REEs on a dry, whole sample basis and 1331 ppm on a dry, ash basis. The ash content of the representative sample was 18.2% on a dry basis. The distribution of the REEs shown in Figure 1 indicates that nearly 24% of the REEs are permanent magnet elements (Nd, Pr, Sm, and Dy) while scandium represents 3.1% of the total, which is valued as a component in high performance aluminum alloy. The high ash-based REE content value provided an indication that the use of grinding to liberate the associated mineral matter could potentially provide a means for significant pre-concentration of the REEs prior to the leaching process while producing a marketable clean coal product.

The 150 × 9 mm middling bulk sample was crushed using a laboratory jaw crusher and subsequently ground in a hammer mill to achieve a top particle size of 1 mm. The bulk sample was split into multiple representative samples by riffling the hammer mill product. For each test, a representative sample was pulverized to a top size of 80 mesh (177 μm) in a smaller hammer mill and then ground in an attrition mill at a pre-selected rotation speed and retention time to achieve the desired particle size reduction. The solids concentration used in the attrition mill was 40 g/L. The ground material was treated in an 8-L Denver conventional flotation system to recover clean coal material using diesel fuel No. 2 as the collector and 4-methyl-2-pentanol as the frother. The chemical concentrations varied from test-to-test based on the feed particle size distribution. As shown in Figure 2, a rougher stage of treatment was followed by three cleaner stages to ensure maximum recovery of the liberated mineral matter and to achieve an ultraclean coal product. The tailings material collected from the coal recovery process was further treated by a second flotation step using octanohydroxamic acid ($C_8H_{17}NO_2$) at a dosage of 1.5 kg/t and a pH value of 9.5 to remove acid consuming materials (e.g., calcite) [17]. No additional frother was used in the mineral flotation stage due to the frothing property of octanohydromaxic acid. A rougher-cleaner treatment process was used for the alkali mineral rejection to maximize recovery of non-floatable material to the tailings stream. The final tailings material was the material used as feed for the leaching studies.

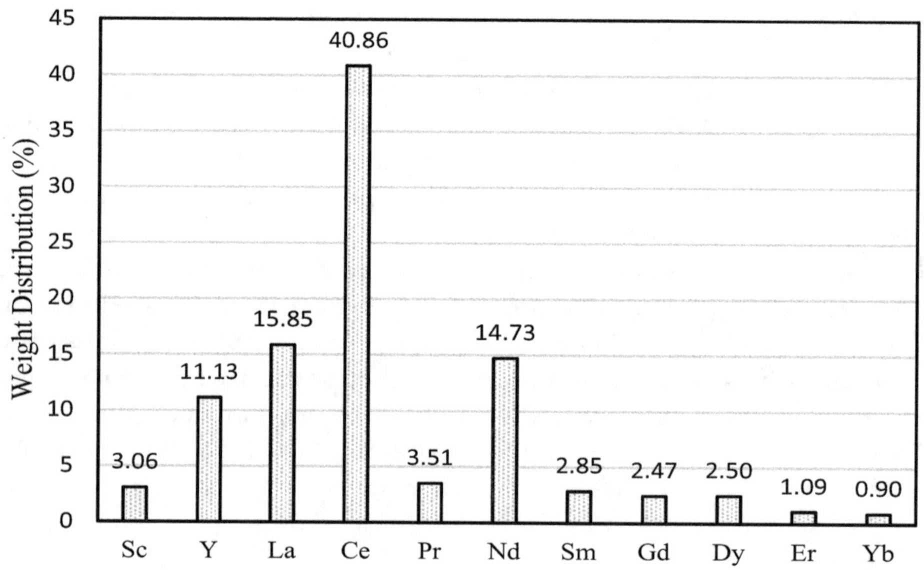

Figure 1. Rare earth element distribution in the nominal 1.4 × 1.6 Specific Gravity (SG) middling fraction of Fire Clay coal as collected from a secondary dense medium vessel product stream.

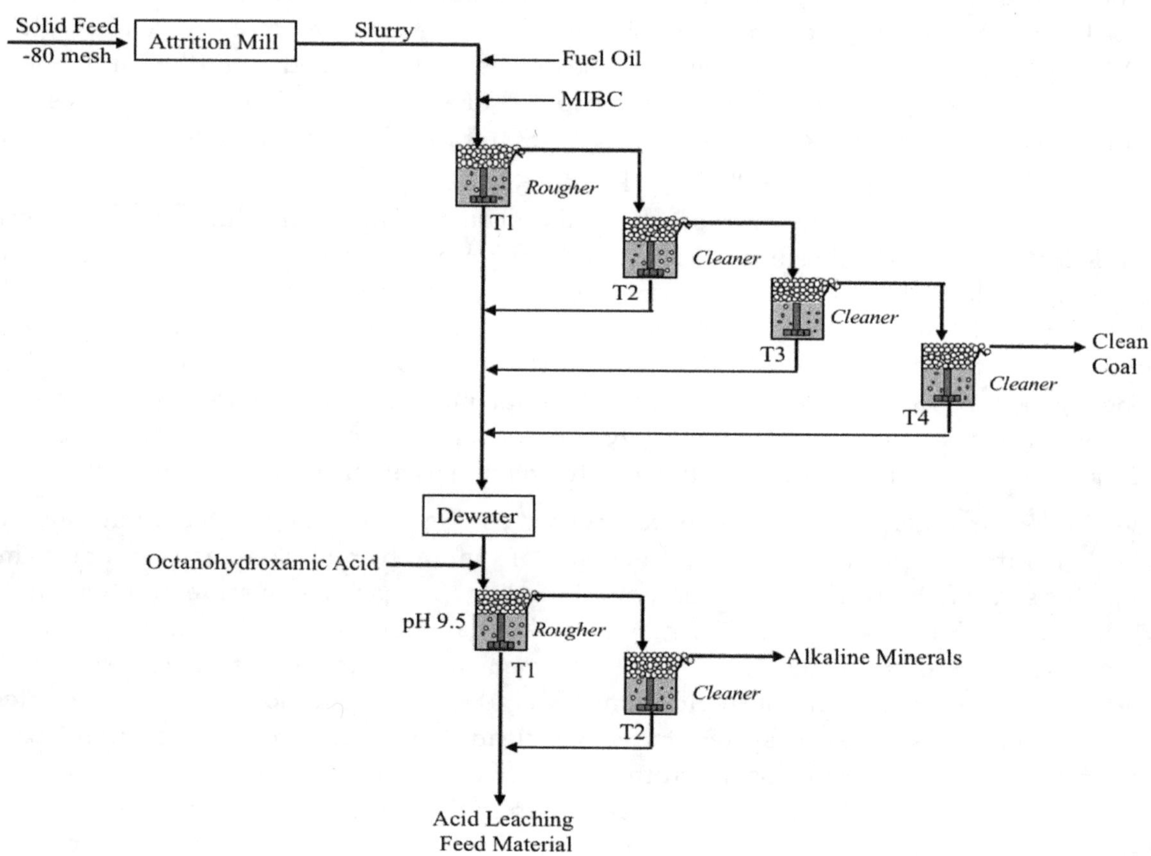

Figure 2. Sample preparation of the Fire Clay middlings for the leaching studies.

2.2. *Methodology*

Leaching experiments were conducted in a triple neck round bottom flask with the middle neck connected to a total reflux condenser, which ensured that a constant volume was maintained. The reactor was placed in a water bath where the solution was heated using an immersion heater and a precise temperature control system to maintain the temperature throughout the duration of the test. Agitation was provided using a magnetic stir at a speed that could be varied up to around 1200 rpm. The leaching experiments were conducted using deionized water and trace metal grade acid (purity > 99.99%).

The investigation involved the evaluation of the following parameters on leaching recovery and kinetics: i) lixiviant type (i.e., H_2SO_4, HCl, and HNO_3), ii) acid concentration (i.e., 0.1M, 0.5M, 1M, and 2M etc.), iii) solid-to-liquid ratio (i.e., S/L = 1/100, 1/50, 1/10, 1/5, etc.), and iv) solution temperature (i.e., 25 °C, 40 °C, 50 °C, 60 °C, and 75 °C). To assess leaching kinetics, samples were collected at time intervals established from the initial start of the test, i.e., 1 min, 3 min, 5 min, 10 min, 20 min, 30 min, 60 min, 90 min, and 120 min. Micro-filter (0.45 μm) plunger syringes were used to separate the leachate from the solids to immediately stop the solid–liquid reaction. The final solid residue was filtered and washed with deionized water.

The REE contents in addition to other elements of interest in the leachate and solid residue samples were determined using inductively coupled plasma optical emission spectrometry (ICP-OES). The results were used to calculate elemental and overall recovery of the REEs. Solid loss, REE content in the leach solid residue and leachate solution are presented in the supplementary materials (Table S1) that is accessible online. Leach recovery represents the amount of material in the test feed source that was solubilized into solution during the leaching process, which was quantified using the following expression:

$$\text{Leach recovery } (\%) = \frac{c_L * V_L}{c_L * V_L + c_{SR} * m_{SR}} \times 100\% \tag{1}$$

in which c_L is the elemental concentration in the leachate solution (μg/mL); V_L the volume of the analyzed leachate solution (mL); c_{SR} the elemental concentration in solid residue (μg/g); and m_{SR} the weight of solid residue (g).

2.3. *Analytical Methods*

REE content was determined by digestion and analysis of the resultant solution in an ICP-OES. The solid sample preparation procedure followed the ASTM D6357 method for ashing and digestion of coal and refuse samples with modifications made to the digestion to allow for use of a digestion block apparatus. The ICP-OES unit was calibrated using a standard solution identified as VHG-SM68 multi standard, which contained 48 elements. The REE recovery of these check standards was maintained at +/− 10% relative standard deviation (RSD). A duplicate sample was chosen at random and run through the entire process to verify repeatability at the frequency of not less than one every 40 samples. A certified coal ash sample (1633b) was utilized to ensure the digestion procedure and as a reference standard for peak selection. Three standard sample were repeated with each batch of digestion and the standard deviations for the rare earth elements are: <2% for Ce, Dy, Er, Eu, Gd, Ho, La, Lu, Nd, Pr, Sc, Th, Y, and Yb; <5% for Sm and Tm; <15% for Tb.

X-ray diffraction (XRD) analyses were conducted on feed samples using an Advance D8 instrument produced by the Bruker Company. The scanning was performed from 10° to 70° with a stepwise increase of 0.02° and a scanning speed of 0.5°/min. The XRD spectra were analyzed to estimate concentrations of major mineral components using the EVA software developed by the same company.

3. Results and Discussions

3.1. Particle Size Effect

A reduction in particle size may provide two significant benefits, i.e., (1) liberation of the clay particles and other mineral matter which exposes more surface area and exchangeable REEs for lixiviants to interact and extract the RE ions and/or (2) liberation of nano-sized RE minerals and RE oxides that may be dissolved in acid solutions. Acid leaching on finer size material can also provide faster kinetic rates and higher efficiency for REE extraction. On the negative side, by reducing the particle size to a micron level, the newly generated surface area is increased exponentially which escalates the consumption of hydrogen ions by dissolving more contaminate metal ions. As such, selectivity may be reduced, thereby elevating the cost of leaching and downstream concentration processes. Additional issues are the higher cost of energy for grinding and difficulties associated with thickening and dewatering ultrafine solid residuals.

To assess the effect of particle size on leaching performance, representative samples of the Fire Clay middlings material were ground for different lengths of time before the de-carbonization step to generate samples having a range of 80% passing sizes (P80). Acid leaching tests were conducted using 1.2 mol/L sulfuric acid solution with a solid concentration of 10 g/L at 75 °C. The REE recovery was quantified using test data generated after leaching for five hours and Equation (1).

The results shown in Table 1 indicated that reducing the particle size liberated mineral matter containing higher concentrations of REE. For example, by reducing the particle size to a P80 size of 150 μm (80 mesh top size) in the feed, the flotation tailings material contained 444 ppm of total REEs with a P80 size of 32 μm. The REE concentration of 444 ppm reflects the content of the coarser mineral matter dispersed in the middling particles. By grinding for greater lengths of time resulting in more applied energy, the P80 size was reduced to sub-micron level and the tailing material generated by flotation nearly doubled to 719 ppm of total REEs. This finding indicates that the finest mineral matter dispersed within the Fire Clay coal has the highest concentration of REEs. A previous study found that REEs in the Fire Clay coal were strongly associated with micro-dispersed kaolinite which may be liberated and released through size reduction [9].

Table 1. Liberation of REEs from the de-carbonized Fire Clay middlings material.

Attrition Mill		P80		Flotation		TREE Concentration		Acid Leaching	
Grinding Time (min)	Rotor Speed (rpm)	Flotation Feed (micron)	Flotation Tailing (micron)	Tailing Yield (%)	Ash Content (%)	Ash Basis (ppm)	Whole Mass Basis (ppm)	Solid Loss (%)	TREE Recovery (%)
Feed					18.2	1331	242		
0	0	150	32.0	3.7	89.1	499	444	11.9	71.2
2.5	150	45	14.5	7.4	85.9	581	499	9.9	74.8
5	200	37	10.2	9.5	87.1	609	531	9.2	79.0
20	200	13	8.7	12.8	90.8	685	622	16.2	83.7
40	200	11	7.3	14.2	91.0	768	699	17.2	84.3
60	200	10	6.5	13.9	91.4	771	705	20.4	83.6
90	200	9	5.0	13.6	90.9	788	716	24.2	82.3
120	200	8	0.9	15.4	91.0	791	719	27.9	82.0

TREE: total rare earth elements.

Reducing particle size resulted in a significant increase in leach recovery from 71.2% to 84.3% over the range of P80 sizes. As such, the size reduction increased the amount of REEs reporting in the leach feed and increased the percentage of the REEs being recovered through leaching. These two positive outcomes suggest that the REEs associated with micro-dispersed mineral matter in the Fire Clay middlings are more concentrated and more easily extractable by leaching relative to the coarser grain fractions. In addition, the finer mineral matter is, in general, more soluble as indicated by an increase in the amount of solid loss during the leaching process. As much as 20% of the solids in the finest sample tested was dissolved under the standard leaching conditions, which may reflect both the solubility of the mineral matter and surface area exposure. A negative impact is an increase in the amount of contaminates in the leachate due to the elevated level of dissolved solids.

Based on the liberation test results, 20 min of grinding time was selected to generate the acid leach feed material used in this study. A flow sheet of the sample preparation process is shown in Figure 3 along with weight yield and content data for each feed and product stream. The decarbonization step resulted in a high-quality clean coal product containing around 7% ash-forming material while the reject material was nearly pure mineral matter as indicated by an ash content of 90.81%. The mineral flotation rougher-cleaner treatment resulted in 0.74% of the total feed reporting to the concentrate having an ash content of 75.12%. The lower ash content in the flotation concentration was an indicator of calcite flotation ($CaCO_3$). The float product also contained 741 ppm of TREEs, which may be due to RE mineral flotation resulting from the use of octanohydroxamic acid. The flotation tailing material produced from the two stages of flotation represented 12.1% of the feed and contained 615 ppm of TREEs and 90.67% ash-forming material. The REE upgrade is 2.54:1 starting from a feed content of 242 ppm. This material was used as the acid leach feed in the subsequent leaching studies presented in this paper.

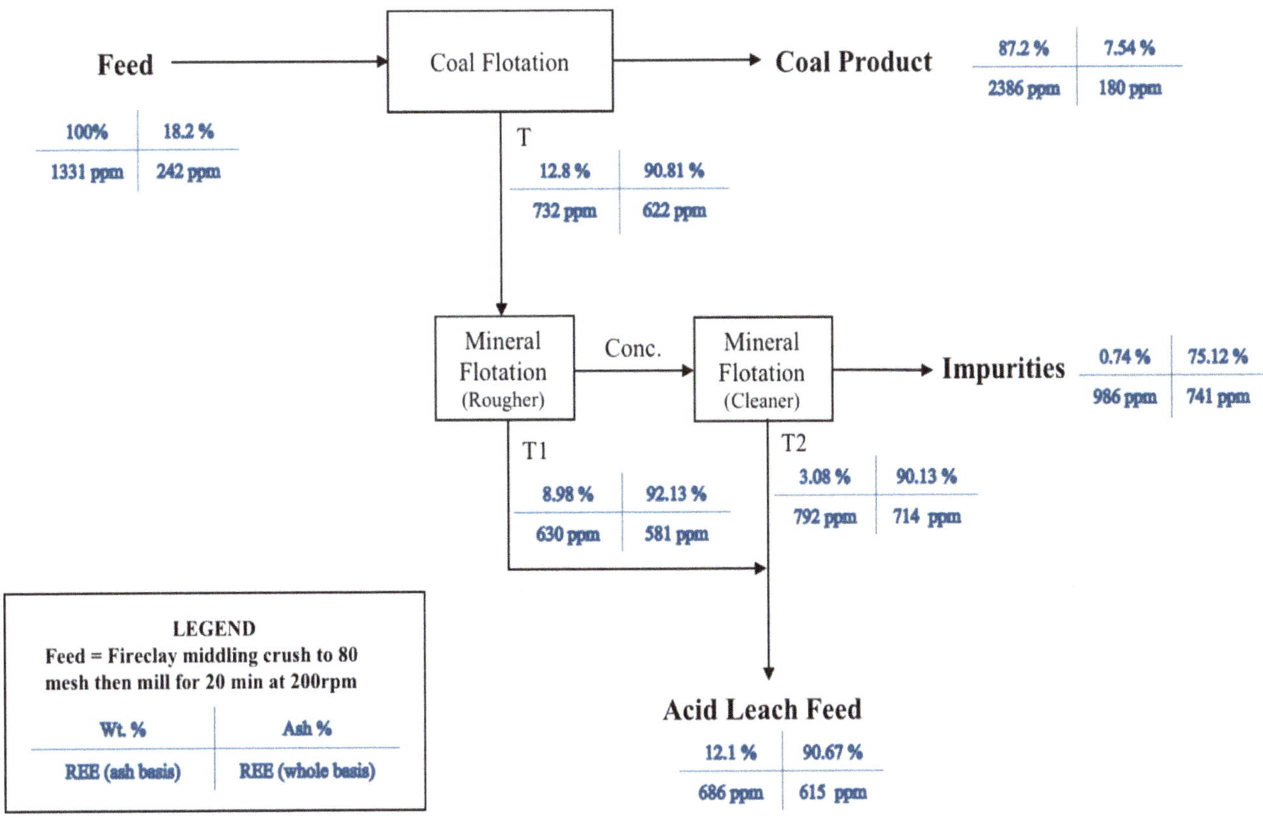

Figure 3. Schematic of sample preparation for the acid leach feed material using coal and mineral flotation.

Five representative samples of the acid leach feed were analyzed to assess the repeatability of the ICP-OES. The average TREE value was 607 ± 18 ppm (2.97% variation) with thorium content of 41 ± 0.6 ppm (1.46% variation). The Ce content accounted for 42% of the total REEs as shown in Figure 4a. The Fire Clay material was rich in light REEs (i.e., Ce, La, Pr, Nd, Sm, Eu, Gd, Sc) as indicated by a content of 534 ppm or 88.0% of the total. Yttrium was the most abundant heavy REE (i.e., Y, Tb, Dy, Ho, Er, Tm, Yb, and Lu) with a concentration of 47 ppm. The major minerals present in the sample were quartz, kaolinite, illite, and muscovite as shown in the XRD plot in Figure 4b.

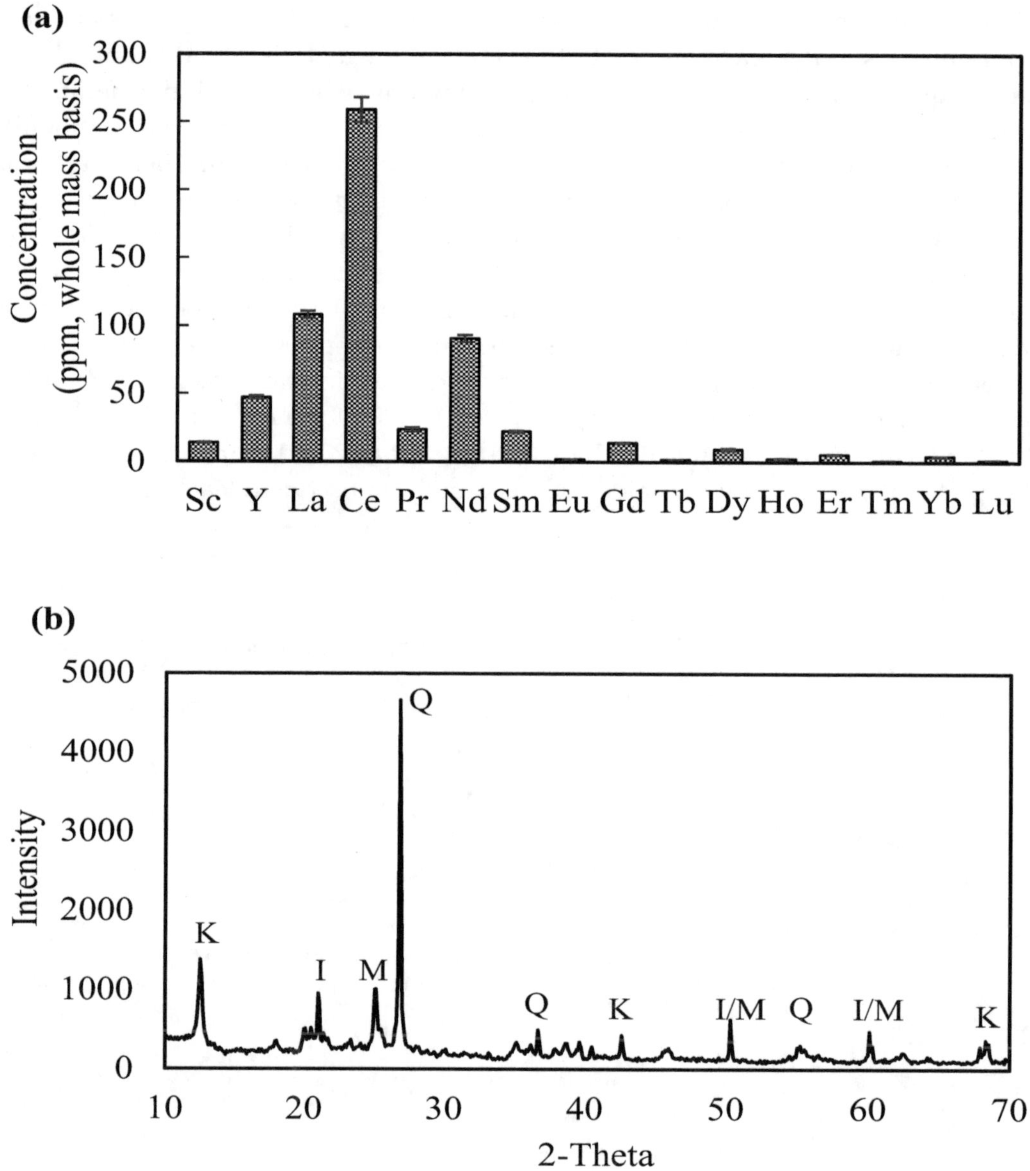

Figure 4. The composition of the acid leaching feed material used in this study on the basis of (**a**) rare earth content and (**b**) mineralogy as determined by X-ray diffraction analysis (Q—quartz, K—kaolinite, I—illite, M—muscovite).

3.2. Effect of Major Variables on REE Leaching

3.2.1. Effect of Acid Type

The lixiviant type affects the REE leaching characteristics by changing the solution speciation stabilities due to the existence of various anions in varying concentrations. Sulfate ions were reported to have a higher coordination ability with rare earths than chloride ions even in high monovalent concentration solutions [18]. Leaching experiments were conducted using different inorganic acids at an acid concentration of 1M, solid/liquid ratio of 10 g/L, and a temperature of 75 °C. Sulfuric acid (H_2SO_4), hydrochloric acid (HCl), and nitric acid (HNO_3) were used to study the effect on REE leaching recovery and reaction rate as shown in Figure 5. Total REE (TREEs) recovery values of 80%, 76%, and 74% were achieved after 3 h of leaching using HCl, HNO_3, and H_2SO_4 solution, respectively. The pH of the leachate solutions at the end of the tests were 0.105, 0.113, and 0.112, respectively.

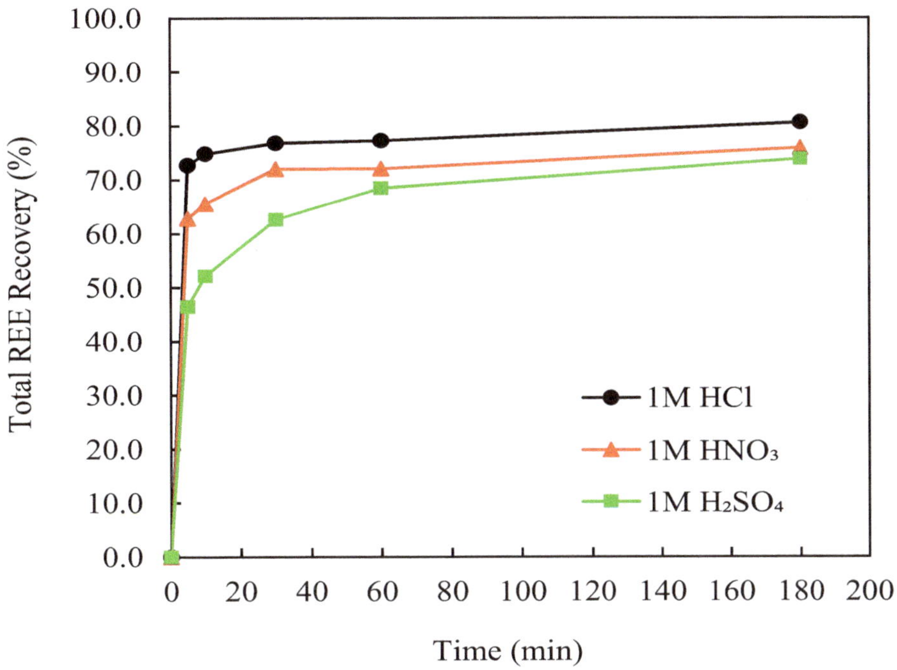

Figure 5. Effect of acid type on the leaching recovery of total rare earth elements contained in the Fire Clay coal middlings (75 °C, 530 rpm, solid-to-liquid ratio (S/L) = 10 g/L, d_{80} = 8.7 μm).

Hydrochloric acid provided the fastest leaching rate, which achieved 73% recovery after the first 5 min of leaching, and slowly reached equilibrium after 3 h. Nitric acid also provided fast leaching rate within the first 30 min. Sulfuric acid was the least effective under the leaching conditions and provided the slowest leaching rate. This finding was likely due to the fact that sulfate ions have a higher coordination ability with rare earths than chloride ions even in high monovalent concentration solutions [19].

The coal-based leachate contained high concentrations of trivalent ions that may coordinate with sulfate ions resulting in depression of the rare earth-sulfate coordination. In addition, sulfuric acid requires two steps of dissociation reaction to release H^+ into solution whereas hydrochloric acid and nitric acid dissociates more rapidly into solution. Viscosity of the sulfuric acid solution is another factor that could have resulted in the slower reaction rate as the wetting rate of the solid particle surfaces is reduced when the solution viscosity is high. Despite the negative aspects of sulfuric acid, the lixiviant is still considered a viable lixiviate due to its relatively low cost and the negative aspects of the other lixiviants including the volatility of hydrochloric acid and the decomposability of nitric acid under 75 °C [20].

3.2.2. Stirring Speed Effect

Stirring speed affects the thickness of the film layer surrounding a solid particle suspended in the lixiviate solution. A high stirring speed creates an enhanced shear rate in solution which reduces the film layer thickness thereby increasing the mass transfer rate through the film diffusion layer [21]. The effect of stirring speed was evaluated at 300 rpm, 530 rpm, 760 rpm, and 900 rpm as shown in Figure 6. The leaching condition included 1M H_2SO_4 solution and a solid/liquid ratio of 10 g/L at 75 °C. The test results indicated that a stirring speed of 300 rpm did not provide sufficient agitation due to inadequate suspension of the slurry based on visual observations, while stirring speeds of 530 rpm to 900 rpm provided nearly equal kinetics. The recovery achieved using a 900-rpm stirring speed was slightly lower than that obtained at 760 rpm. A stirring speed of 530 rpm was established as an adequate value for the standard test conditions.

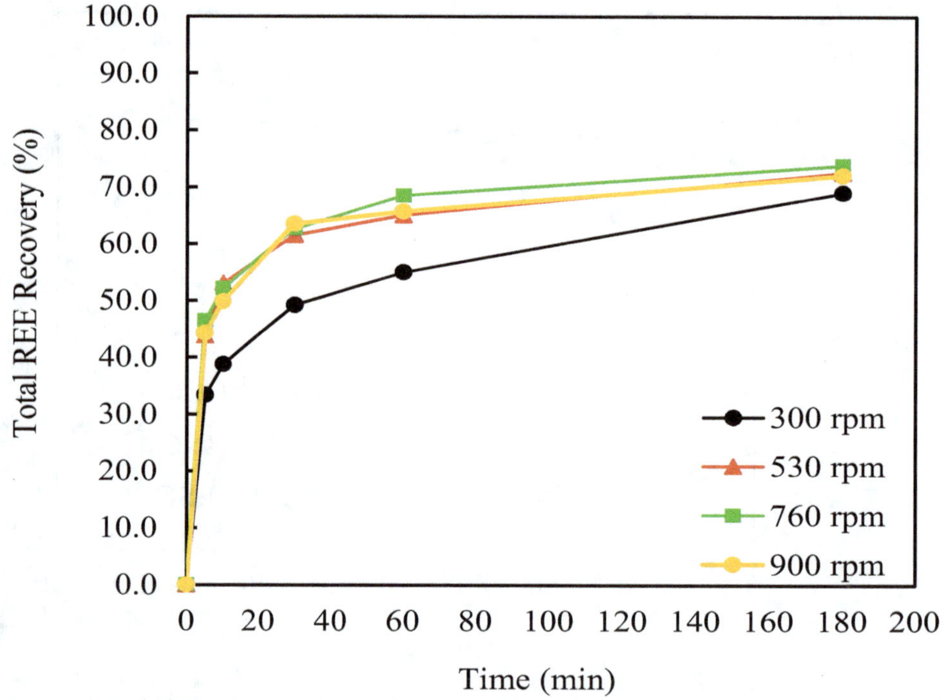

Figure 6. Effect of stirring speed on the leaching recovery of total rare earth elements contained in the Fire Clay coal middlings (75 °C, 1 M H_2SO_4, S/L = 10g/L, d_{80} = 8.7 μm).

3.2.3. Solid-to-Liquid Ratio Effect

The solid-to-liquid (S/L) ratio establishes to the stochiometric ratio of reactants, which directly affects the reaction equilibrium. The effect of the S/L ratio on rare earth leaching recovery was investigated in the range of 10 g/L to 200 g/L while maintaining the other parameters constant at 75 °C, 1 M H_2SO_4, and 530 rpm. The association between reactants decreased with an increase in the solid/liquid ratio, which resulted in a decrease in the extraction rate as shown in Figure 7. Leach recovery was reduced from 74% to 40% after increasing the S/L ratio from 10 g/L to 200 g/L. The magnitude of the recovery reduction is not commonly observed in other metal leaching operations. In the metallic copper leaching process, the leaching reaction was more effective when the Cu^{2+} concentration in solution was higher due to Cu^{2+} reacting with metallic Cu to Cu^{+} [22]. This type of reaction mechanism does not occur in a REE solution since the REEs exist mostly as a compound. Niobium leaching from titanium oxide residues did not show any effect from the S/L ratio on leaching recovery [21]. However, Li et al. (2013) reported on a rare earth concentrate leaching study that found the S/L ratio to have a negative effect on the leaching of a rare earth concentrate when the ratio was higher than 100 g/L [23]. Therefore, the solid/liquid ratio effect varies from source-to-source in different leaching environments.

3.2.4. Effect of Acid Concentration

The effect of sulfuric acid concentration on leaching recovery was studied using 0.1 M, 0.5 M, 1 M, and 2 M acid concentrations using the standard values for temperature, stirring speed, and solid-to-liquid ratio. The initial acid concentrations of 0.1 M, 0.5 M, 1 M, and 2 M resulted in ending pH values of 1.04, 0.38, 0.11, and −0.25, respectively, after 3 h of leaching. As shown in Figure 8, the total REE recovery increased substantially from 40% to 74% by elevating acid concentration from 0.1 to 1 M. However, from 1 M to 2 M, the increase was marginal at around 2.5 absolute percentage points. The optimal acid concentration was selected to be 1 M since higher concentrations of acid did not provide a significant increase in recovery of REEs.

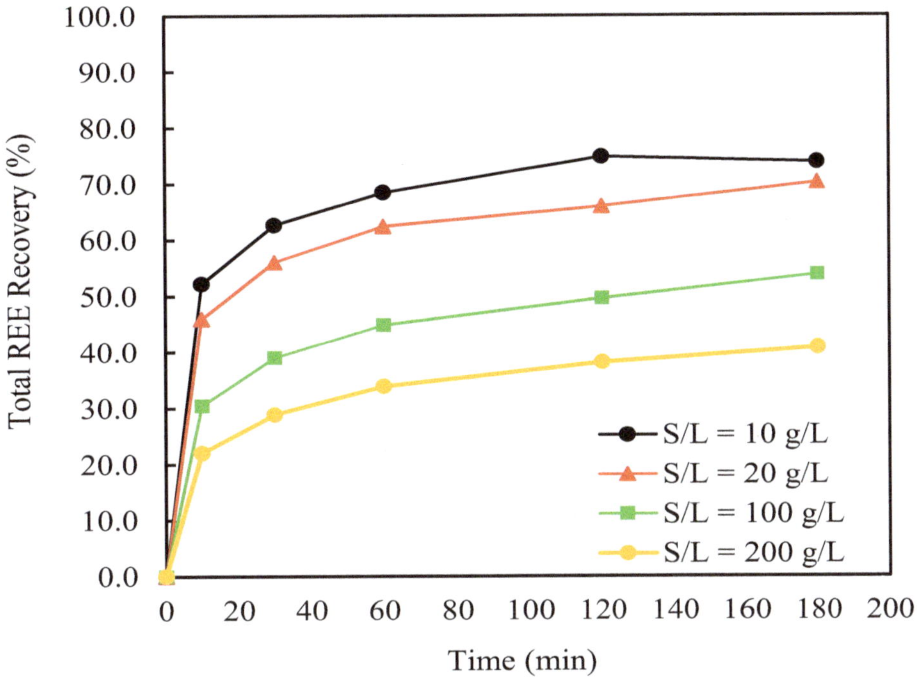

Figure 7. Effect of solid-to-liquid ratio on the leaching recovery of total rare earth elements contained in the Fire Clay coal middlings (75 °C, 1 M H_2SO_4, 530 rpm, D_{80} = 8.7 μm).

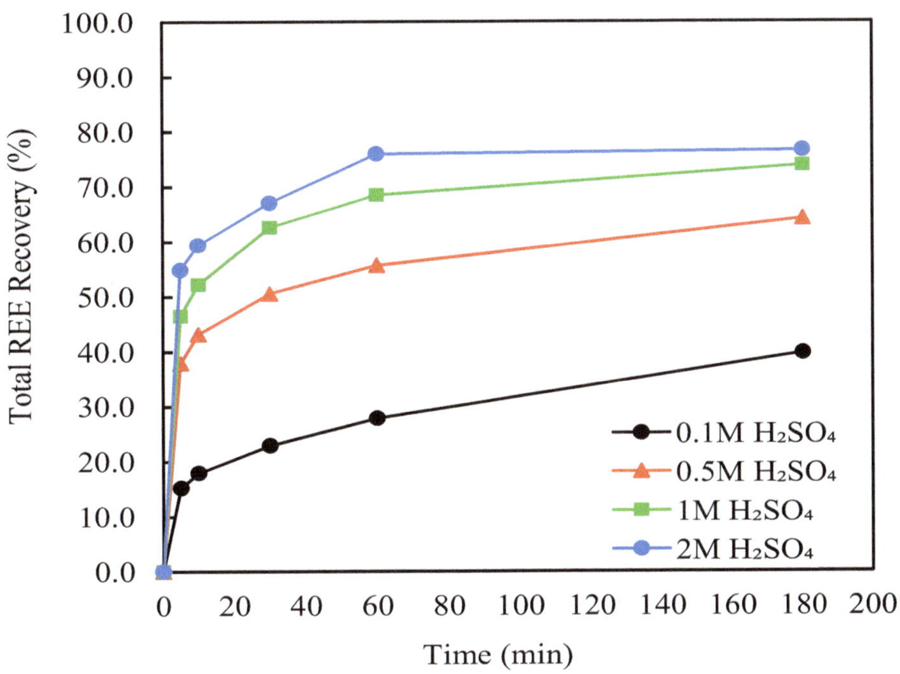

Figure 8. Effect of sulfuric acid solution concentration on the leaching recovery of total rare earth elements contained in the Fire Clay coal middlings (75 °C, 530 rpm, S/L = 10g/L, D_{80} = 8.7 μm).

The effect of acid concentration on individual RE leaching recovery is shown in Figure 9. Recovery of the light REEs significantly increased when acid concentration was elevated from 0.1 M to 1 M and very little improvement was realized when using a 2 M acid solution. Scandium was the least sensitive to acid concentration. The remaining un-leached portion of Sc is likely associated with a mineral structure that requires a higher level energy to break down. Yang et al. (2018) improved the scandium leaching recovery from 31% to 74% after treated by roasting with no chemical additives at 750 °C for 2 h from a coal-based material [24].

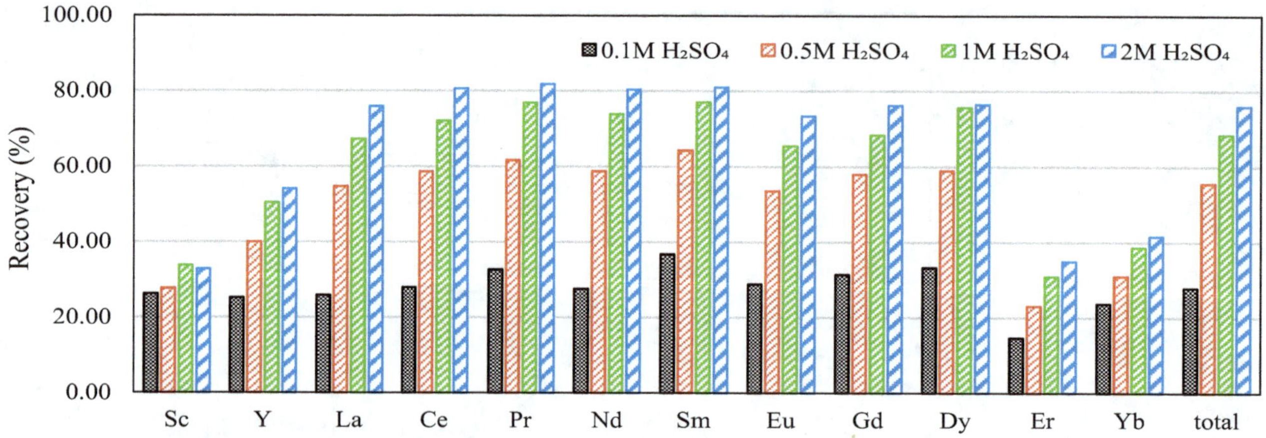

Figure 9. Effect of sulfuric acid solution concentration on individual rare earth element leaching recovery (2 h, 75 °C, 530 rpm, S/L = 10 g/L, D_{80} = 8.7 μm).

3.2.5. Effect of Temperature

A leaching process that is mainly controlled by a diffusion process is more dependent on mixing conditions whereas temperature has a more significant effect on chemical reaction controlled processes [25]. The effect of temperature on REE leaching using 1 M H_2SO_4 was investigated using a stirring speed of 530 rpm and S/L ratio of 10 g/L for 2 h. Samples were taken over shorter time increments due to the relatively fast kinetics during the first 20–30 min. Figure 10 shows that REE leach recovery increased significantly with an elevation in leaching temperature. When the temperature was increased from 298 K (25 °C) to 348 K (75 °C), leaching recovery increased from 35% to 75% after 2 h of leaching. The data suggests the existence of a relatively fast leaching process during the first 20 min followed by a slow process. As such, two or more reaction mechanisms may be occurring when leaching the coal source.

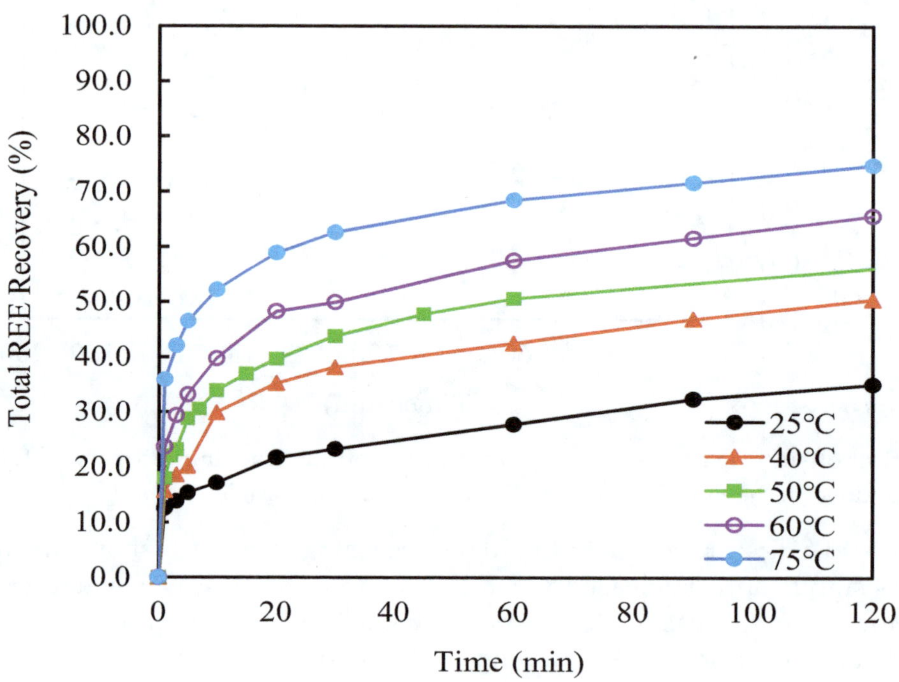

Figure 10. Effect of leaching reaction temperature on the leaching recovery of total rare earth elements contained in the Fire Clay coal middling (1 M H_2SO_4, 530 rpm, S/L = 10g/L, D_{80} = 8.7 μm).

The effect of temperature on individual REEs is shown in Figure 11. Most of the light REEs (i.e., La, Ce, Pr, Nd, Sm) appeared to be very sensitive to temperature, which indicated that the leaching mechanism of light REEs was mostly chemical reaction controlled. The recovery of Ce, Pr, and Nd increased from 36%, 39%, and 36% to 79%, 84%, and 80%, respectively, by increasing the temperature from 25 °C to 75 °C. The heavy REEs and scandium recovery improved with higher temperature, but the increase was not as significant. Scandium recovery rose from 29% to 36%. For the recovery of elements that were relatively insensitive to temperature, the activation energy is generally low and more likely to be a result of a diffusion controlled process [26].

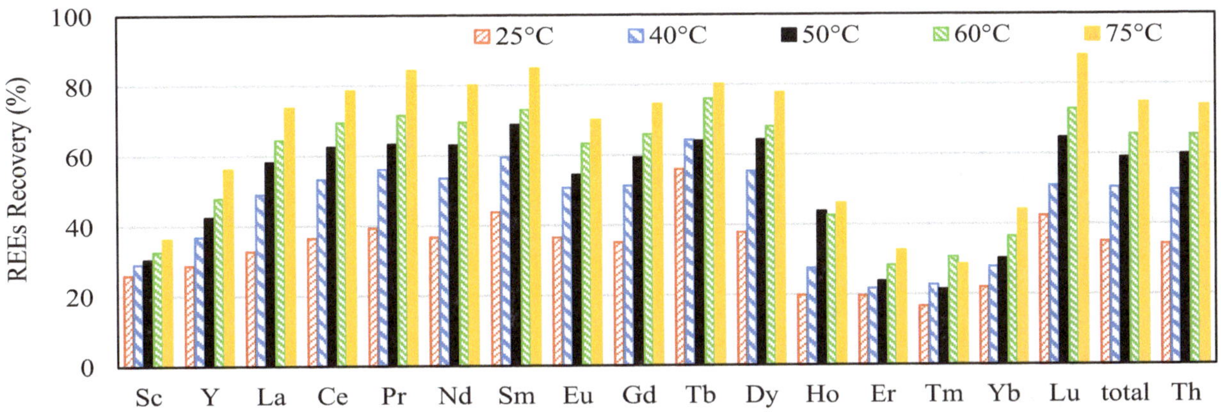

Figure 11. Effect of leaching reaction temperature on the leaching recovery of individual rare earth element (1 M H_2SO_4, 530 rpm, S/L = 10g/L, D_{80} = 8.7 μm, retention time of 120 min).

3.2.6. Morphology

The morphology of the Fire Clay coal middling particles before and after sulfuric acid leaching was studied using SEM. As shown in Figure 12a, the feed material consisted of heterogeneous particles comprised of mostly quartz and clay, which agrees well with the XRD analysis shown in Figure 12b. After 2 h of leaching at 50 °C, the particles were found to have a porous structure on the surface with a micro pore structure as shown in Figure 12b. After 2 h of leaching at 75 °C, the porous structure on some of the particle surfaces appeared larger in size as shown in Figure 12c. The images show no visible layer on the surface instead of porous structure due to dissolution. Therefore, the diffusion process in this reaction may be the results of interfacial transfer of the products and the reagent diffusion through the porous structure of solid particles.

Figure 12. SEM images of particles found in (**a**) leaching feed material; (**b**) solid residue after 2 h leaching at 50 °C; (**c**) solid residue after 2 h leaching at 75 °C (1 M H_2SO_4, 530 rpm, S/L = 10 g/L, D_{80} = 8.7 μm).

The leaching process involved several simultaneous reactions due to the mineral composition and the variety of REE associations. The REEs were found to exist in crystallized structures (mostly silicates and phosphate compounds), which usually require decomposition to be extracted under the current leaching conditions. A small portion of the REEs are present as RE ion substitution form in clays whereas most are associated with soluble RE containing minerals. Based on the results shown in Section 3.2.5, extraction of light REEs in this coal source was more sensitive to temperature thus the light REEs were more likely to be mineral associated, whereas the heavy REEs extraction was more independent to temperature thus more likely to be soluble metal oxides and adsorbed ions onto clay minerals.

3.3. Kinetic Analysis

The leaching process is classified as a fluid–particle heterogenous reaction in which a liquid reacts with a solid by contacting and transforms the solid into a product. A solid particle that reacts with a liquid and shrinks in size during the reaction can be described by a shrinking core model. The reaction is a five-step process, i.e., (1) diffusion through the film layer, (2) diffusion through the product layer, (3) chemical reaction on the surface, (4) product diffusion through the product layer and (5) product diffusion through the film layer to the solution. The slowest step is known as the rate determining process. The activation energy of a certain leaching step can be quantified by selecting the most accurate rate equation to represent the reactions [25].

A variety of rate equations have been developed and reported in literature that describe the leaching rate process [25,27,28]. Among the equations, the rate equation (Equation (2)) developed by Crank–Ginstling–Brounshtein, which describes the mass transfer across product layer, fits the experimental data well, i.e.,

$$k_d t = \left[1 - \frac{2}{3}\alpha - (1-\alpha)^{\frac{2}{3}}\right] \tag{2}$$

here α is the fraction that reacted; k_d is the kinetic constant.

The Crank–Ginstling–Brounshtein equation was used to linearize the extraction fraction (α) among all the temperatures using the experimental data for the first 20 min of leaching and the following 20–120 min of the reaction as shown in Figure 13. The correlation coefficient values (R^2) and the corresponding slopes (k) of the plots are listed in Table 2. Rate constants were calculated and the Arrhenius plots of ln(k) versus 1/K are as shown in Figure 14 for the two leaching stages. The activation energy determined for the first 20 min was 36 kJ/mol and 27 kJ/mol for the following 20–120 min of leaching. The activation energy values for both leaching periods were close to the energy barrier that is typically used to identify a diffusion controlled or chemical reaction controlled process, which is around 20 kJ/mol [26].

Table 2. Correlation coefficients of diffusion-controlled kinetics models at different temperatures for total REEs.

Temperature		**Diffusion 1–20 min**			**Diffusion 20–120 min**		
°C	**K**	**k**	**a**	**R^2**	**k**	**a**	**R^2**
25	298	0.0002	0.0017	0.9949	0.0001	0.0036	0.996
40	313	0.0007	0.0022	0.9648	0.0002	0.0129	0.9977
50	323	0.0009	0.0046	0.9616	0.0003	0.0196	0.9402
60	333	0.0014	0.0068	0.9919	0.0004	0.0262	0.9934
75	348	0.0019	0.0189	0.9627	0.0005	0.0487	0.9796

Note: k denotes the slope of the regression line, and a denotes the intercept of the regression line.

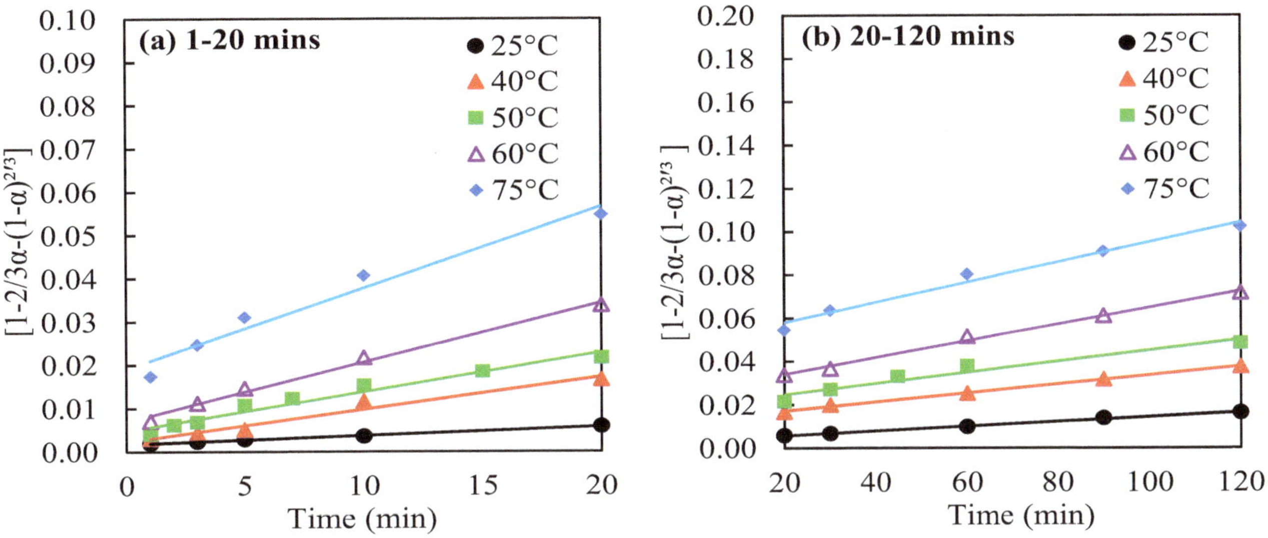

Figure 13. Kinetic modeling of total REEs recovery during the (**a**) first 20 min, and (**b**) 20–120 min of leaching at various temperatures for the Fire Clay middlings (1 M H_2SO_4, 530 rpm, S/L = 10 g/L, D_{80} = 8.7 μm, retention time of 120 min).

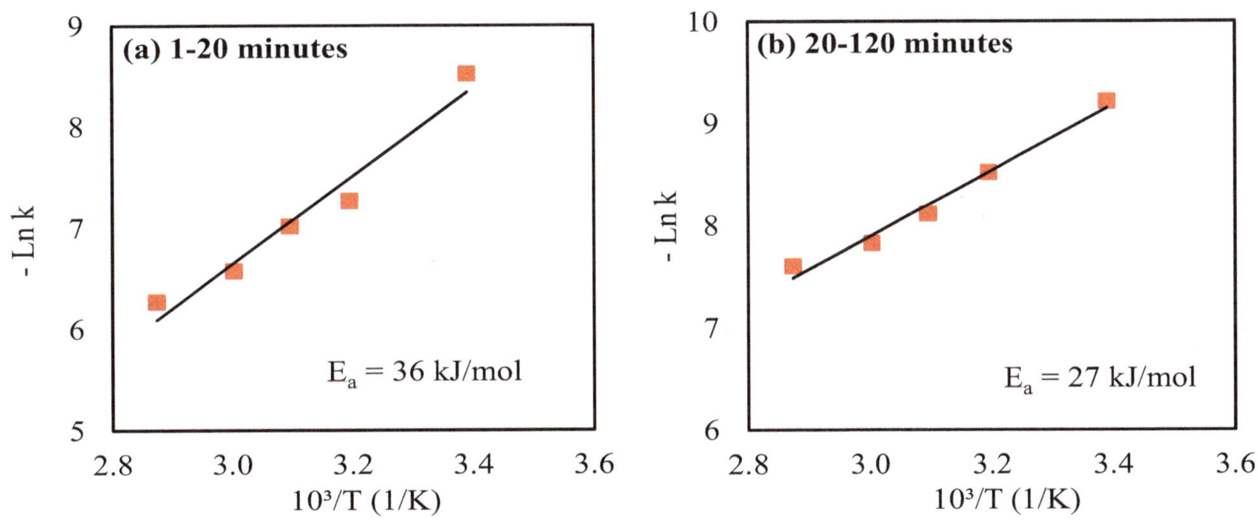

Figure 14. Arrhenius plot for the total REEs leached from the Fire Clay coal middlings during the (**a**) first 20 min, and (**b**) 20–120 min of leaching (1 M H_2SO_4, 530 rpm, S/L = 10g/L, d_{80} = 8.7 micron).

Since the coal tailing material is a heterogenous material that contains a number of potential modes of occurrence of REEs, the leaching process is not a single reaction. The resulting requirement for activation energy is a combination of the various forms of REEs. In addition, the material contains both calcite and pyrite among other soluble minerals that create a complex solution environment where the localized pH elevation on the solid particle surface could cause a product layer to be formed. The interfacial transfer of product through the porous structure of the solid particles requires high activation energies as reported by Li et al. (2010 and 2013), which can be as high as 40 kJ/mol [23,29].

To support the hypothesis, the activation energies for light and heavy REE groups were calculated using the data provided in Tables 3 and 4, respectively. The activation energy values for leaching the light REEs over the first 20 min and the period between 20 and 120 min were 41.8 kJ/mol and 28.1 kJ/mol, respectively. On the other hand, the activation energy values for the leaching of heavy REEs for the first 20 min and the 20–120 min of reaction were 24.2 kJ/mol and 26.1 kJ/mol, respectively. These values indicate that the leaching of the light REEs during the initial stage is more of a chemical reaction followed by the formation of a product layer and a reduced activation energy. The activation energy required for leaching the heavy REEs during the initial stage was significantly lower than that of the light REEs. This finding implies that the major rate controlling mechanism for heavy REEs leaching is diffusion.

Table 3. Correlation coefficients of diffusion-controlled kinetics models at different temperatures for light REEs.

Temperature		Diffusion 1–20 min			Diffusion 20–120 min		
°C	K	k	a	R^2	k	a	R^2
25	298	0.0002	0.0016	0.9975	0.0001	0.0034	0.9971
40	313	0.0005	0.0017	0.9963	0.0002	0.0115	0.9931
50	323	0.001	0.0042	0.9712	0.0003	0.0214	0.9542
60	333	0.0015	0.0068	0.9929	0.0004	0.0278	0.9928
75	348	0.0021	0.0198	0.9648	0.0005	0.0571	0.9888

Note: k denotes the slope of the regression line, and a denotes the intercept of the regression line.

Table 4. Correlation coefficients of diffusion-controlled kinetics models at different temperatures for heavy REEs.

Temperature		Diffusion 1–20 min			Diffusion 20–120 min		
°C	K	k	a	R^2	k	a	R^2
25	298	0.0002	0.0028	0.9593	0.00005	0.0049	0.9634
40	313	0.0004	0.0032	0.9943	0.00008	0.104	0.98
50	323	0.0005	0.0051	0.939	0.00009	0.0127	0.9704
60	333	0.0007	0.0068	0.9803	0.0002	0.016	0.9972
75	348	0.0008	0.013	0.941	0.0002	0.0251	0.9857

Note: k denotes the slope of the regression line, and a denotes the intercept of the regression line.

It was noted that the reaction kinetics was extremely fast within the first 1 min of the reaction. A possible explanation is that, due to the different modes of occurrence of REEs in coal-based material, the easy-to-leach REEs was instantaneously released at the beginning of the leaching process, and the hard-to-leach fraction controlled the reaction rate. The mode of occurrence of REEs can be categorized into different forms (i.e., ion exchange form, carbonate form, metal oxide form, acid soluble form, and insoluble form) with different levels of activation energy needed for extraction [24]. Zhang and Honaker (2020) studied the REE mode of occurrence in coal using the sequential chemical extraction method and quantified the REEs associated with each mode [30]. The REEs associated with ion exchangeable form and carbonates are likely to be released instantaneously at the beginning of the extraction process under the leaching conditions of the current study.

4. Conclusions

The Fire Clay coal is an excellent source for REE extraction due to its elevated REE contents and high leaching recovery values. Reducing the particle size of the middings material prior to de-carbonization resulted in the liberation of the associated micro-dispersed mineral matter that contained significantly higher concentrations of REEs. Decreasing the P80 particle size from 38 μm to 5 μm nearly doubled the concentration of REEs in the tailings of the flotation steps, which was the material used a feed for the leaching tests. The major minerals present in the sample were quartz, kaolinite, illite, and muscovite. Exponentially increasing the particle surface area through grinding resulted in higher solid dissolution which elevates the consumption of hydrogen ions and the concentration of contaminant metal ions.

The type of inorganic acid does not affect leaching recovery significantly but has an impact on the initial leaching rate. The mixing condition is sufficient at rotational speed values above 500 rpm.

Furthermore, solid concentration and acid concentration have a significant effect on leaching recovery of REEs. Scanning electron microscope (SEM) images showed no visible product layer or coating product on the particle surfaces. Therefore, the diffusion process in this reaction may be contributed by the interfacial transfer of the products and the reagent diffusion through the porous structure of solid particles.

The kinetic data obtained from leaching over a range of temperatures suggested that the leaching process follows the diffusion control shrinking core model. The activation energy determined from test data obtained over a range of temperatures using 1 M sulfuric acid was 36 kJ/mol for the first 20 min of reaction time and 27 kJ/mol for the leaching period between 20 and 120 min. The leaching of light REEs during the initial stage is much greater than the typical energy barrier of 20 kJ/mol, which indicates that a mixed control mechanism may occur as a result of several heterogenous materials leaching simultaneously. The energy required for later stage of leaching reduced significantly. The activation energy required for leaching the heavy REEs during the initial stage was significantly lower than that of the light REEs, which implies that the major rate controlling mechanism for heavy REE leaching is the diffusion.

Author Contributions: Conceptualization, X.Y. and R.Q.H.; methodology, X.Y. and R.Q.H.; software, X.Y.; validation, X.Y.; formal analysis, X.X.; investigation, X.Y. and R.Q.H.; resources, R.Q.H.; data curation, X.Y.; writing—original draft preparation, X.Y.; writing—review and editing, R.Q.H.; visualization, X.Y.; supervision, R.Q.H.; project administration, R.Q.H.; funding acquisition, R.Q.H. All authors have read and agreed to the published version of the manuscript.

Acknowledgments: This material is based upon work supported by the Department of Energy under Award Number DE-FE0029900. Disclaimer: This report was prepared as an account of work sponsored by an agency of the United States Government. Neither the United States Government nor any agency thereof, nor any of their employees, makes any warranty, express or implied, or assumes any legal liability or responsibility for the accuracy, completeness, or usefulness of any information, apparatus, product, or process disclosed, or represents that its use would not infringe privately owned rights. Reference herein to any specific commercial product, process, or service by trade name, trademark, manufacturer, or otherwise does not necessarily constitute or imply its endorsement, recommendation, or favoring by the United States Government or any agency thereof. The views and opinions of authors expressed herein do not necessarily state or reflect those of the United States Government or any agency thereof.

References

1. Kumari, A.; Panda, R.; Jha, M.K.; Lee, J.Y.; Kumar, J.R.; Kumar, V. Thermal treatment for the separation of phosphate and recovery of rare earth metals (REMs) from Korean monazite. *J. Ind. Eng. Chem.* **2015**, *21*, 696–703. [CrossRef]
2. Chi, R.; Zhang, X.; Zhu, G.; Zhou, Z.A.; Wu, Y.; Wang, C.; Yu, F. Recovery of rare earth from bastnasite by ammonium chloride roasting with fluorine deactivation. *Miner. Eng.* **2004**, *17*, 1037–1043. [CrossRef]
3. Jha, M.K.; Kumari, A.; Panda, R.; Rajesh Kumar, J.; Yoo, K.; Lee, J.Y. Review on hydrometallurgical recovery of rare earth metals. *Hydrometallurgy* **2016**, *165*, 2–26. [CrossRef]
4. Binnemans, K.; Jones, P.T.; Blanpain, B.; Van Gerven, T.; Yang, Y.; Walton, A.; Buchert, M. Recycling of rare earths: A critical review. *J. Clean. Prod.* **2013**, *51*, 1–22. [CrossRef]
5. Wu, G.; Li, J.; Xu, Z. Triboelectrostatic separation for granular plastic waste recycling: A review. *Waste Manag.* **2013**, *33*, 585–597. [CrossRef]
6. Peelman, S.; Sun, Z.H.I.; Sietsma, J.; Yang, Y. *Chapter 21—Leaching of Rare Earth Elements: Review of Past and Present Technologies*; Rare Earths Industry, Elsevier Inc.: Delft, The Netherlands, 2016; ISBN 9780128023280. [CrossRef]
7. Lin, R.; Bank, T.L.; Roth, E.A.; Granite, E.J.; Soong, Y. International Journal of Coal Geology Organic and inorganic associations of rare earth elements in central Appalachian coal. *Int. J. Coal Geol.* **2017**, *179*, 295–301. [CrossRef]

8. Lin, R.; Howard, B.H.; Roth, E.A.; Bank, T.L.; Granite, E.J.; Soong, Y. Enrichment of rare earth elements from coal and coal by-products by physical separations. *Fuel* **2017**, *200*, 506–520. [CrossRef]
9. Zhang, W.; Yang, X.; Honaker, R.Q. Association characteristic study and preliminary recovery investigation of rare earth elements from Fire Clay seam coal middlings. *Fuel* **2018**, *215*, 551–560. [CrossRef]
10. Honaker, R.Q.; Zhang, W.; Yang, X.; Rezaee, M. Conception of an integrated flowsheet for rare earth elements recovery from coal coarse refuse. *Miner. Eng.* **2018**, *122*, 233–240. [CrossRef]
11. Honaker, R.; Groppo, J.; Bhagavatula, A.; Rezaee, M.; Zhang, W. *Recovery of Rare Earth Minerals and Elements from Coal and Coal Byproducts*; International Coal Preparation Conference: Louisville, KY, USA, 2016.
12. Arbuzov, S.I.; Maslov, S.G.; Finkelman, R.B.; Mezhibor, A.M.; Ilenok, S.S.; Blokhin, M.G.; Peregudina, E.V. Modes of occurrence of rare earth elements in peat from Western Siberia. *J. Geochem. Explor.* **2018**, *184*, 40–48. [CrossRef]
13. Finkelman, R.B.; Palmer, C.A.; Wang, P. Quantification of the modes of occurrence of 42 elements in coal. *Int. J. Coal Geol.* **2018**, *185*, 138–160. [CrossRef]
14. Dai, S.; Graham, I.T.; Ward, C.R. A review of anomalous rare earth elements and yttrium in coal. *Int. J. Coal Geol.* **2016**, *159*, 82–95. [CrossRef]
15. Riley, K.W.; French, D.H.; Farrell, O.P.; Wood, R.A.; Huggins, F.E. Modes of occurrence of trace and minor elements in some Australian coals. *Int. J. Coal Geol.* **2012**, *94*, 214–224. [CrossRef]
16. Hower, J.C.; Ruppert, L.F.; Eble, C.F. Lanthanide, yttrium, and zirconium anomalies in the Fire Clay coal bed, Eastern Kentucky. *Int. J. Coal Geol.* **1999**, *39*, 141–153. [CrossRef]
17. Zhang, W.; Honaker, R.; Groppo, J. *Fundamental Study of the Monazite Calcite Flotation Separation*; Canadian Institute of Mining, Metallurgy and Petroleum: Montrel, QC, Canada, 2016.
18. Xiao, Y.F.; Chen, Y.Y.; Feng, Z.Y.; Huang, X.W.; Huang, L.; Long, Z.Q.; Cui, D.L. Leaching characteristics of ion-adsorption type rare earths ore with magnesium sulfate. *Trans. Nonferrous Met. Soc. China Engl. Ed.* **2015**, *25*, 3784–3790. [CrossRef]
19. Xiao, Y.; Feng, Z.; Huang, X.; Huang, L.; Chen, Y.; Liu, X.; Wang, L.; Zhiqi, L. Recovery of rare earth from the ion-adsorption type rare earths ore: II. Compound leaching. *Hydrometallurgy* **2016**, *163*, 83–90. [CrossRef]
20. Yu, J.; Guo, Z.; Tang, H. Dephosphorization Treatment of High Phosphorus Oolitic Iron Ore by Hydrometallurgical Process and Leaching Kinetics. *ISIJ Int.* **2013**, *53*, 2056–2064. [CrossRef]
21. Makanyire, T.; Jha, A.; Sutcliffe, S. Kinetics of hydrochloric acid leaching of niobium from TiO2 residues. *Int. J. Miner. Process.* **2016**, *157*, 1–6. [CrossRef]
22. Wang, Z.; Guo, S.; Ye, C. Leaching of copper from metal powders mechanically separated from waste printed circuit boards in chloride media using hydrogen peroxide as oxidant. *Procedia Environ. Sci.* **2016**, *31*, 917–924. [CrossRef]
23. Li, M.; Zhang, X.; Liu, Z.; Hu, Y.; Wang, M.; Liu, J.; Yang, J. Kinetics of leaching fluoride from mixed rare earth concentrate with hydrochloric acid and aluminum chloride. *Hydrometallurgy* **2013**, *140*, 71–76. [CrossRef]
24. Yang, X.; Werner, J.; Honaker, R.Q. Leaching of rare Earth elements from an Illinois basin coal source. *J. Rare Earths* **2019**, *37*, 312–321. [CrossRef]
25. Levenspiel, O. *Chemical Reaction Engineering-3rd edition*; John Wiley & Sons, Inc.: Hoboken, NJ, USA, 1999; Volume 38, ISBN 9780471530169.
26. Free, M.L. *Hydrometallurgy: Fundamentals and Applications*; John Wiley & Sons, Inc.: Hoboken, NJ, USA, 2013; ISBN 978-1-118-23077-0.
27. Salmi, T.; Grenman, H.; Warna, J.; Murzin, D.Y. New modelling approach to liquid-solid reaction kinetics: From ideal particles to real particles. *Chem. Eng. Res. Des.* **2013**, *91*, 1876–1889. [CrossRef]
28. Dickinson, C.F.; Heal, G.R. Solid–liquid diffusion controlled rate equations. *Thermochim. Acta* **1999**, *340–341*, 89–103. [CrossRef]
29. Li, M.; Wei, C.; Qiu, S.; Zhou, X.; Li, C.; Deng, Z. Kinetics of vanadium dissolution from black shale in pressure acid leaching. *Hydrometallurgy* **2010**, *104*, 193–200. [CrossRef]
30. Zhang, W.; Honaker, R. Characterization and recovery of rare earth elements and other critical metals (Co, Cr, Li, Mn, Sr, and V) from the calcination products of a coal refuse sample. *Fuel* **2020**, *267*, 117236. [CrossRef]

11

Grammatikopoulosite, NiVP, a New Phosphide from the Chromitite of the Othrys Ophiolite, Greece

Luca Bindi [1], Federica Zaccarini [2,*], Elena Ifandi [3,4], Basilios Tsikouras [3], Chris Stanley [5], Giorgio Garuti [2] and Daniela Mauro [6]

1 Dipartimento di Scienze della Terra, Università degli Studi di Firenze, I-50121 Florence, Italy; luca.bindi@unifi.it
2 Department of Applied Geological Sciences and Geophysics, University of Leoben, A-8700 Leoben, Austria; giorgio.garuti1945@gmail.com
3 Faculty of Science, Physical and Geological Sciences, Universiti Brunei Darussalam, BE 1410 Gadong, Brunei Darussalam; selena.21@windowslive.com (E.I.); basilios.tsikouras@ubd.edu.bn (B.T.)
4 Department of Geology, Section of Earth Materials, University of Patras, 265 00 Patras, Greece
5 Department of Earth Sciences, Natural History Museum, London SW7 5BD, UK; c.stanley@nhm.ac.uk
6 Dipartimento di Scienze della Terra, Università degli Studi di Pisa, I-56126 Pisa, Italy; daniela.mauro@dst.unipi.it
* Correspondence: federica.zaccarini@unileoben.ac.at

Abstract: Grammatikopoulosite, NiVP, is a new phosphide discovered in the podiform chromitite and hosted in the mantle sequence of the Othrys ophiolite complex, central Greece. The studied samples were collected from the abandoned chromium mine of Agios Stefanos. Grammatikopoulosite forms small crystals (from 5 μm up to about 80 μm) and occurs as isolated grains. It is associated with nickelphosphide, awaruite, tsikourasite, and an undetermined V-sulphide. It is brittle and has a metallic luster. In plane-polarized light, it is creamy-yellow, weakly bireflectant, with measurable but not discernible pleochroism and slight anisotropy with indeterminate rotation tints. Internal reflections were not observed. Reflectance values of mineral in air (R_1, R_2 in %) are: 48.8–50.30 at 470 nm, 50.5–53.5 at 546 nm, 51.7–55.2 at 589 nm, and 53.2–57.1 at 650 nm. Five spot analyses of grammatikopoulosite give the average composition: P 19.90, S 0.41, Ni 21.81, V 20.85, Co 16.46, Mo 16.39, Fe 3.83, and Si 0.14, total 99.79 wt %. The empirical formula of grammatikopoulosite—based on Σ(V + Ni + Co + Mo + Fe + Si) = 2 apfu, and taking into account the structural results—is $(Ni_{0.57}Co_{0.32}Fe_{0.11})_{\Sigma 1.00}(V_{0.63}Mo_{0.26}Co_{0.11})_{\Sigma 1.00}(P_{0.98}S_{0.02})_{\Sigma 1.00}$. The simplified formula is (Ni,Co)(V,Mo)P and the ideal formula is NiVP, which corresponds to Ni 41.74%, V 36.23%, P 22.03%, total 100 wt %. The density, calculated on the basis of the empirical formula and single-crystal data, is 7.085 g/cm^3. The mineral is orthorhombic, space group *Pnma*, with a = 5.8893(8), b = 3.5723(4), c = 6.8146(9) Å, V = 143.37(3) Å^3, and Z = 4. The mineral and its name have been approved by the Commission of New Minerals, Nomenclature and Classification of the International Mineralogical Association (IMA 2019-090). The mineral honors Tassos Grammatikopoulos, geoscientist at the SGS Canada Inc., for his contribution to the economic mineralogy and mineral deposits of Greece.

Keywords: grammatikopoulosite; phosphide; chromitite; Agios Stefanos mine; Othrys; ophiolite; Greece

1. Introduction

Natural phosphides are very rare phases, representing only 3% of the minerals approved by the International Mineralogical Association (IMA) ([1] and references therein). Most of these natural phosphides have been discovered and described in meteorites. Only recently have several new phosphides been discovered in terrestrial rocks, most of which are associated with the pyrometamorphic rocks of the Hatrurim Formation, Southern Levant [2]. Phosphides are documented in podiform chromitites hosted in the mantle sequence of ophiolite complexes of Greece and Russia [3–6] but are very rare. The phosphide found in the Russian chromitite was found in situ, in contact with serpentine of the altered interstitial silicates. The phosphides associated with the Greek chromitite were documented in heavy concentrates. Recently, the new mineral tsikourasite ($Mo_3Ni_2P_{1+x}$ ($x < 0.25$) was discovered in heavy mineral concentrates from a chromitite from the Othrys ophiolite (Greece)) [1]. Further investigation of the same mineral concentrates led to the discovery of a second new mineral. Quantitative chemical analysis and crystal structure proved that the studied phase is a new phosphide, characterized by the simplified formula (Ni,Co)(V,Mo)P and by the ideal formula NiVP, which corresponds to Ni 41.74%, V 36.23%, P 22.03%, total 100 wt %. The mineral is orthorhombic, space group *Pnma*, with $a = 5.8893(8)$, $b = 3.5723(4)$, $c = 6.8146(9)$ Å, and $Z = 4$. The mineral and its name have been approved by the Commission of New Minerals, Nomenclature and Classification of the International Mineralogical Association (IMA 2019-090). The mineral honors Tassos Grammatikopoulos (b. 1966), a geoscientist at the SGS Canada Inc., for his contribution to the investigation of economic mineralogy and mineral deposits of Greece. Holotype material is deposited in the Mineralogical Collection of the Museo di Storia Naturale, Università di Pisa, Via Roma 79, Calci (Pisa, Italy), under catalogue number 19,911.

2. Geological Background and Occurrence of Grammatikopoulosite

Grammatikopoulosite was discovered in a heavy mineral concentrate, which was prepared from chromitite specimens hosted in the mantle sequence of the Mesozoic Othrys ophiolite, central Greece (Figure 1A–C).

Othrys ophiolite is a complete but dismembered suite (Mirna Group) and consists of three structural units: the uppermost succession with variably serpentinized peridotites, which is structurally bounded by an ophiolite mélange; the intermediate Kournovon dolerite, including cumulate gabbro and local rhyolite; and the lower Sipetorrema Pillow Lava unit including also basaltic flows, siltstones, and chert. The Mirna Group constituted multiple inverted thrust sheets [7], which were eventually obducted onto the Pelagonian Zone during the Late Jurassic–Early Cretaceous [8–14]. Three types of basalts with different geochemical signatures have been described: (i) alkaline within-plate (WPB), (ii) normal-type mid-ocean ridge (N-MORB), and (iii) low-K tholeiite (L-KT). A biostratigraphic investigation indicated that radiolarites associated with N-MORB were deposited in the Middle and Late Triassic. Radiolarites deposited over the L-KT basalts are Early Carnian–Middle Norian–Late Norian in age [15]. N-MORB erupted during the Middle–Late Triassic period. The L-KT basalts erupted during the Middle–Late Triassic period, in part simultaneously with N-MORB, in the ocean–continent transition zone, close to the rifted continental margin. Finally, the alkaline WPB are interpreted to have formed in oceanic seamounts or in the ocean–continent transition zone adjacent to the rifted continental margin [15].

Electron microprobe analyses of the assemblages of these chromitites revealed that the studied spinel-supergroup minerals are magnesiochromites [1,5,6]. However, spinel mineral chemistry is rather heterogeneous with Cr_2O_3 (44.96–51.64 wt %), Al_2O_3 (14.18–20.78 wt %), MgO (13.34–16.84 wt %), and FeO (8.3–13.31 wt %). The calculated Fe_2O_3 ranges from 6.72 to 9.26 wt %. The amounts of MnO (0.33–0.60 wt %), V_2O_3 (0.04–0.30 wt %), ZnO (up to 0.07 wt %), and NiO (0.03–0.24 wt %) exhibit minor variations. The TiO_2 content is low (0.03–0.23 wt %), in agreement with the typical value reported for the mantle-hosted podiform chromitites. The Cr/(Cr + Al) ratios of the investigated magnesiochromite are lower than those of chromites formed in the supra-subduction zone (SSZ) and those related to boninitic melt [1,5,6].

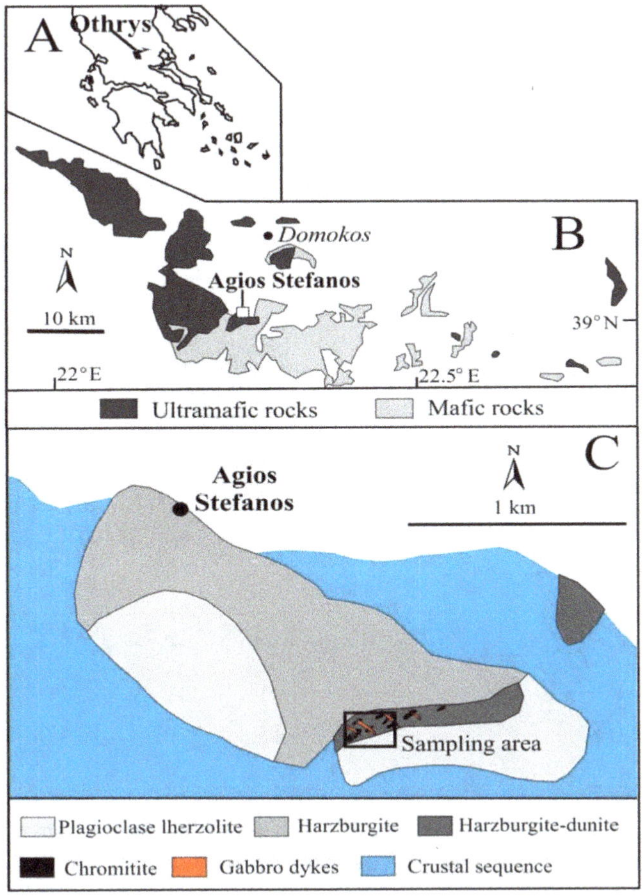

Figure 1. Location of the Othrys complex in Greece: (**A**) general geological map of the Othrys ophiolite showing the location of the Agios Stefanos chromium mine; (**B**) and (**C**) detailed geological setting of the Agios Stefanos area (modified after [5,10,16,17]).

3. Analytical Methods

The processing and recovery of the heavy minerals were carried out by treating about 10 kg of massive chromitite at SGS Mineral Services, Canada, following the procedure described by several authors [1,5,6,18]. Approximately 10 kg of chromitite were collected in the abandoned mine of Agios Stefanos. Subsequently, they were crushed and blended to generate a homogenous composite sample. About 500 g of the composite sample were riffled and stage crushed to a P80 (80% passing) of 75 µm. After, the sample was treated with heavy liquid (the density of the heavy liquids was 3.1 g/cm^3) to separate the heavy minerals. This procedure produced a heavy and light mineral concentrate. The heavy fraction was selected and processed with a superpanner. This method is designed for small amounts of sample and is closely controlled, leading to very effective separation. It consists of a tapering triangular deck with a "V" shape cross section. The table reproduces the concentrating action of a gold pan. Initially, the sample is swirled to stratify the minerals. Then, the heaviest minerals settle to the bottom and are deposited on the deck surface. The less dense material moves towards the top, overlying the heavy minerals. The operation of the deck is then changed to a rapid reciprocal motion, with an appropriate "end-knock" at the up-slope end of the board, and a steady flow of wash water is introduced. The "end-knock" forces the heavy minerals to migrate to the up-slope end of the deck. The wash water carries the light minerals to move to the narrower, down-slope end of the deck. The heaviest fractions were split into the heaviest fraction (tip) followed by a less dense fraction (middling). The "tip" and the "middlings" of the superpanner are the densest fractions and included liberated grains of chromite, sulphides, alloys, and phosphides, while the lighter tail consisted of a particle mixture of chromite and silicates. No source of contamination is likely during sample collection and subsequent treatment. The heavy minerals were prepared in epoxy blocks, and then polished for mineralogical examination. Quantitative chemical analyses and acquisition of back-scattered electron images of grammatikopoulosite were performed with a JEOL JXA-8200 electron microprobe, installed in the E. F. Stumpfl laboratory, Leoben University, Austria, operating in Wavelength Dispersive Spectrometry (WDS) mode. Major and minor elements were determined at a 20 kV accelerating voltage and a 10 nA beam current, with 20 s as the counting time for the peak and 10 s for the backgrounds. The beam diameter was about 1 μm in size. For the WDS analyses, the following lines and diffracting crystals were used: P, S, Si = ($K\alpha$, PETJ), V, Fe, Co, Ni = ($K\alpha$, LIFH), and Mo = ($L\alpha$, PETJ). The following standards were selected: synthetic Ni_3P for Ni and P, molybdenite for S and Mo, synthetic Fe_3P for Fe, synthetic metallic vanadium for V, skutterudite for Co, quartz for Si, and chromite for Cr. The ZAF correction method was applied. Automatic corrections were performed for interferences P-Mo, Cr-V, and Mo-S. Representative analyses of grammatikopoulosite are listed in Table 1.

Table 1. Chemical data (wt % of elements) for grammatikopoulosite.

Sample	P	S	Ni	V	Co	Mo	Fe	Si	Total
VP40-1	20.38	0.41	21.98	21.02	16.33	16.72	3.82	0.14	100.79
VP40-2	19.83	0.42	21.70	20.48	16.66	16.36	3.83	0.14	99.41
VP40-3	19.65	0.39	21.72	20.73	16.51	16.35	3.86	0.13	99.33
VP40-4	19.65	0.40	21.95	21.05	16.37	16.31	3.85	0.13	99.71
VP40-5	20.01	0.41	21.69	20.98	16.45	16.20	3.78	0.16	99.67
average	19.90	0.41	21.81	20.85	16.46	16.39	3.83	0.14	99.79

Single-crystal and powder X-ray diffraction data were collected at the University of Florence using a Bruker D8 Venture equipped with a Photon II CCD detector, with graphite-monochromatized Mo$K\alpha$ radiation (λ = 0.71073 Å) using a crystal fragment hand-picked from the polished section under a reflected light microscope (cif file, see Supplementary Materials). The crystal (about 80 μm in size) was carefully and repeatedly washed in acetone. It did not show any other visible phase attached to the surface. Single-crystal X-ray diffraction intensity data were integrated and corrected for standard Lorentz polarization factors with the software package *Apex3* [19,20].

The reflectance measurements on grammatikopoulosite were carried out using a WTiC standard and a J&M TIDAS diode array spectrophotometer at the Natural History Museum of London, UK.

4. Physical and Optical Properties

More than 30 grains of grammatikopoulosite were found in the studied polished sections. Grammatikopoulosite occurs as anhedral to subhedral grains. Most of them were less than 10 μm in size and only two of them were up to about 80 μm. Grammatikopoulosite consists of single (Figure 2A) or poly-phase grains associated with other minerals, such as tsikourasite, nickelphosphide, awaruite, and a V-sulphide, which likely represents another new mineral (Figure 2B,C). Compositions of these minerals have been reported in previous papers [5,6]. In plane-polarized light, grammatikopoulosite is creamy-yellow, weakly bireflectant, with measurable but not discernible pleochroism and slight anisotropy with indeterminate rotation tints. Internal reflections were not observed. Reflectance values of the mineral in air (R in %) are reported in Table 2 and in Figure 3. Density was not measured because of the small amount of available material and the presence of fine intergrowths with awaruite. The calculated density is equal to 7.085 g·cm^{-3}, based on the empirical composition and unit–cell volume refined from single-crystal XRD data.

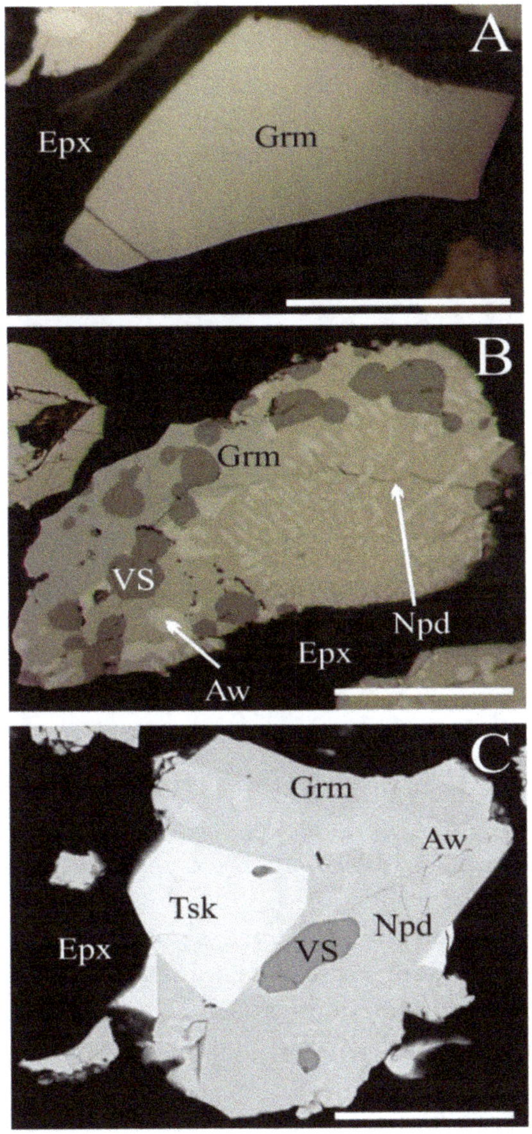

Figure 2. Digital image in reflected plane-polarized light (**A,B**) and a back-scattered electron image (**C**) showing grammatikopoulosite from the chromitite of Agios Stefanos. Abbreviations: Grm = grammatikopoulosite, Tsk = tsikourasite, VS = V-sulphide, Aw = awaruite, Npd = nickelphosphide, Epx = epoxy. Scale bar = 50 microns.

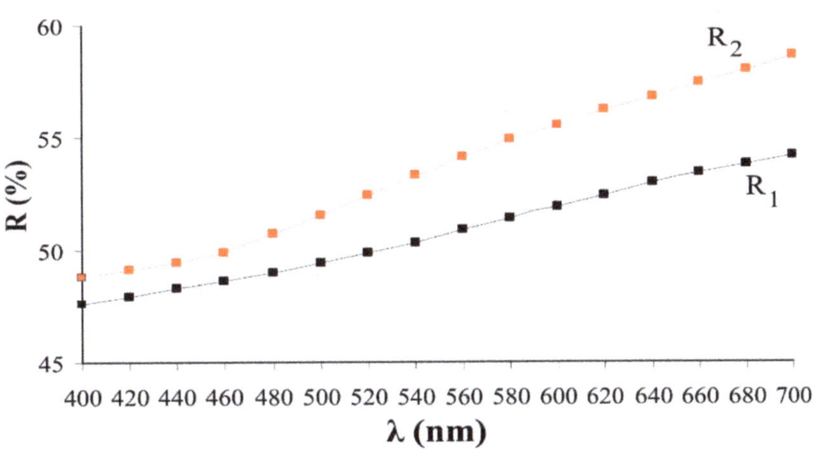

Figure 3. Reflectance data for grammatikopoulosite.

Table 2. Reflectance values for grammatikopoulosite. The values required by the Commission on Ore Mineralogy are given in bold.

λ nm	R_1	λ nm	R_2
400	47.6	400	48.8
420	47.9	420	49.1
440	48.3	440	49.4
460	48.6	470	49.9
470	**48.8**	**470**	**50.3**
480	49.0	480	50.7
500	49.4	500	51.5
520	49.9	520	52.4
540	50.3	540	53.3
546	**50.5**	**546**	**53.5**
560	50.9	560	54.1
580	51.4	580	54.9
589	**51.7**	**589**	**55.2**
600	51.9	600	55.5
620	52.4	620	56.2
640	53.0	640	56.8
650	**53.2**	**650**	**57.1**
680	53.8	680	58.0
700	54.2	700	58.6

5. Chemical Composition and X-Ray Crystallography

Chemical composition and X-ray data reveal that the empirical formula of grammatikopoulosite, based on Σ(V + Ni + Co + Mo + Fe + Si) = 2 apfu and taking into account the structural results (see below), is $(Ni_{0.57}Co_{0.32}Fe_{0.11})_{\Sigma 1.00}(V_{0.63}Mo_{0.26}Co_{0.11})_{\Sigma 1.00}(P_{0.98}S_{0.02})_{\Sigma 1.00}$. The simplified formula is (Ni,Co)(V,Mo)P and the ideal formula is NiVP, which corresponds to the composition of Ni = 41.74%, V = 36.23%, and P 22.03% (total 100 wt %).

A small grammatikopoulosite grain (about 80 μm in size) was handpicked from the polished section under a reflected light microscope and mounted on a 5 μm diameter carbon fiber, which was, in turn, attached to a glass rod in preparation for the single-crystal X-ray diffraction study. The fragment consists of crystalline grammatikopoulosite associated with minor, fine-grained polycrystalline awaruite. A total of 493 unique reflections were collected up to $2\theta = 80.33°$. The mineral is orthorhombic, space group *Pnma*, with $a = 5.8893(8)$, $b = 3.5723(4)$, $c = 6.8146(9)$ Å, and Z = 4. Given the similarity in unit–cell values and space groups, the structure was refined starting from the atomic coordinates reported for allabogdanite [21] using the software Shelxl-97 [22]. The site occupancy factor (s.o.f.) at the two cation sites was allowed to vary (Ni vs. Mo) using scattering curves for neutral atoms taken from the International Tables for Crystallography [23], leading to 27.5 and 28.4 e^- for *M*1 and *M*2 sites, respectively. The P site was found to be fully occupied (P vs. structural vacancy) by phosphorous (site scattering = 15.0 e^-) and fixed accordingly. Taking into account the site distribution observed in florenskyite [24], allabogdanite [21], and andreyivanovite [25], and in most Co_2Si-structure-type synthetic compounds [26], we assigned Ni, Fe, and Co to fill the *M*1 site (i.e., $Ni_{0.57}Co_{0.32}Fe_{0.11}$) and all the other elements (i.e., $V_{0.63}Mo_{0.26}Co_{0.11}$) to the *M*2 site. The mean electron numbers calculated with such a site distribution were identical (27.36 and 28.38 e^-) to those obtained in the refinement. For this reason, the subsequent cycles of refinement were run with the above constrained site populations, yielding an $R_1 = 0.0276$ for 465 reflections with $Fo > 4\sigma(Fo)$ and $R_1 = 0.0291$ for all the 493 independent reflections and 19 parameters. Refined atomic coordinates and isotropic displacement parameters are given in Table 3, whereas selected bond distances are reported in Table 4.

X-ray powder diffraction data for grammatikopoulosite (Table 5) were obtained with a Bruker D8 Venture equipped with a Photon II CCD detector, with graphite-monochromatized Cu$K\alpha$ radiation (λ = 1.54138 Å). The least squares refinement gave the following values: $a = 5.8088(2)$, $b = 3.5993(2)$, $c = 6.8221(3)$ Å, and $V = 142.634(8)$ Å^3. The calculated powder diffraction pattern obtained using the site occupancies and atomic coordinates is reported in Table 3.

Table 3. Atoms, site occupancies, atom coordinates, and isotropic displacement parameters (Å^2) for grammatikopoulosite.

Atom	Site Occupancy	x/a	y/b	z/c	U_{iso}
$M1$	$Ni_{0.57}Co_{0.32}Fe_{0.11}$	0.35709(6)	$\frac{1}{4}$	0.93703(5)	0.00578(10)
$M2$	$V_{0.63}Mo_{0.26}Co_{0.11}$	0.47087(6)	$\frac{1}{4}$	0.33109(5)	0.00595(9)
P	$P_{1.00}$	0.23639(12)	$\frac{1}{4}$	0.62449(10)	0.00547(13)

Table 4. Selected bond distances (Å) for grammatikopoulosite.

Atoms	Bond Distance
$M1$–P	2.2453(8)
$M1$–P (×2)	2.2639(5)
$M1$–P	2.2728(8)
$M1$–$M1$ (×2)	2.6000(6)
$M1$–$M2$ (×2)	2.7280(5)
$M1$–$M2$ (×2)	2.7487(5)
$M1$–$M2$	2.7677(5)
$M1$–$M2$	2.7696(5)
$M2$–P	2.4299(8)
$M2$–P (×2)	2.5008(6)
$M2$–P (×2)	2.5811(6)
$M2$–$M1$ (×2)	2.7280(5)
$M2$–$M1$ (×2)	2.7487(5)
$M2$–$M1$	2.7677(5)
$M2$–$M1$	2.7696(5)
$M2$–$M2$ (×2)	2.9339(6)

Table 5. Observed and calculated X-ray powder diffraction data (d in Å) for grammatikopoulosite. The strongest observed reflections are given in bold.

Indices	1		2	
hkl	d_{obs}	I_{obs}	d_{calc}	I_{calc}
101	4.43	10	4.4559	14
002	-	-	3.4073	5
102	**2.950**	**20**	**2.9493**	**19**
111	**2.785**	**25**	**2.7872**	**24**
201	2.699	5	2.7031	5
112	**2.273**	**60**	**2.2743**	**65**
210	2.269	10	2.2722	8
201	2.230	10	2.2279	9
211	**2.157**	**100**	**2.1555**	**100**
103	**2.118**	**25**	**2.1194**	**27**
013	1.915	15	1.9168	14
301	1.888	10	1.8864	12
113	1.824	15	1.8227	20
020	**1.784**	**20**	**1.7861**	**21**
004	1.702	10	1.7036	10
302	1.700	15	1.7010	22
213	1.608	10	1.6065	7
114	1.489	5	1.4878	6
303	1.482	5	1.4853	6
400	1.470	5	1.4723	6
123	1.367	10	1.3658	7
322	1.233	10	1.2318	9
314	1.211	5	1.2106	6
215	1.170	10	1.1688	9
511	1.102	5	1.1038	6
513	1.005	5	1.0035	5

1 = observed diffraction pattern; 2 = calculated diffraction pattern obtained with the atomic coordinates reported in Table 3 (only reflections with $I_{rel} \geq 4$ are listed).

6. Description of the Structure and Relations to Other Species

In the structure of grammatikopoulosite (Figure 4), *M*1 links four P atoms and eight *M*2 (Figure 5—left), whereas *M*2 links five P, six *M*1, and two *M*2 (Figure 5—right). The metal–phosphorous distances are much shorter in the *M*1 coordination sphere than in that of *M*2 (Table 4). Interestingly, if only the *M*–P distances are considered in the coordination polyhedra of the *M* atoms, $M1P_4$ tetrahedra (Figure 6) forming corner-sharing chains along the **b**-axis or $M2P_5$ square pyramids forming zig-zag chains along the **a**-axis (Figure 7) can be observed.

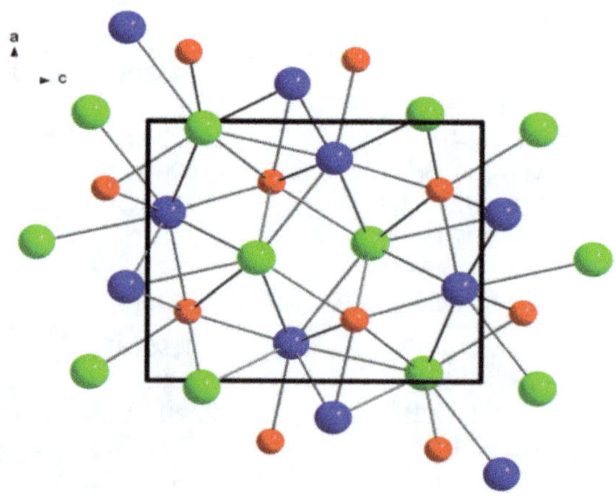

Figure 4. The crystal structure of grammatikopoulosite projected down [10]. Blue, green, and red circles refer to *M*1, *M*2, and P, respectively.

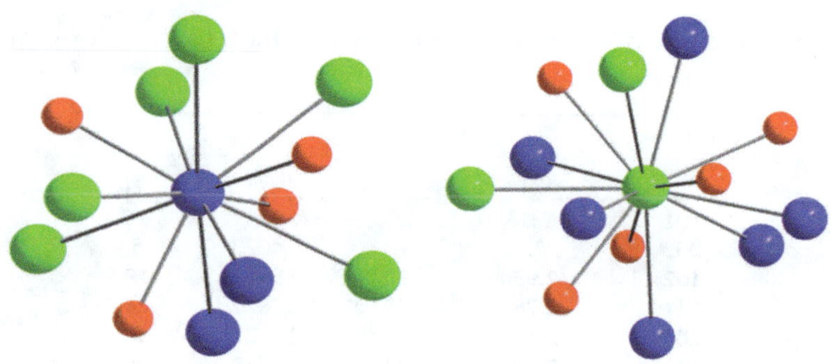

Figure 5. Coordination environment of *M*1 (left) and *M*2 (right) sites in the crystal structure of grammatikopoulosite. Colors and symbols as in Figure 4.

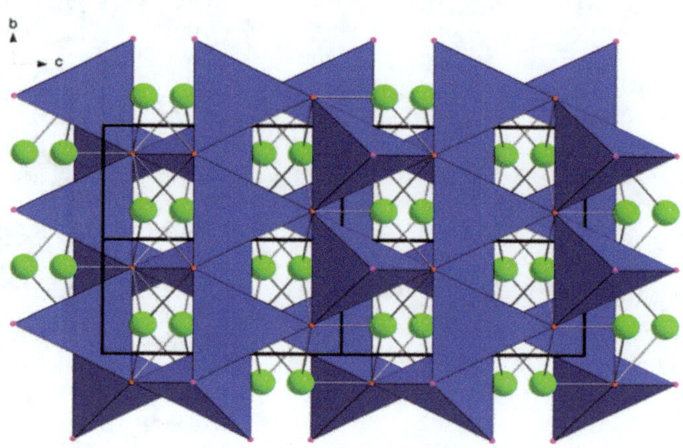

Figure 6. Polyhedral representation of the $M1P_4$ tetrahedra in the crystal structure of grammatikopoulosite. Colors and symbols as in Figure 4.

Figure 7. Polyhedral representation of the $M2P_5$ distorted pyramids in the crystal structure of grammatikopoulosite. Colors and symbols as in Figure 4.

It was reported that the departure from the hexagonal, high-temperature *P*-62*m* structure (barringerite structure) was linked to some ordering between the metals at the *M* sites. This was not immediately evident in both allabogdanite [21] and andreyivanovite [25] because of the close site scattering of Fe, Ni, and Cr. On the contrary, the ordering is evident if we take into consideration florenskyite [24] and grammatikopoulosite with ideal formulae FeTiP and NiVP, respectively. Ti and V preferentially enter the larger *M*2 site in agreement with their larger metallic radius (1.47 and 1.35 Å, respectively [27]).

Grammatikopoulosite does not correspond to any valid or invalid unnamed mineral [28] and it belongs to the group of natural phosphides with the Co_2Si orthorhombic *Pnma* structure, i.e., florenskyite [24], allabogdanite [21], and andreyivanovite [25].

7. Discussion and Genetical Implications

Grammatikopoulosite, similar to tsikourasite, is natural in origin despite the fact that both phases were found in heavy mineral concentrates and not in situ [1]. This assumption is supported by the following observations: during the sampling, the only tool used to collect the chromitite was a steel hammer; the concentrates were obtained in a laboratory that uses a grinder manufactured by TM Engineering, with media Alloy 1 composed of Cr and Mo, and does not contain Ni, V, Co, Fe, or P; the used material is very hard and is recognized to be best for applications where abrasion and impact is the norm; and the chemical composition of grammatikopoulosite indicates crystallization under reducing conditions, similar to what has been previously proposed for other phosphides found in ophiolitic chromitites [1,3–6]. According to literature data, the crystallization temperature of most of the terrestrial phosphides is generally comprised between 700 °C and 1150 °C [4]. However, a conclusive model to explain the origin of grammatikopoulosite cannot be provided yet, since it may have a terrestrial origin or represent fragments of a meteorite. The following models, which imply the crystallization in a terrestrial environment from low to high temperatures, can be suggested for the origin of grammatikopoulosite: (i) alteration of chromitite during sub-oceanic or on-land serpentinization at temperatures between 400 °C and 150 °C, which is a typical reducing process [29]; (ii) reaction of the chromitite with high-temperature reducing fluids in the mantle; and (iii) interaction of the chromitite and their host serpentinized peridotites with a surface lightning strike. Phosphide minerals occurring in ophiolitic chromitites in Othrys and Gerakini–Ormylia (Greece), as well as in Alapaevsk (Russia), are closely associated with highly reducing phases such as awaruite, typically forming during serpentinization [1,3–6]. Likewise, grammatikopoulosite is also associated with awaruite and other reducing phases, likely suggesting a low-temperature origin during serpentinization. This hypothesis is strongly supported by the discovery of a Ni-phosphide in situ in the serpentine-rich matrix of the Alapaevsk chromitite of Russia [3]. Recently, Etiope et al. [30,31] have shown that the Agios Stefanos chromitite is the source rock of abiotic CH_4 measured in the west Othrys springs. These authors argued that CH_4 formation took place at temperatures below 150 °C via Sabatier reaction during oceanic and on-land serpentinization. This observation provides further evidence that reducing conditions can be achieved at a low temperature during the alteration of mantle-derived rocks. Alternatively, Xiong et al. [32] suggested that reducing conditions can be achieved in mantle rocks, including chromitites, in the shallow lithosphere due to the interaction of mantle-derived fluids enriched in CH_4 and H_2 with basaltic melt. However, no phosphides are reported in the ultra-reduced, high-temperature mineralogical assemblage described by Xiong et al. [32]. Natural phosphides and other ultra-reduced minerals have been rarely described in few fulgurites [33,34] but they are very common accessory phases in meteorites [21,24,25,35–39] or. The rare and localized occurrence of the Othrys phosphides may provide some support to the two last models. However, the probability to have collected a fragment of a meteorite or a fulgurite in the Othrys ophiolite during the sampling of the studied chromitite seems improbable, and for the discrimination of the meteoritic materials, a more detailed isotopic study may be needed. To conclude, an indisputable genetic model to explain how grammatikopoulosite is crystallized still needs further investigation.

Author Contributions: L.B. and F.Z. wrote the manuscript; F.Z. and L.B. performed the chemical analyses and the diffraction experiments, respectively; E.I. and B.T. provided the concentrate sample and information on the sample provenance and petrography of Othrys chromitite; G.G. and D.M. discussed the chemical data; and C.S. obtained the optical data. All authors provided support in the data interpretation and revised the manuscript. All authors have read and agreed to the published version of the manuscript.

Acknowledgments: The authors acknowledge Ritsuro Miyawaki, Chairman of the CNMNC and its members for helpful comments on the submitted new mineral proposal. We are very grateful to the four anonymous referees for constructive comments that improved the manuscript. Many thanks are due to the editorial staff of *Minerals* and to the consulted referees. This paper is dedicated to the memory of our friend and colleague Demetrios.

References

1. Zaccarini, F.; Bindi, L.; Ifandi, E.; Grammatikopoulos, T.; Stanley, C.; Garuti, G.; Mauro, D. Tsikourasite, $Mo_3Ni_2P_{1+x}$ ($x < 0.25$), a new phosphide from the chromitite of the othrys ophiolite, Greece. *Minerals* **2019**, *9*.

2. Britvin, S.N.; Murashko, M.N.; Vapnik, Y.; Polekhovsky, Y.S.; Krivovichev, S.V. Earth's phosphides in levant and insights into the source of Archean prebiotic phosphorus. *Sci. Rep.* **2015**, *5*. [CrossRef]

3. Zaccarini, F.; Pushkarev, E.; Garuti, G.; Kazakov, I. Platinum-group minerals and other accessory phases in chromite deposits of the Alapaevsk ophiolite, Central Urals, Russia. *Minerals* **2016**, *6*, 108. [CrossRef]

4. Sideridis, A.; Zaccarini, F.; Grammatikopoulos, T.; Tsitsanis, P.; Tsikouras, B.; Pushkarev, E.; Garuti, G.; Hatzipanagiotou, K. First occurrences of Ni-phosphides in chromitites from the ophiolite complexes of Alapaevsk, Russia and GerakiniOrmylia, Greece. *Ofioliti* **2018**, *43*, 75–84.

5. Ifandi, E.; Zaccarini, F.; Tsikouras, B.; Grammatikopoulos, T.; Garuti, G.; Karipi, S.; Hatzipanagiotou, K. First occurrences of Ni-V-Co phosphides in chromitite of Agios Stefanos mine, Othrys ophiolite, Greece. *Ofioliti* **2018**, *43*, 131–145.

6. Zaccarini, F.; Ifandi, E.; Tsikouras, B.; Grammatikopoulos, T.; Garuti, G.; Mauro, D.; Bindi, L.; Stanley, C. Occurrences of of new phosphides and sulfide of Ni, Co, V, and Mo from chromitite of the Othrys ophiolite complex (Central Greece). *Per. Ital. Mineral.* **2019**, 88. [CrossRef]

7. Smith, A.G.; Rassios, A. The evolution of ideas for the origin and emplacement of the western Hellenic ophiolites. *Geol. Soc. Am. Spec. Pap.* **2003**, *373*, 337–350.
8. Hynes, A.J.; Nisbet, E.G.; Smith, G.A.; Welland, M.J.P.; Rex, D.C. Spreading and emplacement ages of some ophiolites in the Othris region (eastern central Greece). *Z. Deutsch Geol. Ges.* **1972**, *123*, 455–468.
9. Smith, A.G.; Hynes, A.J.; Menzies, M.; Nisbet, E.G.; Price, I.; Welland, M.J.; Ferrière, J. The stratigraphy of the Othris Mountains, eastern central Greece: A deformed Mesozoic continental margin sequence. *Eclogue Geol. Helv.* **1975**, *68*, 463–481.
10. Rassios, A.; Smith, A.G. Constraints on the formation and emplacement age of western Greek ophiolites (Vourinos, Pindos, and Othris) inferred from deformation structures in peridotites. In *Ophiolites and Oceanic Crust: New Insights from Field Studies and the Ocean Drilling Program*; Dilek, Y., Moores, E., Eds.; Geological Society of America: Boulder, CO, USA, 2001; pp. 473–484.
11. Barth, M.G.; Mason, P.R.D.; Davies, G.R.; Drury, M.R. The Othris Ophiolite, Greece: A snapshot of subduction initiation at a mid-ocean ridge. *Lithos* **2008**, *100*, 234–254. [CrossRef]
12. Barth, M.; Gluhak, T. Geochemistry and tectonic setting of mafic rocks from the Othris Ophiolite, Greece. *Contrib. Mineral. Petrol.* **2009**, *157*, 23–40. [CrossRef]
13. Dijkstra, A.H.; Barth, M.G.; Drury, M.R.; Mason, P.R.D.; Vissers, R.L.M. Diffuse porous melt flow and melt-rock reaction in the mantle lithosphere at a slow-spreading ridge: A structural petrology and LA-ICP-MS study of the Othris Peridotite Massif (Greece). *Geochem. Geophys. Geosyst.* **2003**, *4*. [CrossRef]
14. Magganas, A.; Koutsovitis, P. Composition, melting and evolution of the upper mantle beneath the Jurassic Pindos ocean inferred by ophiolitic ultramafic rocks in East Othris, Greece. *Int. J. Earth Sci.* **2015**, *104*, 1185–1207. [CrossRef]
15. Bortolotti, V.; Chiari, M.; Marcucci, M.; Photiades, A.; Principi, G.; Saccani, E. New geochemical and age data on the ophiolites from the Othrys area (Greece): Implication for the Triassic evolution of the Vardar ocean. *Ofioliti* **2008**, *33*, 135–151.
16. Economou, M.; Dimou, E.; Economou, G.; Migiros, G.; Vacondios, I.; Grivas, E.; Rassios, A.; Dabitzias, S. Chromite deposits of Greece. In *Chromites, UNESCO's IGCP197 Project Metallogeny of Ophiolites*; Petrascheck, W., Karamata, S., Eds.; Theophrastus Publ. S.A.: Athens, Greece, 1986; pp. 129–159.
17. Garuti, G.; Zaccarini, F.; Economou-Eliopoulos, M. Paragenesis and composition of laurite from chromitites of Othrys (Greece): Implications for Os-Ru fractionation in ophiolite upper mantle of the Balkan Peninsula. *Mineral. Depos.* **1999**, *34*, 312–319. [CrossRef]
18. Tsikouras, B.; Ifandi, E.; Karipi, S.; Grammatikopoulos, T.A.; Hatzipanagiotou, K. Investigation of platinum-group minerals (PGM) from Othrys chromitites (Greece) using superpanning concentrates. *Minerals* **2016**, *6*, 94. [CrossRef]
19. Bruker. *APEX3*; Bruker AXS Inc.: Madison, WI, USA, 2016; Available online: https://www.bruker.com/products/x-ray-diffraction-and-elemental-analysis/single-crystal-x-ray-diffraction/sc-xrd-software/apex3.html (accessed on 31 January 2020).
20. Bruker. *SAINT and SADABS*; Bruker AXS Inc.: Madison, WI, USA, 2016. Available online: https://www.bruker.com/products/x-ray-diffraction-and-elemental-analysis/single-crystal-x-ray-diffraction/sc-xrd-software/apex3.html (accessed on 31 January 2020).
21. Britvin, S.N.; Rudashevskii, N.S.; Krivovichev, S.V.; Burns, P.C.; Polekhovsky, Y.S. Allabogdanite, $(Fe,Ni)_2P$, a new mineral from the Onello meteorite: The occurrence and crystal structure. *Am. Mineral.* **2002**, *87*, 1245–1249. [CrossRef]
22. Sheldrick, G.M. A short history of SHELX. *Acta Crystallogr.* **2008**, *A64*, 112–122. [CrossRef]
23. Wilson, A.J.C. *International Tables for Crystallography: Mathematical, Physical, and Chemical Tables*; International Union of Crystallography: Chester, UK, 1992; Volume 3.
24. Ivanov, A.V.; Zolensky, M.E.; Saito, A.; Ohsumi, K.; Yang, S.V.; Kononkova, N.N.; Mikouchi, T. Florenskyite, FeTiP, a new phosphide from the Kaidun meteorite, Locality: Kaidun chondritic meteorite, South Yemen. *Am. Mineral.* **2000**, *85*, 1082–1086. [CrossRef]
25. Zolensky, M.; Gounelle, M.; Mikouchi, T.; Ohsumi, K.; Le, L.; Hagiya, K.; Tachikawa, O. Andreyivanovite: A second new phosphide from the Kaidun meteorite. *Am. Mineral.* **2008**, *93*, 1295–1299. [CrossRef]
26. Fruchart, R.; Roger, A.; Sénateur, J.P. Crystallographic and magnetic properties of solid solutions of the phosphides M2P, M = Cr, Mn, Fe, Co, and N. *J. Appl. Phys.* **1969**, *40*, 1250–1257. [CrossRef]
27. Wells, A.F. *Structural Inorganic Chemistry*, 5th ed.; Clarendon Press: Oxford, UK, 1984; p. 1288.

28. Smith, D.G.W.; Nickel, E.H. A system for codification for unnamed minerals: Report of the subcommittee for unnamed minerals of the IMA commission on new minerals, nomenclature and classification. *Can. Mineral.* **2007**, *45*, 983–1055. [CrossRef]
29. Malvoisin, B.; Chopin, C.; Brunet, F.; Matthieu, E.; Galvez, M.E. Low-temperature Wollastonite formed by carbonate reduction: A marker of serpentinite redox conditions. *J. Petrol.* **2012**, *53*, 159–176. [CrossRef]
30. Etiope, G.; Tsikouras, B.; Kordella, S.; Ifandi, E.; Christodoulou, D.; Papatheodorou, G. Methane flux and origin in the Othrys ophiolite hyperalkaline springs, Greece. *Chem. Geol.* **2013**, *347*, 161–174. [CrossRef]
31. Etiope, G.; Ifandi, E.; Nazzari, M.; Procesi, M.; Tsikouras, B.; Ventura, G.; Steele, A.; Tardini, R.; Szatmari, P. Widespread abiotic methane in chromitites. *Sci. Rep.* **2018**, *8*, 8728. [CrossRef]
32. Xiong, Q.; Griffin, W.L.; Huang, J.X.; Gain, S.E.M.; Toledo, V.; Pearson, N.J.; O'Reilly, S.Y. Super-reduced mineral assemblages in "ophiolitic" chromitites and peridotites: The view from Mount Carmel. *Eur. J. Mineral.* **2017**, *29*, 557–570. [CrossRef]
33. Pasek, M.A.; Hammeijer, J.P.; Buick, R.; Gull, M.; Atlas, Z. Evidence for reactive reduced phosphorus species in the early Archean ocean. *Proc. Natural Acad. Sci. USA* **2013**, *110*, 100089–100094. [CrossRef]
34. Ballhaus, C.; Wirth, R.; Fonseca, R.O.C.; Blanchard, H.; Pröll, W.; Bragagni, A.; Nagel, T.; Schreiber, A.; Dittrich, S.; Thome, V.; et al. Ultra-high pressure and ultra-reduced minerals in ophiolites may form by lightning strikes. *Geochem. Perspec. Lett.* **2017**, *5*, 42–46. [CrossRef]
35. Buseck, P.R. Phosphide from meteorites: Barringerite, a new iron-nickel mineral. *Science* **1969**, *165*, 169–171. [CrossRef]
36. Britvin, S.N.; Kolomensky, V.D.; Boldyreva, M.M.; Bogdanova, A.N.; Krester, Y.L.; Boldyreva, O.N.; Rudashevsky, N.S. Nickelphosphide $(Ni,Fe)_3P$—The nickel analogue of schreibersite. *Zap. Vserossi. Mineral. Obschch.* **1999**, *128*, 64–72.
37. Ma, C.; Beckett, J.R.; Rossman, G.R. Monipite, MoNiP, a new phosphide mineral in a Ca-Al-rich inclusion from the Allende meteorite. *Am. Mineral.* **2014**, *99*, 198–205. [CrossRef]
38. Pratesi, G.; Bindi, L.; Moggi-Cecchi, V. Icosahedral coordination of phosphorus in the crystal structure of melliniite, a new phosphide mineral from the Northwest Africa 1054 acapulcoite. *Am. Mineral.* **2006**, *91*, 451–454. [CrossRef]
39. Skala, R.; Cisarova, I. Crystal structure of meteoritic schreibersites: Determination of absolute structure. *Phys. Chem. Mineral.* **2005**, *31*, 721–732. [CrossRef]

12

Effect of Sulfuric Acid Baking and Caustic Digestion on Enhancing the Recovery of Rare Earth Elements from a Refractory Ore

Rina Kim [1,*], Heechan Cho [2], Jinan Jeong [2], Jihye Kim [3], Sugyeong Lee [4], Kyeong Woo Chung [1], Ho-Sung Yoon [1] and Chul-Joo Kim [1]

1 Resources Recovery Research Center, Mineral Resources Division, Korea Institute of Geoscience and Mineral Resources (KIGAM), Daejeon 34132, Korea; case7@kigam.re.kr (K.W.C.); hsyoon@kigam.re.kr (H.-S.Y.); cjkim@kigam.re.kr (C.-J.K.)

2 Department of Energy Resources Engineering, Seoul National University, Seoul 08826, Korea; hccho@snu.ac.kr (H.C.); krenos@snu.ac.kr (J.J.)

3 Department of Chemical Engineering and Applied Chemistry, University of Toronto, Toronto, ON M5S 3E5, Canada; jhye.kim@mail.utoronto.ca

4 Robert M. Buchan Department of Mining, Queen's University, Kingston, ON K7L 3N6, Canada; sugyeong.lee@queensu.ca

* Correspondence: rkim@kigam.re.kr

Abstract: To improve the recovery of rare earth elements (REEs) from a refractory ore, this study investigated two different chemical decomposition methods, namely sulfuric acid baking and caustic digestion, with their respective leaching processes. The studied lateritic ore contained goethite (FeOOH) as a major constituent with REEs scattered around and forming submicron grains of phosphate minerals, such as apatite and monazite. Therefore, despite the substantially high content of REEs (3.4% total rare earth oxide), the normal acidic leaching efficiency of REEs reached only 60–70%. By introducing sulfuric acid baking and caustic digestion, the REE-leaching efficiency was significantly improved. After sulfuric acid baking at 2.0 acid/solid ratio and 200 °C for 2 h, the leaching efficiency reached 97–100% in the subsequent water-leaching. When the ore was digested with a solid/liquid ratio of 100 g/L in a 30 wt% NaOH solution at 115 °C and 300 rpm for 3 h, the REE-leaching efficiency of 99–100% was attained at 80 °C using a 3.0 M HCl solution. The correlation between the REE and the Fe-leaching was determined. The improvements in REE-leaching in both methods were mostly attributed to the mineral phase and crystallinity changes of Fe-bearing minerals due to the ore pretreatments. Such findings were also supported by X-ray diffraction and scanning electron microscopy analyses.

Keywords: REE; sulfuric acid baking; caustic digestion; acid leaching; water leaching

---

1. Introduction

The rare earth elements (REEs) comprise 17 elements in the periodic table, namely the 15 lanthanide elements (La–Lu) plus Sc and Y. These 17 elements are generally divided into two groups: La–Eu (Nos. 57–63) are light REEs (LREEs), whereas Gd–Lu (Nos. 64–71) and Y (No. 39) are heavy REEs (HREEs). Sc is not classified in any of the two groups. The REEs are used in various high-tech industries, e.g., the manufacturing of computers, cell phones, optical glasses, phosphors, hybrid cars and batteries, because of their unique electrical and magnetic properties [1–4].

REEs can be recovered from different minerals. More than 250 rare earth minerals have been discovered to date [1], and the representative minerals are monazite ($REEPO_4$), bastnasite ($REECO_3F$) and xenotime (YPO_4). Monazite is an REE phosphate mineral that is very stable and difficult to leach under mild conditions. In contrast, bastnasite is an REE fluorocarbonate. When bastnasite is used as an REE resource, the fluorocarbonate should be converted first into an oxide or a sulfate to facilitate the succeeding leaching process. Xenotime is an yttrium phosphate mineral, which also has a very stable form; thus, similar to monazite, it is also difficult to leach under mild conditions. Other REE minerals include apatite, cheralite, eudialyte, loparite, phosphorite, secondary monazite and spent uranium solutions [4]. REEs are also found in ion-adsorbed clays; they are physically adsorbed onto the clays and are mostly HREEs [5–7].

Although there are several REE minerals in nature, only monazite, bastnasite and xenotime are commercially utilized [8]. General processing methods have been developed for these well-known REE minerals: acid baking (mostly by sulfuric acid) followed by water-leaching can be applied to treat them [8–13]. The REE phosphate minerals, i.e., monazite and xenotime, can also be effectively pretreated using a NaOH solution, and then, the converted REE hydroxide compounds are mostly leached in a weak acid solution [8,12,14,15]. Bastnasite can be leached by hydrochloric or sulfuric acid after a roasting process to decompose fluorine and carbonate [12]. The REEs from ion-adsorbed clays can be leached using an ammonium sulfate solution, the mechanism of which is similar to that of the ion exchange process [5–7].

However, the described methods areeffective only when the minerals arerecovered as high-grade concentrates and when the REE minerals exist as major minerals in the ores. Furthermore, the minerals contained within an ore vary depending on the mine; therefore, different processing methods need to be applied based on the ore type. Some researchers studied the leaching behavior of apatite containing considerable quantities of REEs. Habashi [16] reviewed the REE recovery from Kola apatite, and Jorjani et al. [17] conducted a leaching study using an apatite concentrate from the Chadormalu Plant in Iran. For the two studies, nitric acid was used as the leaching agent to prevent gypsum production. In addition, REEs in apatite can generally be recovered as a byproduct from calcium sulfate sludge from the wet processing of phosphoric acid [18,19]. REEs contained in apatite can also be treated by a two-step leaching method using hydrochloric acid as a leaching agent [20]. In the first step, approximately 80% of Ca and 60% of P are removed with almost no REEs loss, and the REEs are recovered at the second step with less impurities. Several metallurgical studies of refractory REE ores have also been conducted. In Brazil, Neumann and Medeiros [21] tried to separate the REE minerals from Araxá REE ore by physical methods; however, they decided to omit physical separation due to the complex mineralogy, fine REE crystals and particles having strong intergrowth with the gangue mineral. In a pilot plant, Araxá REE ore was introduced into direct leaching after sulfuric acid baking [15]. The REE ores in Chuktukon deposits from Russia showed similar mineralogical properties as the Araxá ores and direct chemical processing was also considered, i.e., nitric acid-leaching in an autoclave [22]. The processes mentioned above are summarized in Table 1.

As indicated in Table 1, appropriate processing methods based on the ore properties have to be developed to handle gangue minerals as well as REE minerals. This study investigated the enhancement in REE recovery from a refractory ore containing 3.4% total rare earth oxide (TREO) by applying two different chemical decomposition methods followed by leaching processes, i.e., sulfuric acid baking—water-leaching and caustic digestion–acid-leaching.

Table 1. Rare earth elements (REEs) recovery metallurgical processes.

REE Mineral Type	Origin	Method	REE-Leaching Efficiency	Reference
Monazite	Unknown	H_2SO_4 baking (180–250 °C)/ Water-leaching (70 °C)	85–100%	Berry et al. [9]
Monazite	Australia	H_2SO_4 baking (250 °C)/ Acid-leaching (0.9 M H_2SO_4, 20–25 °C)	96%	Demol et al. [10]
Red mud	Greece	H_2SO_4 baking (700 °C)/ Water-leaching (room temperature)	>80% (Sc 60%)	Borra et al. [11]
Apatite residue	Iran	H_2SO_4 baking (270 °C)/ Water-leaching	95%	Soltani et al. [13]
Bastnasite, monazite	China	Caustic digestion ($NaOH + Ca(OH)_2$ roasting, 700 °C)/ Acid washing (0.5 M HCl + 0.05 M citric acid, 40 °C)/ Acid-leaching (3 M HCl + 0.5 M $AlCl_3$, 70 °C)	86%	Huang et al. [14]
Monazite, xenotime, bastnasite	Kyrgyz Republic	Caustic digestion (Na_2CO_3 fusion)/Acid-leaching (HNO_3)	–	Buchanan et al. [15]
Apatite, allanite, chevkinite, titanite	Turkey	Caustic digestion (600 °C)/ Water–HCl-leaching	–	Buchanan et al. [15]
Ion-adsorbed clays	Unknown	Acidic-leaching (0.5 M $(NH_4)_2SO_4$, pH 3 by H_2SO_4, 25 °C)	80–90%	Moldoveanu and Papangelakis [7]
Apatite	Iran	Acid-leaching (60% HNO_3, 60 °C)	59–74%	Jorjani et al. [17]
Apatite	South Africa	Acid-leaching (1.0 M HNO_3, 0.5 M $Ca(NO_3)_2$)	85%	Preston et al. [18]
Apatite	Mongolia	Two-step acid-leaching (1st: 1.0 M HCl/2nd: 2.0 M HCl)	92–94%	Kim et al. [20]

2. Materials and Methods

2.1. Ore Characteristics

The ore used in this study was a lateritic ore containing goethite as a major mineral. Figure 1 shows an X-ray diffraction (XRD) analysis of the REE ore, where the goethite, ettringite, muscovite, maghemite, and apatite peaks can be observed. However, no REE mineral peaks appeared because their concentrations were much lower than those of Fe-bearing minerals, and the REEs were scattered throughout the ore. REE-containing particles were observed through a scanning electron microscopy (SEM) analysis, as displayed in Figure 2. In the subsequent energy-dispersive X-ray (EDX) analysis (Figure 3), Ce oxide and mixed REE grains were observed, and some REEs were likely to substitute Ca in the apatite lattice. In addition, considering the high intensity of REEs and P in the mixed REE particle presented in Figure 3c, the presence of an REE phosphate mineral, i.e., monazite, was also expected.

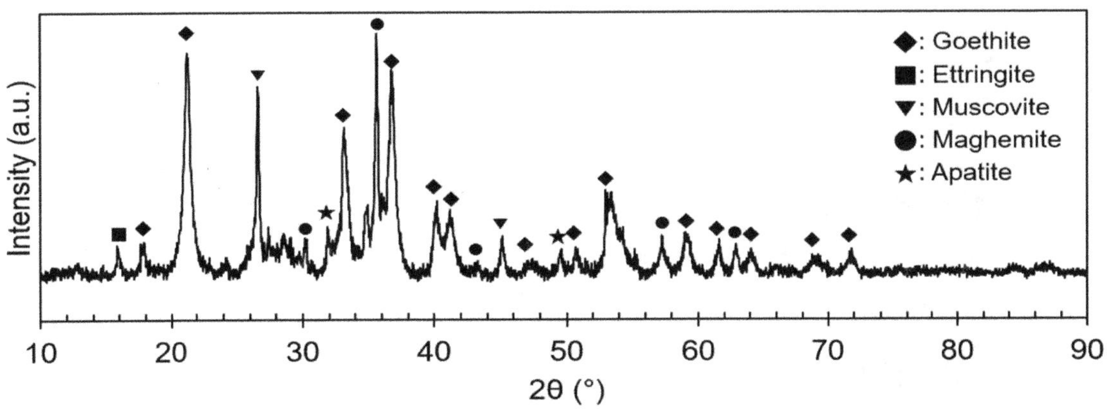

Figure 1. XRD pattern of the REE ore.

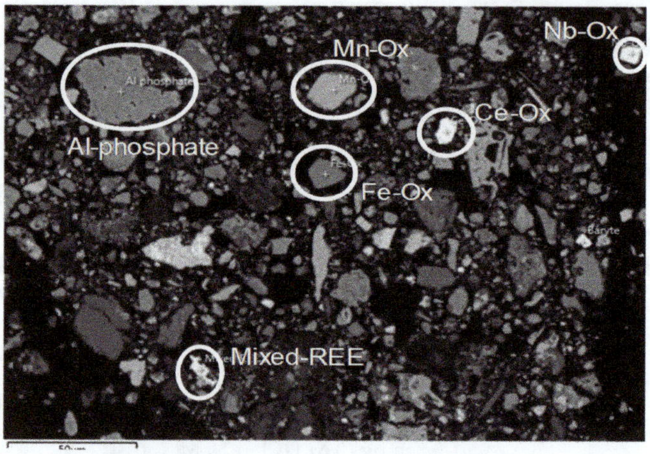

Figure 2. SEM image of the REE ore.

Because the ore was lateritic, the iron content was significantly high (more than 50% as Fe_2O_3) and 10% of manganese was present as an oxide as presented in Table 2. The TREO grade was 3.4%, which is relatively high as an REE ore. Ce was the most abundant element, followed by La, Nd and Y (Table 2). The ore was obtained as a ground material, and the 80% passing size was 30 μm (Figure 4).

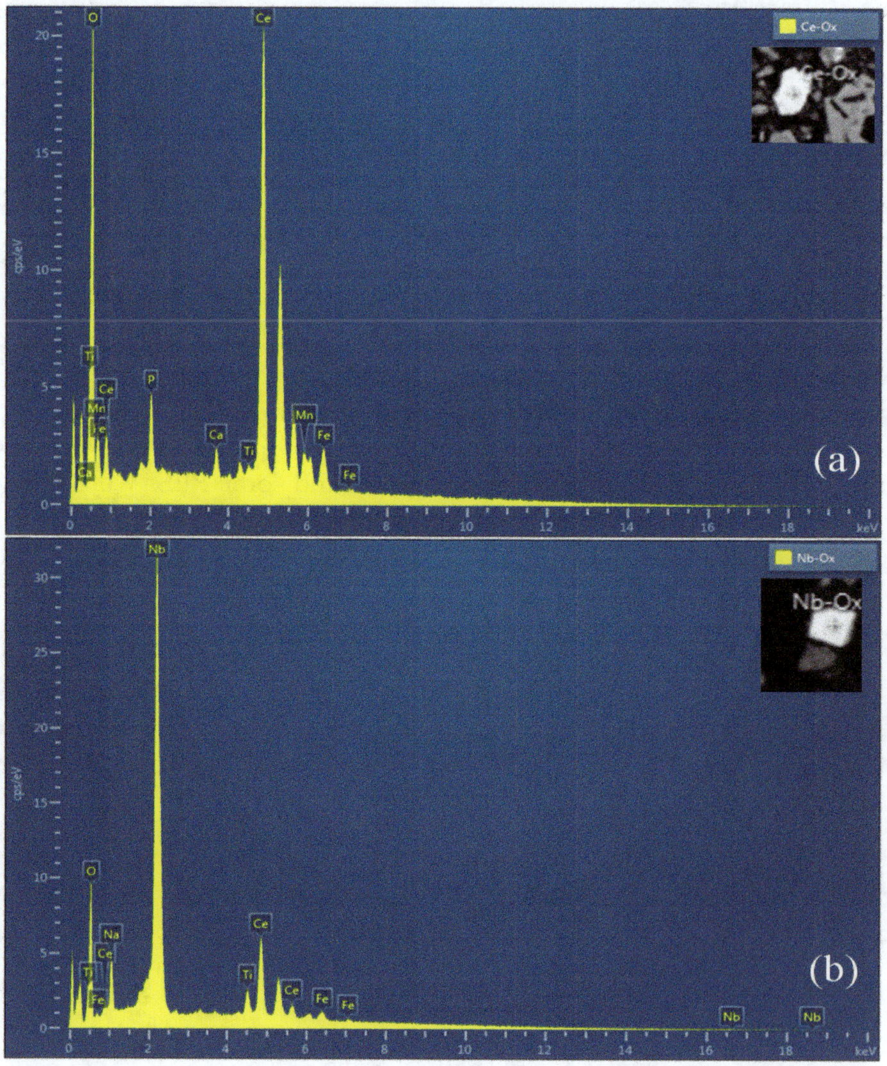

Figure 3. *Cont.*

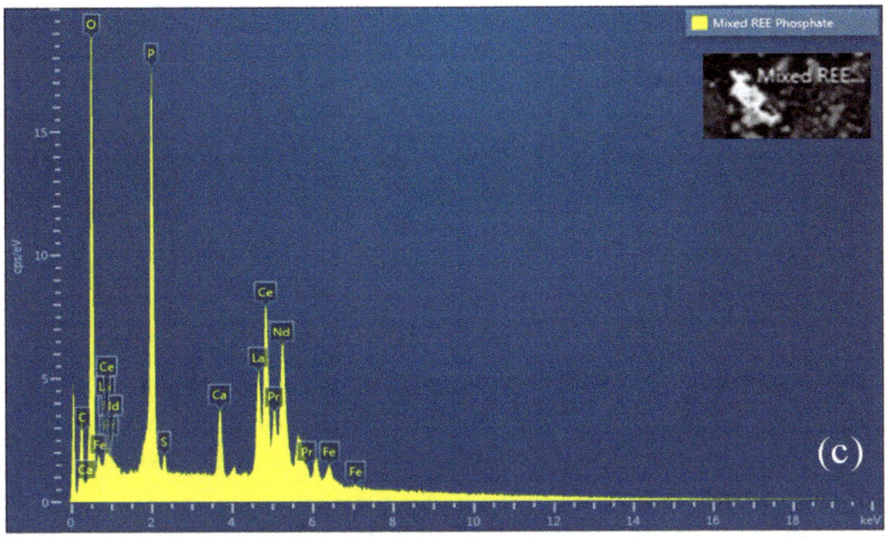

Figure 3. EDX spectra of REE-containing grains: (**a**) Ce oxide grain; (**b**) Nb oxide grain, which is host to minor Ce; (**c**) mixed REE oxide grain with Ca and P.

Table 2. Chemical composition of the REE ore.

Formula	Fe_2O_3	MnO	SiO_2	P_2O_5	Al_2O_3	TiO_2	CaO	BaO	CeO_2	MgO	La_2O_3
wt%	50.1	9.9	7.8	6.9	6.7	4.4	3.6	2.2	1.7	1.2	0.9
Formula	K_2O	SrO	Na_2O	Nd_2O_3	SO_3	Nb_2O_5	ZnO	Cl	ZrO_2	Y_2O_3	MoO_3
wt%	0.8	0.7	0.6	0.6	0.5	0.4	0.3	0.2	0.2	0.2	0.07

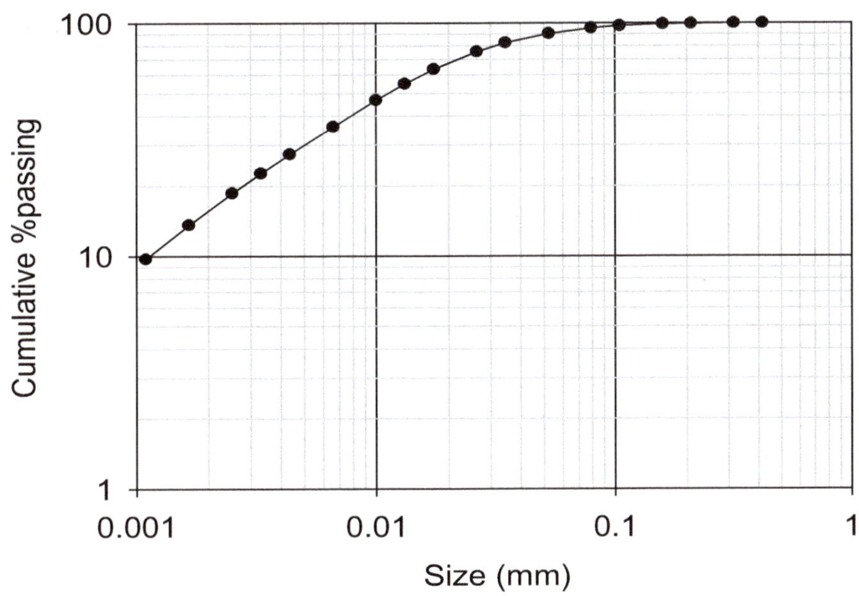

Figure 4. Cumulative particle size distribution of the ground REE ore (P_{80} of 30 μm).

2.2. Preliminary Leaching

Generally, REE oxide can dissolve at a slightly acidic pH (pH of less than 4.0–5.0), and REE phosphate dissolves at a highly acidic pH of approximately −1.0 [20,23], respectively. The purpose of preliminary leaching was to determine if the REE minerals in the ore could really dissolve at those pH ranges. The suggested pH ranges can be achieved by hydrochloric or nitric acid concentrations of 2.0 M or less. Using sulfuric acid for the preliminary tests was avoided because it can cause the REEs to co-precipitate with Ca minerals and interfere with the analysis of the REE-leaching behavior.

In the preliminary leaching, 1.0–3.0 M of nitric acid was used as a leaching reagent as it has less effect on the REE-leaching behavior since nitrate anions rarely react with any metal cations. The leaching temperature was varied as 25, 50 and 80 °C, with leaching times of 10 h, 3 h and 3 h, respectively. All leaching tests were conducted with a pulp density of 10% (*w/w*). The leaching efficiency and final metal recovery were calculated as follows, respectively:

$$\text{Leaching efficiency}(\%) = \frac{C_L \times V}{M_R \times m_R + C_L \times V} \times 100 \quad (1)$$

$$\text{Final recovery}(\%) = \frac{C_L \times V}{M_F \times m_F} \times 100 \quad (2)$$

Here, C_L is the metal concentration dissolved in the leaching solution (mg/L), V is the volume of the leaching solution (L), M_R and M_F are the metal contents in the leaching residue and ore (mg/g), respectively, and m_R and m_F are the amounts of the leaching residue and the ore used in the leaching test (g), respectively.

The leaching solution was sampled at regular intervals, and the samples were analyzed using inductively coupled plasma optical emission spectrometry (ICP-OES) to measure the metal concentration in the solution. The leaching residue and the ore were first fused by NaOH, with an ore to NaOH ratio of 1:10 (*w/w*) at 400 °C for 1 h. The fused products were then dissolved in aqua regia to measure the metal contents in the residue and the ore. They were also analyzed by ICP-OES. This solution analysis procedure was applied to all leaching tests in this study.

2.3. Sulfuric Acid Baking–Water-Leaching

To decompose the ore to a more soluble form, sulfuric acid baking was applied to the ore treatment. Various acid/solid ratios (*w/w*) were tested, i.e., 0.5, 1.0 and 2.0, and the ore and acid mixture was baked in a muffle furnace at 200 °C for 2 h. The baked ore was subsequently introduced to the water-leaching step. Water-leaching was conducted at 25, 50 and 80 °C for 3 h with a pulp density of 10% (*w/w*).

2.4. Caustic Digestion–Acid-Leaching

For caustic digestion, sodium hydroxide was used as a caustic chemical, and NaOH solutions with concentrations of 20, 30, 40 and 50% (*w/w*) were prepared. For the caustic digestion step, the highest temperature without boiling was chosen for each NaOH solution concentration, i.e., 105 °C at 20%, 115 °C at 30%, 120 °C at 40% and 145 °C at 50%. The digestion was conducted for 3 h, and the digested solid was then prepared for the subsequent acid-leaching reaction after solid/liquid separation and washing. The metal contents in the solid before and after digestion were measured by X-ray fluorescence (XRF). The digested solid was leached by hydrochloric acid (1.0–3.0 M) at 80 °C and 10% (*w/w*) pulp density for 3 h.

3. Results and Discussion

3.1. Preliminary Leaching Results

As a baseline, ore-leaching tests in nitric acid solution at different HNO_3 concentrations and temperatures were conducted. In the preliminary leaching, the leaching efficiency was equal to the final recovery as it was a single-step process.

Prior to the leaching tests, Eh–pH diagrams of REEs were considered. According to the Eh–pH diagrams (Figure 5), REE(III) hydroxide can simply dissolve in an acidic pH region. Cerium shows a slightly different characteristic because it can exist as two valence states, i.e., +3 and +4. Ce(III) hydroxide behaves in a similar manner to other REE(III) hydroxides; however, if it oxidizes to a higher valence state, Ce(IV), its leaching can be inhibited under mild acidic conditions. The leaching behavior of Ce(IV) will be discussed thoroughly in later sections.

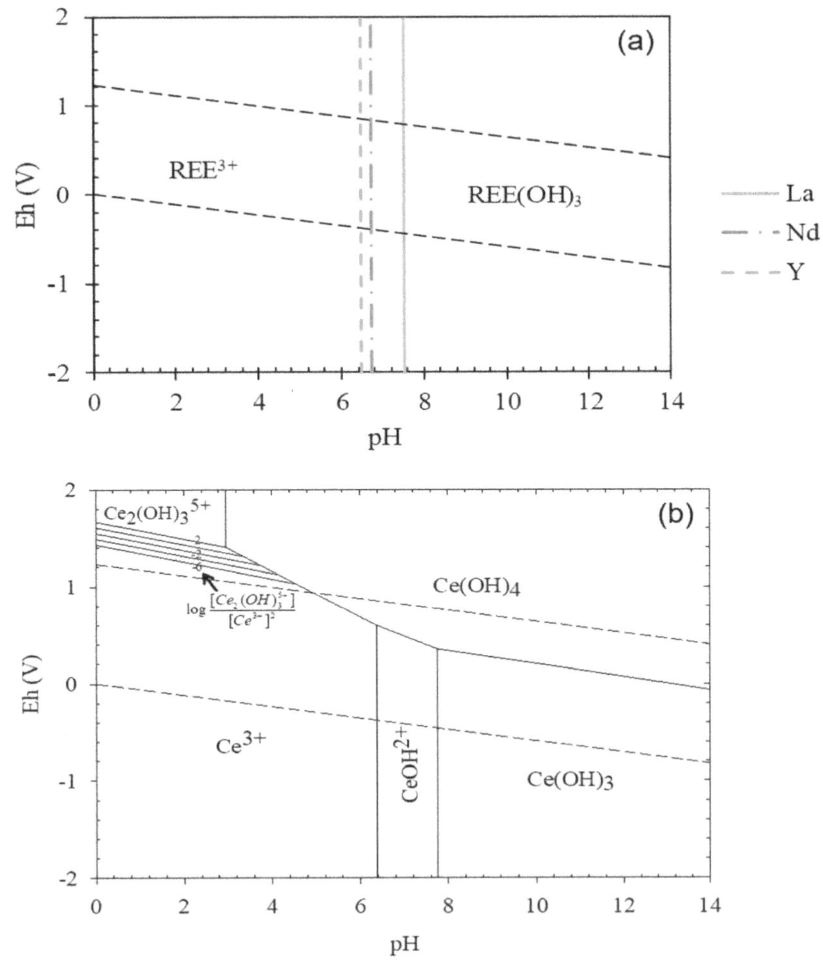

Figure 5. Eh–pH diagrams of REE systems: (**a**) (La, Nd and Y)–OH; (**b**) Ce–OH ([Ce], [La], [Nd], [Y] = 0.01 M).

As depicted in Figure 6 and Table 3, the REE-leaching efficiency generally increased with increasing acid concentration from 1.0 to 2.0 M at 25 °C (Ce = 8.4–17.3%, La = 10.2–18.6%, Nd = 13.8–26.3% and Y = 23.2–31.3%) and temperature from 25 to 80 °C in 2.0 M HNO_3 solution (Ce = 17.3–43.6%, La = 18.6–49.9%, Nd = 26.3–57.3% and Y = 31.3–48.2%). At 3.0 M HNO_3 and 80 °C, the REE recovery further improved to 63.0% Ce, 61.4% La, 72.3% Nd and 55.9% Y. However, even though the REE-leaching efficiency was improved by the high acid concentration and temperature, the maximum efficiency only ranged from 61% to 73% and approximately 30–40% of the REEs remained in the solid phase.

The leaching levels of the major impurities, namely Fe, Ca and P, were also examined. The leaching levels of Fe and P were not significant because their maximum values were only 20.0% and 41.2%, respectively, at 3.0 M HNO_3 and 80 °C. However, the Ca-leaching level was as high as 94–100% even at the most moderate condition of 1.0 M HNO_3 and 25 °C. Considering the mineralogy of the ore, Ca was mostly derived from apatite, and the pH of the solution was sufficiently low to dissolve apatite (less than 0.25, Table 3), which is available at pH values between 1.0 and 3.3 [20]. The impurity-leaching behavior will be discussed in more detail in later sections related to the REE-leaching behavior.

In the succeeding residue analysis by SEM-EDX, some REE oxide particles or REE-containing grains were trapped in other mineral matrices (Figure 7). The size of REE grains was in the range between 1.5 and 12 μm, which suggests that further liberation was essential. However, the ore particle size was already very fine at P_{80} of 30 μm; therefore, the authors decided to conduct chemical decomposition methods to free the REE-containing grains from the trapping matrices and improve the leaching recovery.

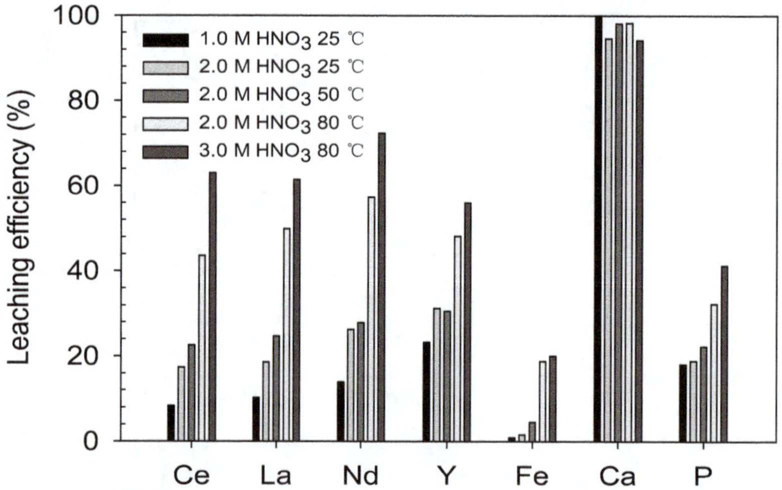

Figure 6. Preliminary leaching results: leaching efficiency (%) of REEs and impurities at different HNO_3 concentrations and temperatures.

Table 3. Preliminary leaching results: leaching efficiency (%) of REEs and impurities at different HNO_3 concentrations and temperatures.

HNO_3 Concentration (M)	Temperature (°C)	Initial pH	Leaching Efficiency (%)						
			Ce	La	Nd	Y	Fe	Ca	P
1.0	25	0.25	8.4	10.2	13.8	23.2	0.9	100.0	18.0
2.0	25	−0.37	17.3	18.6	26.3	31.3	1.6	94.6	18.9
2.0	50	−0.24	22.5	24.7	27.8	30.5	4.5	98.1	22.2
2.0	80	−0.13	43.6	49.9	57.3	48.2	18.8	98.2	32.3
3.0	80	−0.36	63.0	61.4	72.3	55.9	19.9	94.1	41.2

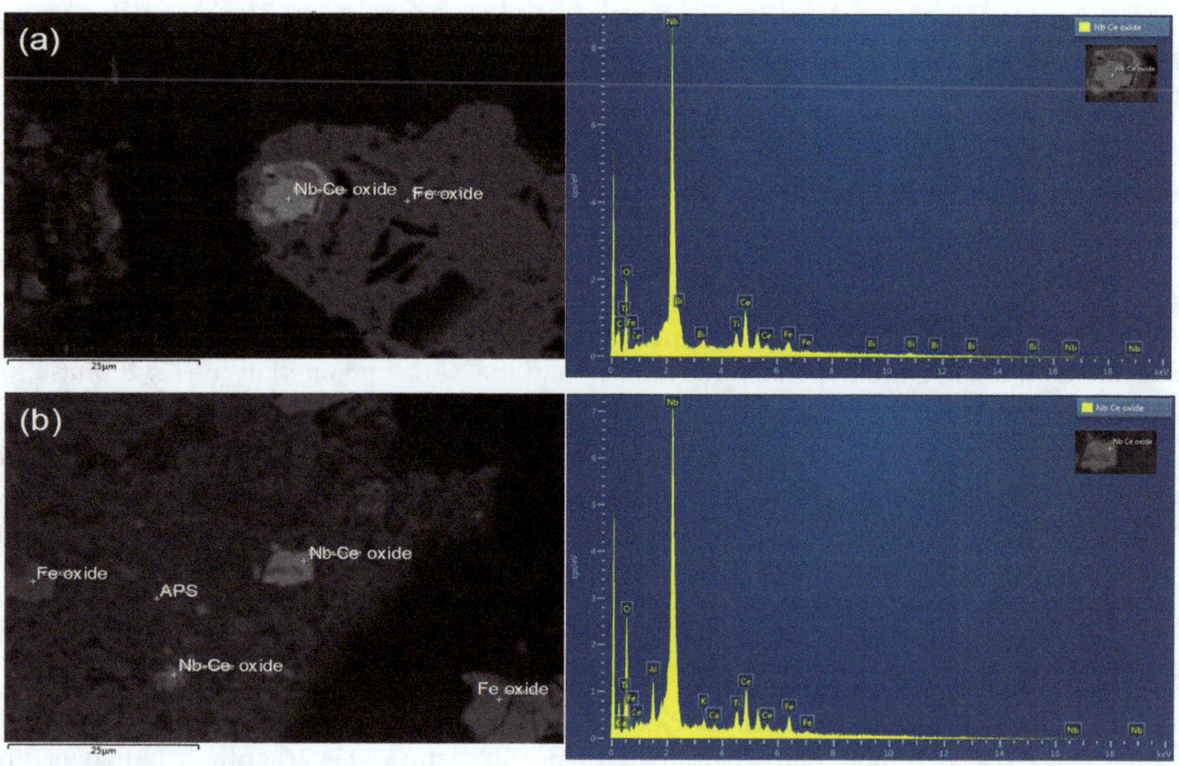

Figure 7. SEM-EDX analysis of preliminary leaching residue (leached at 3.0 M HNO_3 and 80 °C): (**a**) Fe oxide grain trapping Nb–Ce oxide; (**b**) micro-breccia of alumino-phosphate-sulfate (APS) bearing Fe oxide and Nb–Ce oxide.

3.2. Sulfuric Acid Baking–Water-Leaching

3.2.1. Effect of Acid/Solid Ratio

Sulfuric acid baking was expected to decompose the minerals trapping the REEs and the REE phosphate mineral itself (monazite and REE-substituted apatite). Because the entire baking product was sent to the next water-leaching step, it was considered that the leaching efficiency was equal to the final recovery. The detailed recovery and mass balance data are provided in Table S1 of a Supplemental data.

Through the sulfuric acid baking, the expected chemical reactions are likely to be those indicated in Equations (3)–(5). The ore and acid mixture before and after baking are displayed in Figure 8. According to Equation (3) and other previous studies [11,24], FeOOH, an Fe(III)-containing mineral, can only convert to Fe(III) sulfate species during sulfuric acid baking. Similarly, in this study, it was confirmed that most of the FeOOH was converted to $Fe_2(SO_4)_3$ after the acid baking, regardless of the acid amount, at 1.0 and 2.0 acid/solid ratios (Figure 9). Although it was not detected from the XRD analysis, Fe(II) sulfate species seemed to be also formed, because the baked product color was pale green. This implied that a reductive reaction occurred during the baking process, but the mechanism was not clear. The authors assumed that it was the effect of unknown reductive materials in the ore and lack of oxygen sources in the closed muffle furnace.

$$2FeOOH + 3H_2SO_4 \leftrightarrow Fe_2(SO_4)_3 + 4H_2O \tag{3}$$

$$REEPO_4 + H_2SO_4 \leftrightarrow REE_2(SO_4)_3 + H_3PO_4 \tag{4}$$

$$Ca_{10}(PO_4)_6X_2 + 10H_2SO_4 = 10CaSO_4 + 6H_3PO_4 + 2HX\ (X = OH,\ Cl,\ F) \tag{5}$$

Figure 8. The ore and sulfuric acid mixture (**a**) before and (**b**) after baking at 200 °C for 2 h.

In the succeeding water-leaching step, the REE recovery was significantly improved as shown in Figure 10, as the maximum levels were Ce 98.4%, La 99.1%, Nd100% and Y 96.5% after baking at an acid/solid ratio of 2.0. The sulfuric acid concentration in the leaching step was calculated simply from the amount added for the baking step. The corresponding sulfuric acid concentrations in leaching solutions with acid/solid ratios of 0.5, 1.0 and 2.0 were 0.57, 1.13 and 2.27 M, respectively, although some part of the acid was consumed during the baking step. In the preliminary leaching results, when 0.5, 1.0 and 2.0 M of nitric acid were used, the REE-leaching efficiency was significantly less than that of the baked sample. Thus, it can be concluded that the sulfuric acid baking effectively decomposed the minerals in the ore.

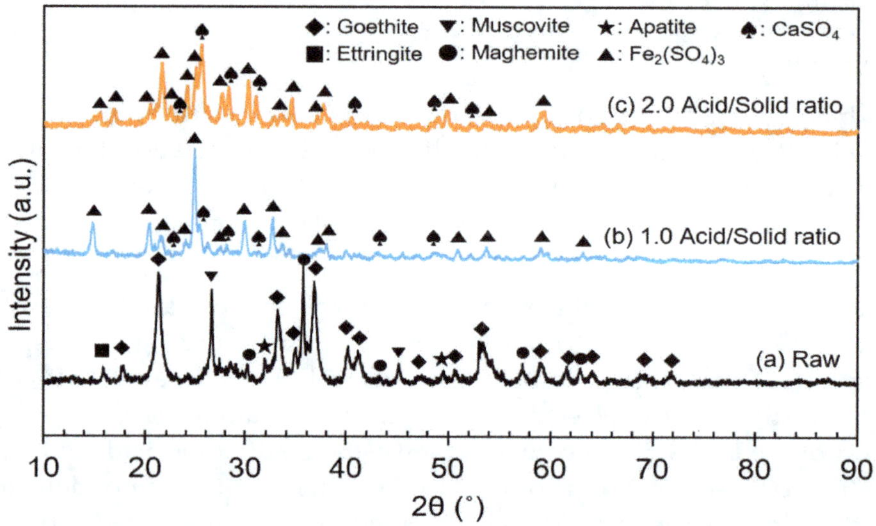

Figure 9. XRD patterns before and after sulfuric acid baking at different acid/solid ratios: (**a**) raw ore; (**b**) acid/solid ratio = 1.0; (**c**) acid/solid ratio = 2.0.

In terms of impurity-leaching, the Fe and P recoveries were considerably increased to 38.0% and 30.3%, 65.9% and 60.6% and 81.1% and 87.5% after baking at 0.5, 1.0 and 2.0 acid/solid ratios, respectively, compared with the maximum Fe and P recoveries of 20.0% and 41.2%, respectively, at 3.0 M nitric acid and 80 °C. This also implied that the minerals were effectively decomposed via the acid baking process. In addition, the Fe recovery was much higher in this study than in the previous one [24], which was approximately 20% at 1.5 acid/ore ratio. This was because the Fe(II) species formed in the current study were more soluble, whereas in the previous study, only less soluble Fe(III) species were present even after sulfuric acid baking.

In contrast, the Ca acted in a different manner from other metals. Its recovery remained at the same level of 57–58% regardless of the acid/solid ratio. This level was 1.62–1.65 times less than the recovery of the preliminary leaching results of 94.1% at 3.0 M nitric acid and 80 °C. Interestingly, the Ca recovery was considerably improved to 61.6, 74.4 and 97.7% after washing the leaching residues of 0.5, 1.0 and 2.0 acid/solid ratios, respectively. This is because calcium can easily form Ca-sulfate precipitates in a leaching solution in the presence of sulfate ions [20,25]; thus, the Ca recovery was not so high initially. After washing the residue with 3–4 times more water than the leaching solution amount, i.e., 300–400 mL, some of the Ca-sulfate precipitates were solubilized. More Ca dissolved at the higher acid/solid ratio because more residual acid remained in the solid residue pore. The precipitation problem will be further discussed in Section 3.2.2.

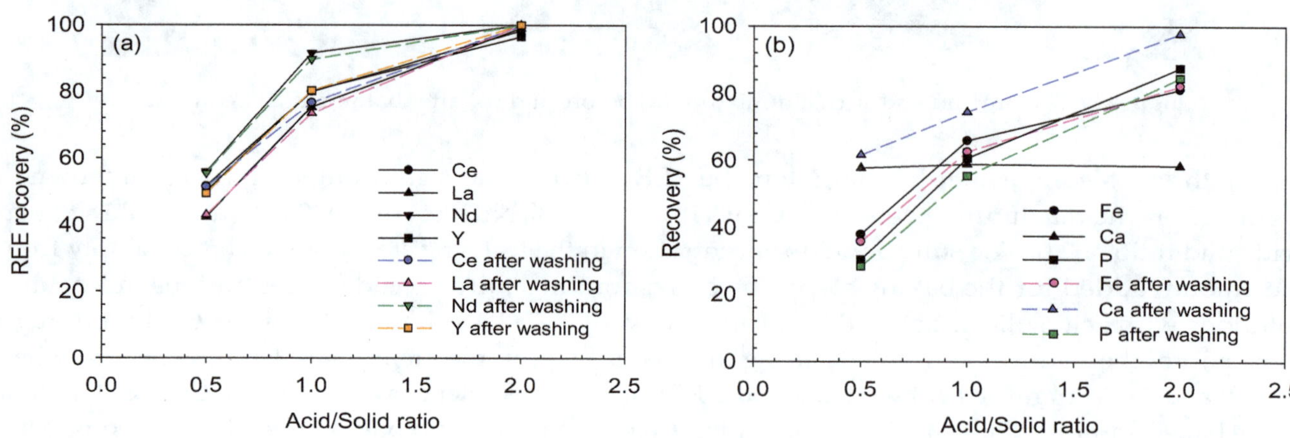

Figure 10. Leaching recoveries of (**a**) REEs and (**b**) impurities after water-leaching of the baked ore at different acid/solid ratios of sulfuric acid baking (baked at 200 °C for 2 h and leached at 25 °C for 3 h).

3.2.2. Effect of Leaching Temperature

The effect of temperature on the leaching process was examined for the ore treated with an acid/solid ratio of 2.0, which was the best condition for the highest recovery. As shown in Figure 11, the REE recovery decreased with increasing temperature from 25 to 80 °C. Between 25 and 50 °C, only a slight decrease in recovery occurred, but the REE recovery significantly decreased to 60–72% at 80 °C, compared with 96–100% at 25 °C. This trend was also observed in the recovery of impurities: the Fe, Ca and P recoveries decreased to 57.9, 51.2% and 73.2% at 80 °C, respectively, from 81.1, 58.3% and 87.5% at 25 °C, respectively.

The decrease in recovery was because of Ca-sulfate precipitation. Calcium can easily form precipitates with sulfate ions owing to its low solubility as presented in Table 4. The solubility of Ca sulfate, i.e., calcium sulfate dihydrate ($CaSO_4{\cdot}2H_2O$, CSD), is only 2.68 g/L at 25 °C (equivalent to 0.62 g/L Ca); then, it increases between 25 and 40 °C, and reduces to 2.59 g/L (0.60 g/L Ca) at 50 °C and 2.30 g/L (0.54 g/L Ca) at 80 °C [26,27]. The REE sulfates also showed a decrease in solubility with increasing temperature, but their solubility concentrations were significantly higher than their concentrations in the leaching solution (1.01 g/L Ce, 0.61 g/L La, 0.36 g/L Nd and 0.09 g/L Y at 2.0 acid/solid ratio and 25 °C). Despite these characteristics, the recoveries of REEs, Fe and P decreased at 80 °C. It seemed that these metals were adsorbed onto CSD and complexly reacted with each other to form co-precipitates. A similar behavior of REE co-precipitation and adsorption onto CSD was also observed in several previous studies [20,25,28]. They showed that the REE removal from the solution occurred at pH regions lower than the thermodynamic REE precipitation pH; thus, the adsorption onto the CSD surface would be a major process.

Moreover, it is known that the solubility of REE phosphates also decreases with increasing temperature [29–31]; therefore, the REE phosphate precipitation also seemed to affect the recoveries of REEs and impurities. After leaching at 80 °C, the recoveries of REEs and impurities were significantly improved by the washing process because of the solubilization of the precipitated CSD, which contained the adsorbed metals.

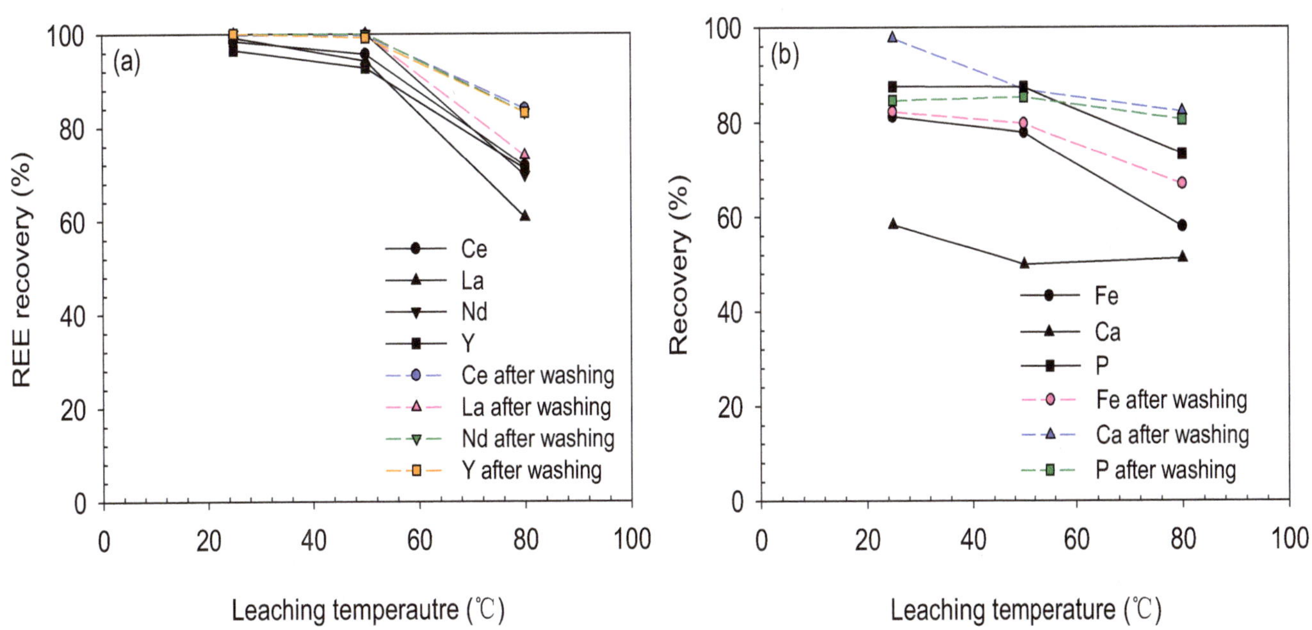

Figure 11. Leaching recoveries of (**a**) REEs and (**b**) impurities after water-leaching of the baked ore at 25–80 °C for 3 h (baked at 2.0 acid/solid ratio and 200 °C for 2 h).

Table 4. Solubilities of Ca sulfate and REE sulfate compounds at 25, 50 and 80 °C (the numbers inside the parentheses are the solubility of Ca and REEs).

Compound	Solubility (g/L)		
	25 °C	50 °C	80 °C
$CaSO_4{\cdot}2H_2O$ [1]	2.68 (0.62)	2.59 (0.60)	2.30 (0.54)
$Ce_2(SO_4)_3$ [2]	85.2 (42.0)	46.9 (23.1)	32.2 (15.9)
$La_2(SO_4)_3$ [2]	20.9 (10.3)	14.7 (7.22)	9.53 (4.68)
$Nd_2(SO_4)_3$ [2]	48.3 (24.2)	28.8 (14.4)	19.7 (9.87)
$Y_2(SO_4)_3$ [2]	61.4 (23.4)	43.1 (16.5)	29.0 (11.1)

[1] Dutrizac, 2002 [26]; [2] Das et al., 2019 [32].

3.3. Caustic Digestion–Acid-Leaching

3.3.1. Caustic Digestion Behavior

The second decomposition method, i.e., caustic digestion, was conducted at different concentrations of NaOH solution, i.e., 20–50 wt%. The significant chemical reactions of caustic digestion are shown in Equations (6)–(9).

$$REEPO_4 + 3NaOH \leftrightarrow REE(OH)_3 + Na_3PO_4 \tag{6}$$

$$Ca_{10}(PO_4)_6X_2 + 20NaOH \leftrightarrow 10Ca(OH)_2 + 6\ Na_3PO_4 + 2NaX\ (X = Cl,\ F) \tag{7}$$

$$Ca_{10}(PO_4)_6(OH)_2 + 18NaOH \leftrightarrow 10Ca(OH)_2 + 6\ Na_3PO_4 \tag{8}$$

$$AlPO_4 + 4NaOH \leftrightarrow Al(OH)_4^- + 4\ Na^+ + PO_4^{3-} \tag{9}$$

It was expected that due to caustic digestion, the Al and P would be dissolved into the solution because they are very soluble in strong basic solutions, whereas the REEs would remain in the solid phase as a hydroxide. Actually, as the NaOH concentration was increased, the REEs were concentrated in the digested product, except Y (Y is slightly amphoteric [33]), and the Al and P contents decreased in the solid phase and finally reached 1.0 wt% and 0.1 wt%, respectively, in the 50 wt% NaOH digested product (Figure 12). Iron is not soluble in basic solutions; thus, it was also concentrated in the digested product.

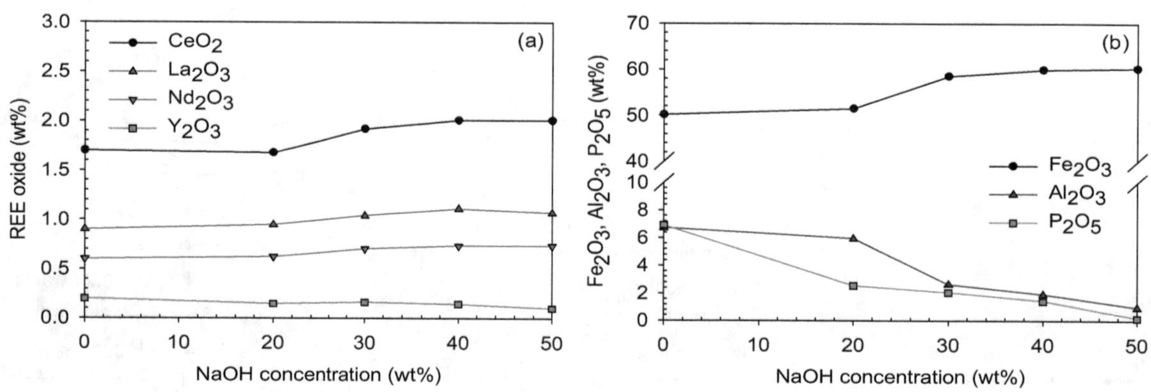

Figure 12. Oxide contents (wt%) of (**a**) REE and (**b**) Fe, Al and P of the digested product at different digesting NaOH concentrations.

Figure 13 shows that the intensity of goethite peaks in the digested product decreased. Representatively, at 2θ value of 21.3°, the peak intensity of goethite was 1019 and 363 before and after the digestion, respectively. Even some of the goethite peaks disappeared after digestion, although the Fe concentration increased. In contrast, the intensities of gangue mineral peaks, such as those of muscovite and ettringite, were relatively improved. These phenomena implied that the crystallinity of the goethite weakened, and it may be possible to free the trapped REE particles. Actually, as indicated in Figure 14, the surface of ore particles were chemically affected and eroded after caustic digestion.

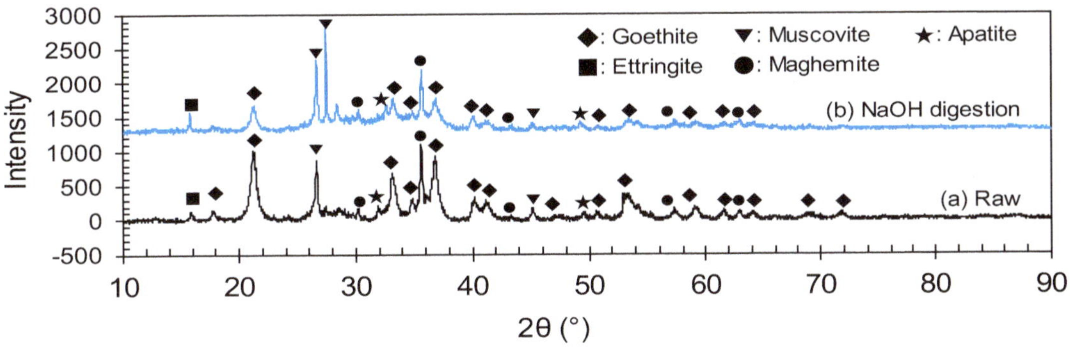

Figure 13. XRD patterns before and after caustic digestion using 50 wt% NaOH solution.

Figure 14. SEM images of the ore (**a**) before and (**b**) after caustic digestion (digestion condition: 50 wt% NaOH and 145 °C for 3 h).

3.3.2. Acid-Leaching Behavior

The caustic digested product was introduced to the succeeding acid-leaching step. During the digestion step, some elements of the ore were lost to the liquid phase, e.g., Al and P; thus, the leaching efficiency and final recovery were different in this test. The leaching efficiency was calculated based on the amount of elements in the digested product, whereas the recovery was based on the amount in the raw ore. The detailed recovery and mass balance data are provided in Table S2 of the Supplemental data.

The effect of acid concentration was initially investigated using 1.0–3.0 M hydrochloric acid. The digested ore was produced using a 50 wt% NaOH solution. As presented in Figure 15, the REE-leaching efficiency generally increased as the acid concentration increased, while Ce behaved in a slightly different manner.

In the case of Ce, its leaching efficiency only reached 12.1% at 1.0 M HCl. It was improved at elevated HCl concentrations of 2.0 and 3.0 M as 75.5% and 79.7%, respectively; however, these values were still lower than those of other REEs at 2.0 and 3.0 M HCl, which were in the range of 87–92% and 94–100%, respectively. It seemed that some part of Ce was converted from Ce(III) to Ce(IV) compound during the digestion step, probably $Ce(OH)_4$, which led to the low recovery of Ce. $Ce(OH)_4$ is less soluble than $Ce(OH)_3$ as the solubility product of $Ce(OH)_4$ is $10^{-51.56}$ and that of $Ce(OH)_3$ is $10^{-26.15}$ [34]. This can be also confirmed from the Eh–pH diagram of the Ce-OH system in Figure 5; $Ce(OH)_3$ can simply be leached at a slightly acidic pH, but $Ce(OH)_4$ needs a highly acidic pH. Although $Ce(OH)_4$ was not detected from the XRD analysis due to its low concentration, it was probable that $Ce(OH)_4$ was formed during the caustic digestion because of the significantly lower Ce-leaching efficiency than those of the other REEs.

The final recoveries of Ce, La and Nd showed almost similar trends with their leaching efficiencies. However, that of Y was significantly low, i.e., 26.2–38.4% at 1.0–3.0 M HCl. This was because Y was dissolved and lost during the digestion step, and only 40% of Y from the raw ore was introduced into the acid-leaching step.

In terms of impurity-leaching, the Fe-leaching efficiency and recovery almost linearly increased as the acid concentration increased. The Al-leaching efficiency ranged from 95% to 100%; hence, its final recovery was similar to that in the digestion step, which was in the range of 10–12%. Phosphorus was not detected in any leaching solutions because it was removed during the digestion step.

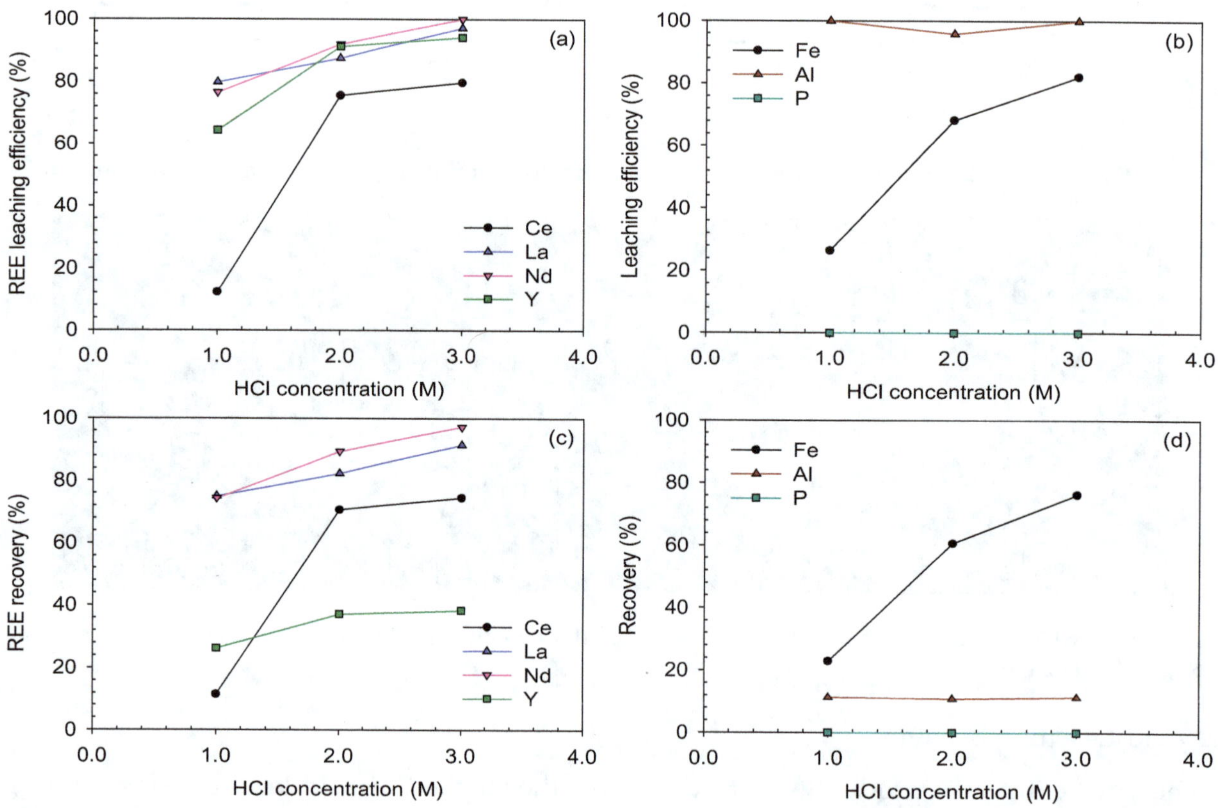

Figure 15. Leaching efficiencies and recoveries after acid-leaching of the caustic digested ore at HCl concentrations of 1.0–3.0 M and 80 °C for 3 h (digested using 50 wt% NaOH solution for 3 h): **(a)** REE and **(b)** Fe-, Al- and P-leaching efficiencies (%); **(c)** REE and **(d)** Fe, Al and P recoveries (%).

Figure 16 illustrates variation of leaching results with the digesting NaOH concentration. In terms of leaching efficiency and recovery, both were maximized at 30 wt% NaOH concentration, and those of La and Nd were not much changed at higher NaOH concentrations. However, Ce showed an evident decrease in the leaching efficiency and recovery at NaOH concentrations higher than 30 wt%, which was caused by the less soluble $Ce(OH)_4$ formation at higher NaOH concentrations. The final recovery of Y also significantly decreased as the NaOH concentration increased from 30 wt% to 50 wt%, although the leaching efficiency remained high at 93–94%. This was because its recovery from the digestion step decreased from 68% to 40% as the NaOH concentration increased from 30 wt% to 50 wt%.

Concerning the impurities, the Fe-leaching efficiency had a similar trend to that of the REEs (except Ce), which increased between 20 wt% and 30 wt% in the range of 74–82%. The Fe recovery was similar to its leaching efficiency, which ranged from 70% to 78%. However, the Al and P recoveries considerably decreased from 64.2% to 11.3% and 27.0% to 0%, respectively, as they were removed from the solution phase during the digestion step. In terms of leaching efficiency, that of P monotonously decreased, while that of Al fluctuated. It seemed that the residual NaOH in the digested product using 40 wt% NaOH solution suppressed the Al-leaching. A small amount of Al remained in the digested product using 50 wt% NaOH solution; thus, the Al-leaching efficiency reached 100%, although a higher NaOH concentration was applied.

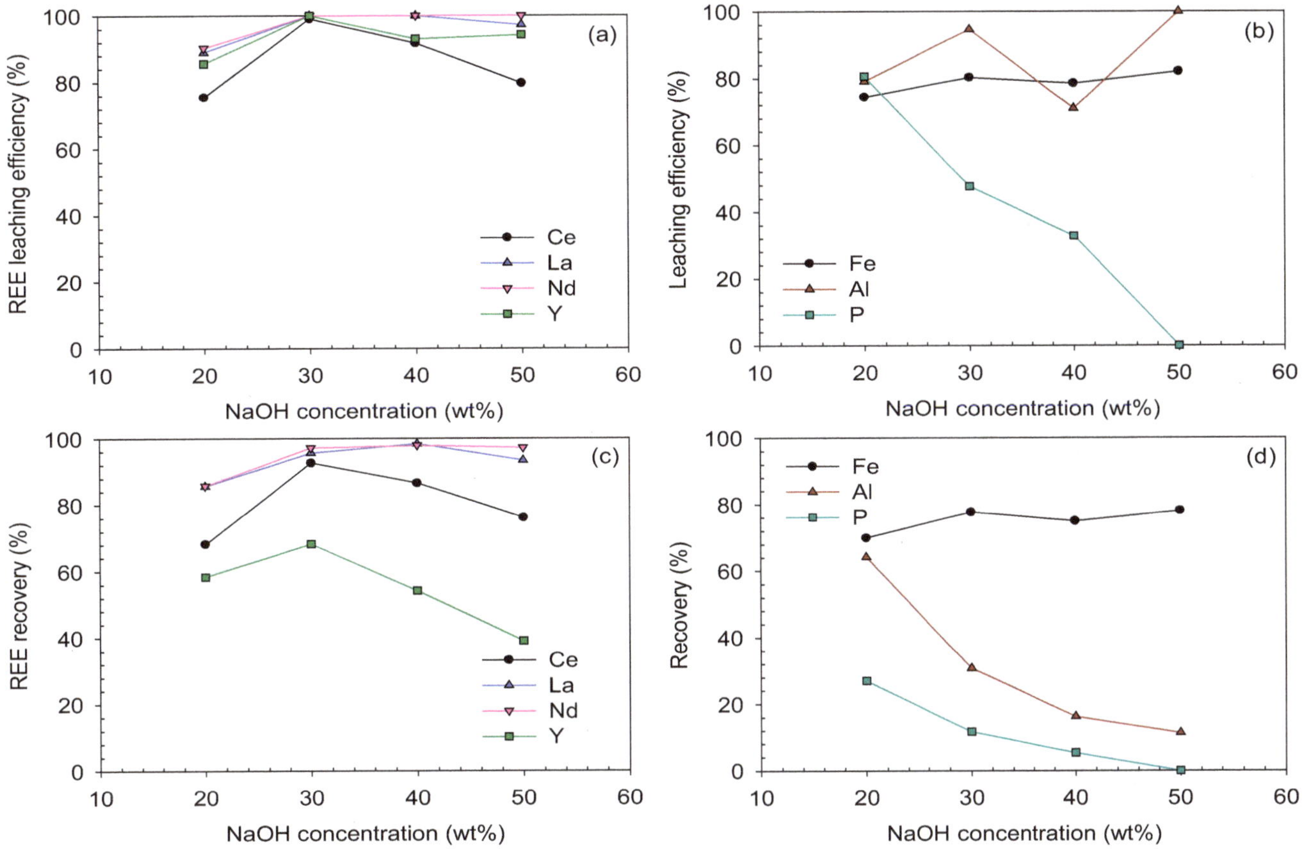

Figure 16. Leaching efficiencies and recoveries after acid-leaching of the caustic digested ore at digesting NaOH concentrations of 20–50 wt% (leached using 3.0 M HCl at 80 °C for 3 h): (**a**) REE and (**b**) Fe-, Al- and P-leaching efficiencies (%); (**c**) REE and (**d**) Fe, Al and P recoveries (%).

3.4. Determination of the Optimum Condition for Analyzing the REE–Fe-Leaching Efficiency Correlation

To statistically analyze the REE-leaching behavior, correlation curves of the REE–Fe-leaching efficiency are plotted in Figures 17 and 18 and the calculated R^2 values are listed in Table 5.

According to the analysis, the REE and Fe-leaching efficiencies were generally positively correlated. Specifically, the correlation was stronger during sulfuric acid baking–water-leaching (because the R^2 values were much closer or even equal to 1.0000) than during caustic digestion–acid-leaching. This means that (i) REE-leaching highly depends on the Fe-bearing mineral decomposition, which traps the REE constituents during the sulfuric acid baking–water-leaching process and (ii) REEs and Fe co-precipitation by CSD affected a good R^2 value; from REEs and Fe recovery trend before/after the residue washing process shown in Figures 10 and 11, their leaching behavior was somewhat governed by CSD formation as both the dissolved REEs and Fe can adsorb and desorb from CSD in a similar extent.

The curves showed a weaker REE–Fe correlation for the caustic digestion–acid-leaching test results, and the Ce–Fe curve based on the NaOH concentration showed the weakest correlation with an R^2 value of 0.1657. This is because of the Ce oxidation to the Ce(IV) species of $Ce(OH)_4$ as discussed earlier. In addition, the La and Nd leaching efficiencies showed weak correlations with that of Fe based on the digestion condition (Figure 18a) because they reached 100% recovery at 80% Fe-leaching efficiency, which is slightly lower than the maximum Fe recovery of 82%. If the points corresponding to the 50 wt% NaOH digestion are removed, the REE–Fe correlations are improved: 0.9998 Ce–Fe, 0.9158 La–Fe, 0.9158 Nd–Fe and 0.9575 Y–Fe. This implies that both the Fe-bearing mineral decomposition and digesting NaOH concentration can considerably affect the succeeding acid-leaching step in the caustic digestion–acid-leaching process. Based on the leaching condition, i.e., HCl concentration, the REE–Fe correlation was satisfactory regardless of the elements (Figure 18b).

Consequently, under the applied decomposition methods and leaching conditions in this study, the REE-leaching was more affected by Fe-leaching during sulfuric acid baking–water-leaching. At the Fe-leaching efficiencies of 68% and 66%, the REE-leaching efficiencies (except Ce) were 88–92% and 75–80% from the caustic digestion–acid-leaching and sulfuric acid baking–water-leaching, respectively. Therefore, caustic digestion was a better decomposition method in terms of REE-leaching, and the REE-leaching was inhibited by CSD co-precipitation in the sulfuric acid baking–water-leaching process as the leaching efficiency was lower than in caustic digestion–acid-leaching at the same corresponding Fe-leaching level.

Overall, as the Fe-leaching reaches the maximum, the REE-leaching efficiency also attains its highest value; among the factors, the Fe-bearing mineral decomposition has the most significant effect on REE-leaching. Thus, considering the REE-leaching efficiency only, the optimum decomposition-leaching conditions are as follows: 2.0 acid/solid ratio sulfuric acid baking +25 °C water-leaching for the sulfuric acid baking–water-leaching process and 30 wt% NaOH digestion + 3.0-M HCl-leaching at 80 °C for the caustic digestion–acid-leaching.

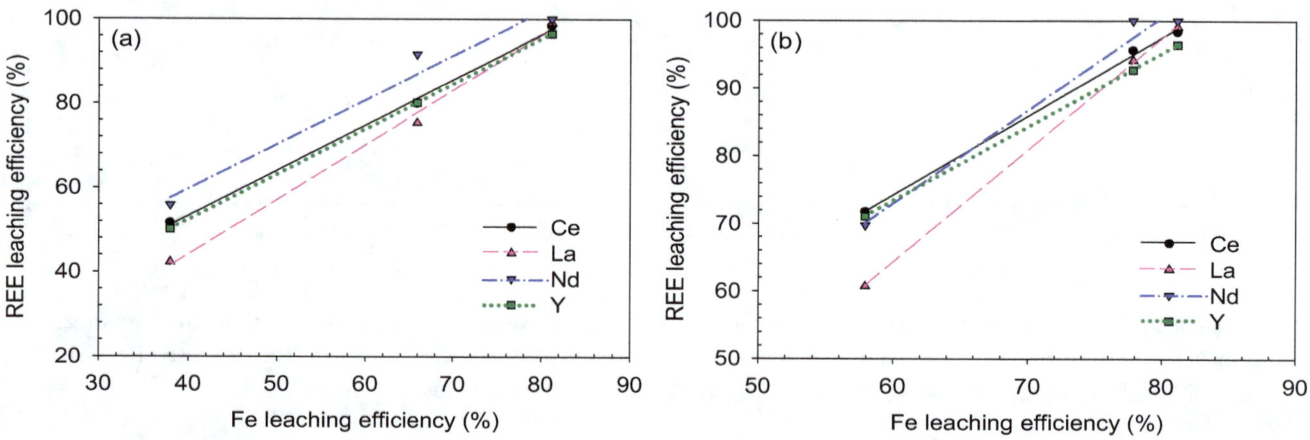

Figure 17. REE–Fe-leaching efficiency correlation for the sulfuric acid baking–water-leaching results based on (**a**) acid/solid ratio and (**b**) leaching temperature.

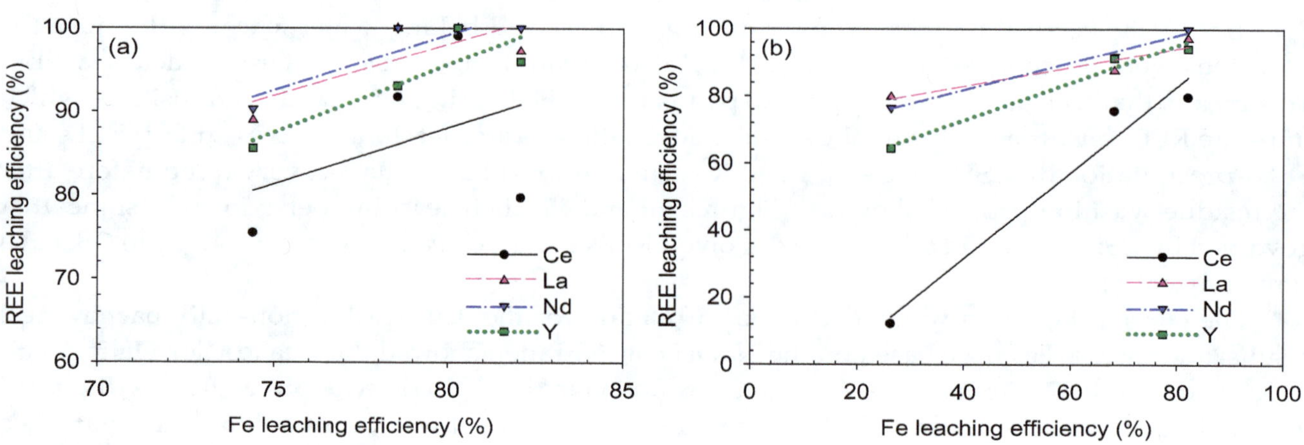

Figure 18. REE–Fe efficiency correlation for the caustic digestion–acid-leaching results based on (**a**) digesting NaOH solution concentration and (**b**) leaching HCl concentration.

Table 5. R^2 values of REE–Fe correlation curves based on the decomposition method and leaching condition.

Correlated Metals	Sulfuric Acid Baking-Water-Leaching		Caustic Digestion–Acid-Leaching	
	Acid/Solid Ratio	Leaching Temperature	NaOH Concentration	HCl Concentration
Ce–Fe	0.9980	0.9984	0.1657	0.9652
La–Fe	0.9946	0.9997	0.5990	0.8849
Nd–Fe	0.9692	0.9824	0.8062	0.9899
Y–Fe	1.0000	1.0000	0.7706	0.9772

4. Conclusions

In this study, the effects of two different chemical decomposition methods, namely sulfuric acid baking and caustic digestion, to enhance the REE recovery from a refractory ore were investigated. The REE-containing particles were scattered throughout the ore, and some of them were even trapped in other minerals such as Fe oxide, although the total REE concentration was relatively high at 3.4% TREO. From the preliminary leaching tests using nitric acid, the maximum REE recovery only reached 61–73%; thus, a significant amount of REEs was still left in the residue.

Using sulfuric acid baking–water-leaching, the REE recovery was improved to 97–100% after the ore was baked at 2.0 acid/solid ratio and 200 °C for 2 h. The REE recovery reduced with increasing leaching temperature because of the CSD solid formation at elevated temperatures and the accompanying REE co-precipitation.

During the caustic digestion reaction, the REEs were concentrated in the solid phase as Al and P were removed to the solution phase by the decomposition reaction. In the succeeding acid-leaching, all REE-leaching efficiencies reached the maximum after 30 wt% NaOH digestion for 3 h and 3.0 M HCl-leaching at 80 °C for 3 h. With increasing NaOH concentration, Ce became oxidized to Ce(IV) species of $Ce(OH)_4$, and its leaching efficiency from the acid-leaching step decreased.

From the REE–Fe correlation analysis, a stronger correlation between the REEs and Fe was observed for the sulfuric acid baking–water-leaching results because of not only Fe-bearing mineral decomposition to iron sulfate, but also CSD precipitation with REEs and Fe. In addition, comparing the sulfuric acid baking–water-leaching and caustic digestion–acid-leaching, the caustic digestion showed a more powerful effect on enhancing the REE-leaching as it resulted to a higher REE-leaching efficiency at the same Fe-leaching level than that of sulfuric acid baking. It can be concluded that in terms of REE-leaching efficiency, as the Fe-leaching reached the maximum, the REE-leaching efficiency also attained its highest value. Furthermore, the Fe-bearing mineral decomposition significantly affected the REE-leaching of the studied ore.

Author Contributions: Conceptualization, R.K. and H.C.; methodology, R.K. and H.C.; validation, R.K., J.J., J.K. and S.L.; formal analysis, R.K., J.J., J.K. and S.L.; investigation, R.K.; resources, H.C., K.W.C., H.-S.Y. and C.-J.K.; data curation, R.K.; writing—original draft preparation, R.K.; writing—review and editing, H.C., K.W.C., H.-S.Y. and C.-J.K.; visualization, R.K.; supervision, H.C.; funding acquisition, H.C. and K.W.C. All authors have read and agreed to the published version of the manuscript.

References

1. Jordens, A.; Cheng, Y.P.; Waters, K.E. A review of the beneficiation of rare earth element bearing minerals. *Miner. Eng.* **2013**, *41*, 97–114. [CrossRef]
2. Massari, S.; Ruberti, M. Rare earth elements as critical raw materials: Focus on international markets and future stategies. *Resour. Policy* **2013**, *38*, 36–43. [CrossRef]
3. Golev, A.; Scott, M.; Erskine, P.D.; Ali, S.H.; Ballantyne, G.R. Rare earths supply chains: Current status, constraints and opportunities. *Resour. Policy* **2014**, *41*, 52–59. [CrossRef]
4. Xie, F.; Zhang, T.A.; Dreisinger, D.; Doyle, F. A critical review on solvent extraction of rare earths from aqueous solutions. *Miner. Eng.* **2014**, *56*, 10–28. [CrossRef]
5. Tian, J.; Chi, R.; Yin, J. Leaching process of rare earths from weathered crust elution-deposited rare earth ore. *Trans. Nonferrous Met. Soc. China* **2010**, *20*, 892–896. [CrossRef]
6. Chen, Z. Global rare earth resources and scenarios of future rare earth industry. *J. Rare Earths* **2011**, *29*, 1–6. [CrossRef]
7. Moldoveanu, G.A.; Papangelakis, V.G. Recovery of rare earth elements adsorbed on clay minerals: II. Leaching with ammonium sulfate. *Hydrometallurgy* **2013**, *131–132*, 158–166. [CrossRef]
8. Sadri, F.; Nazari, A.M.; Ghahreman, A. A review on the cracking, baking and leaching processes of rare earth element concentrates. *J. Rare Earths* **2017**, *35*, 739–752. [CrossRef]
9. Berry, L.; Agarwal, V.; Galvin, J.; Safarzadeh, M.S. Decomposition of monazite concentrate in sulphuric acid. *Can. Metall. Quart.* **2018**, *57*, 422–433. [CrossRef]
10. Demol, J.; Ho, E.; Senanayake, G. Sulfuric acid baking and leaching of rare earth elements, thorium and phosphate from a monazite concentrate: Effect of bake temperature from 200 to 800 °C. *Hydrometallurgy* **2018**, *179*, 254–267. [CrossRef]
11. Borra, C.R.; Mermans, J.; Blanpain, B.; Pontikes, Y.; Binnemans, K.; Gerven, T.V. Selective recovery of rare earths from bauxite residue by combination of sulfation, roasting and leaching. *Miner. Eng.* **2016**, *92*, 151–159. [CrossRef]
12. Jha, M.K.; Kumari, A.; Panda, R.; Kumar, J.R.; Yoo, K.; Lee, J.Y. Review on hydrometallurgical recovery of rare earth metals. *Hydrometallurgy* **2016**, *165*, 2–26. [CrossRef]
13. Soltani, F.; Abdollahy, M.; Petersen, J.; Ram, R.; Javad Koleini, S.M.; Moradkhani, D. Leaching and recovery of phosphate and rare earth elements from an iron-rich fluorapatite concentrate: Part II: Selective leaching of calcium and phosphate and acid baking of the residue. *Hydrometallurgy* **2019**, *184*, 29–38. [CrossRef]
14. Huang, Y.; Dou, Z.; Zhang, T.; Liu, J. Leaching kinetics of rare earth elements and fluoride from mixed rare earth concentrate after roasting with calcium hydroxide and sodium hydroxide. *Hydrometallurgy* **2017**, *173*, 15–21. [CrossRef]
15. Buchanan, J.; Reveley, S.; Forrester, K.; Cummings, A. Review of current rare earth processing practice. In Proceedings of the ALTA 2014, Perth, Australia, 24–31 May 2014.
16. Habashi, F. The recovery of the lanthanides from phosphate rock. *J. Chem. Technol. Biotechnol. Chem. Technol.* **1985**, *35*, 5–14. [CrossRef]
17. Jorjani, E.; Bagherieh, A.H.; Chelgani, S.C. Rare Earth Elements Leaching from Chadormalu Apatite Concentrate: Laboratory Studies and Regression Predictions. *Korean J. Chem. Eng.* **2011**, *28*, 557–562. [CrossRef]
18. Preston, J.S.; Cole, P.M.; Craig, W.M.; Feather, A.M. the recovery of rare earth oxides from a phosphoric acid by-product. Part 1: Leaching of rare earth values and recovery of a mixed rare earth oxide by solvent extraction. *Hydrometallurgy* **1996**, *41*, 1–19. [CrossRef]
19. Wang, L.; Long, Z.; Huang, X.; Yu, Y.; Cui, D.; Zhang, G. Recovery of rare earths from wet-process phosphoric acid. *Hydrometallurgy* **2010**, *101*, 41–47. [CrossRef]
20. Kim, R.; Cho, H.; Han, K.N.; Kim, K.; Mun, M. Optimization of acid leaching of rare-earth elements from mongolian apatite-based ore. *Minerals* **2016**, *6*, 63. [CrossRef]
21. Neumann, R.; Medeiros, E.B. Comprehensive mineralogical and technological characterisation of the Araxá (SE Brazil) complex REE (Nb-P) ore, and the fate of its processing. *Int. J. Miner. Process.* **2015**, *144*, 1–10. [CrossRef]
22. Kuzmin, V.I.; Pashkov, G.L.; Lomaev, V.G.; Voskresenskaya, E.N.; Kuzmina, V.N. Combined approaches for comprehensive processing of rare earth metal ores. *Hydrometallurgy* **2012**, *129–130*, 1–6. [CrossRef]

23. Kim, R.; Cho, H.; Han, K.N. Behavior of anions in association with metal ions under hydrometallurgical environments: Part I–OH^- effect on various cations. *Min. Metall. Explor.* **2014**, *31*, 34–39. [CrossRef]
24. Guo, X.; Li, D.; Park, K.; Tian, Q.; Wu, Z. Leaching behavior of metals from a limonitic nickel laterite using a sulfation–roasting–leaching process. *Hydrometallurgy* **2009**, *99*, 144–150. [CrossRef]
25. Sadri, F.; Kim, R.; Yang, Z.; Ghahreman, A. The effect of calcium sulfate crystallization and the crystal modification on aqueous REE stability in Ca saturated REE-Ca-SO_4-H_2O systems. *Hydrometallurgy* **2018**, *182*, 82–96. [CrossRef]
26. Dutrizac, J.E. Calcium sulphate solubilities in simulated zinc processing solutions. *Hydrometallurgy* **2002**, *65*, 109–135. [CrossRef]
27. Azimi, G.; Papangelakis, V.G.; Dutrizac, J.E. Modelling of calcium sulphate solubility in concentrated multi-component sulphate solutions. *Fluid Phase Equilibr.* **2007**, *260*, 300–315. [CrossRef]
28. Sadri, F.; Kim, R.; Ghahreman, A. Substitution of Calcium with Ce, Nd, Er, and Tb in the Structure of Microcrystals of Calcium Sulfates with Controlled Hydration Water: A Proposed Mechanism. *Cryst. Growth Des.* **2019**, *19*, 2621–2631. [CrossRef]
29. Firsching, F.H.; Brune, S.N. solubility products of the trivalent rare-earth phosphates. *J. Chem. Eng. Data* **1991**, *36*, 93–95. [CrossRef]
30. Poitrasson, F.; Oelkers, E.; Schott, J.; Montel, J. Experimental determination of synthetic $NdPO_4$ monazite end-member solubility in water from 21 °C to 300 °C: Implications for rare earth element mobility in crustal fluids. *Geochim. Cosmochim. Acta* **2004**, *68*, 2207–2221. [CrossRef]
31. Cetiner, Z.S.; Wood, S.A.; Gammons, C.H. The aqueous geochemistry of the rare earth elements. Part XIV. The solubility of rare earth element phosphates from 23 to 150 °C. *Chem. Geol.* **2005**, *217*, 147–169. [CrossRef]
32. Das, G.; Lencka, M.M.; Eslamimanesh, A.; Wang, P.; Anderko, A.; Riman, R.E.; Mavrotsky, A. Rare earth sulfates in aqueous systems: Thermodynamic modeling of binary and multicomponent systems over wide concentration and temperature ranges. *J. Chem. Thermodyn.* **2019**, *131*, 49–79. [CrossRef]
33. Ivano-Emie, B.N.; Nisel'son, L.A.; Ivolgina, A.T. Solubility of yttrium hydroxide in sodium hydroxide solution. *Zhur. Neorg. Khim.* **1961**, *6*, 1483–1484.
34. Joshi, S.; Kulp, E.A.; Fahrenholtz, W.G.; O'Keefe, M.J. Dissolution of cerium from cerium-based conversion coatings on Al 7075-T6 in 0.1 M NaCl solutions. *Corros. Sci.* **2012**, *60*, 290–295. [CrossRef]

13

Factors Controlling the Chromium Isotope Compositions in Podiform Chromitites

Maria Economou-Eliopoulos [1,*], Robert Frei [2] and Ioannis Mitsis [1]

1 Department of Geology and Geoenvironment, University of Athens, 15784 Athens, Greece; mitsis@geol.uoa.gr

2 Department of Geosciences and Natural Resource Management, University of Copenhagen, Øster Voldgade 10, 1350 Copenhagen K, Denmark; robertf@ign.ku.dk

* Correspondence: econom@geol.uoa.gr

Abstract: The application of Cr isotope compositions to the investigation of magmatic and post-magmatic effects on chromitites is unexplored. This study presents and compiles the first Cr stable isotope data (δ^{53}Cr values) with major and trace element, contents from the Balkan Peninsula, aiming to provide an overview of the compositional variations of δ^{53}Cr values in ophiolite-hosted chromitites and to delineate geochemical constraints controlling the composition of chromitites. The studied chromitites exhibit δ^{53}Cr values ranging from −0.184‰ to +0.159‰, falling in the range of so-called "igneous Earth" or "Earth's mantle inventory" with values −0.12 ± 0.11‰ to 0.079 ± 0.129‰ (2sd). A characteristic feature is the slightly positively fractionated δ^{53}Cr values of all chromitite samples from Othrys (+0.043 ± 0.03‰), and the occurrence of a wide range of δ^{53}Cr values spanning from positively, slightly negatively to the most negatively fractionated signatures (Pindos, δ^{53}Cr = −0.147 to +0.009‰; Skyros, δ^{53}Cr = −0.078 to +0.159‰). The observed negative trend between δ^{53}Cr values and Cr/(Cr + Al) ratios may reflect a decrease in the δ^{53}Cr values of chromitites with increasing partial melting degree. Alternatively, it may point to processes related to magmatic differentiation, as can be seen in our data from Mikrokleisoura (Vourinos).

Keywords: chromium isotopes; chromitites; ophiolites; Balkan Pensula

---

1. Introduction

Chromitites in ophiolite complexes are studied extensively, in particular with respect to mantle-lithospheric slab interactions, post magmatic processes and subduction recycling scenarios [1–8]. Chromitites differ in their major and trace element compositions, including their platinum-group element (PGE) content, and also have been shown to exhibit variations in their chromium isotope signatures. Although the PGE content in chromitites is commonly a few 100 s of ppb, a significant PGE-enrichment, reaching tens of ppm has been reported in high-Cr or high-Al type chromitites from small occurrences, either in Pt and Pd (PPGE; incompatible, with $Di > 1$) or in Os, Ir, Ru (IPGE; compatible, $Di < 1$). Primary magmatic compositional trends recorded in chromite may be obliterated later on during secondary processes, in that Platinum-Group Minenerals (PGM) and silicates may have been substantially modified during subsequent subsolidus re-equilibration and/or by reaction with metasomatic fluids, during extended periods of deformation (including ductile asthenosphere mantle flow), and by shallow crustal, brittle deformation. The factors controlling the trace element distribution in chromitites remain unclear [8–20].

Likewise, limited data on the chromium isotope compositions of chromitites have shown isotopically heavier signatures compared to mantle peridotites [2,21–23]. However, the role of magmatic processes on the Cr isotope system during partial melting and magma differentiation is still unexplored, and there is still extensive debate on the reasons for the diversity of PGE contents in chromitites [22–24].

Although $\delta^{53}Cr$ values of chromitites from the Balkan peninsula are still very limited [2], Cr isotope studies have been performed by several authors in the view to evaluate hexavalent chromium (Cr(VI)) contamination in groundwater and rock leachates [25–30]. For example, Cr isotope signatures indicated that Cr(VI) contamination in groundwater from the Friuli Venezia Giulia Region of Italy is related to the oxidation of trivalent (Cr(III)) deposited as a consequence of past industrial activities [25]. The $\delta^{53}Cr$ values and Cr(VI) concentrations in contaminated water from the Evia and Assopos Basins (Greece) are compatible with global data ranges that characterize waters contaminated by both natural processes and anthropogenic activities, and probably delineate potential contamination sources [28]. Also, Cr isotope-based studies have shown oxidative mobilization of Cr(VI) from ultramafic host rocks, and successive back-reduction of the thus mobilized Cr(VI) fractions [29]. Furthermore, the record of a 12-month long time-series of $\delta^{53}Cr$ values in run-off from a small serpentinite-dominated catchment in Central Europe, revealed that the Cr(VI) export flux during winter was significantly higher than during the summer [29]. These recent studies and respectively supporting data, in the light of the widespread distribution of ophiolite complexes in Europe and in orogenic zones elsewhere around the globe, emphasize the need for further more detailed studies addressing the variation of $\delta^{53}Cr$ values in Cr-bearing rocks and ores.

The ophiolite complexes form an important component of the Tethyan metallogenetic belt, extending from the Mirdita zone in Albania to the Vourinos, Pindos and Othrys complexes of the south zone. These ophiolite complexes host large chromite deposits and smaller chromite occurrences, characterized by high-Cr and high-Al, low- and high-IPGE and PPGE compositions, and by PGE-patterns with both negative and positive slopes [10,14,15,31]. Such a compositional diversity in chromitites may provide an opportunity for in detail studies aimed at a better understanding of the role of magmatic processes on the Cr isotope signatures of the respective chromites [21–24]. In our study we present the first systematic chromium isotope data (expressed as $\delta^{53}Cr$ values) of massive chromitites hosted in ophiolite complexes of the Balkan Peninsula, along with trace element contents, including PGE, and combined with scanning electron microscopy/energy-dispersive X-ray spectroscopy (SEM/EDS) identifications. We aim at delineating geochemical constraints that potentially control the elemental composition of these chromitites and their Cr isotope compositions, ultimately with the overall intention to contribute to the origin and genesis of chromite and PGE mineralization and deposits formation.

2. Materials and Methods

2.1. Preparation and Analysis of the Chromitites

For the purpose of the present study, 2.5 to 3.0 kg amounts of chromitite sample mostly from locations of the Balkan Peninsula (Table 1) were crushed into pieces of approximately 0.3 × 0.3 cm with the help of a jaw crusher. The crushed piles were mixed well and divided into four equal portions. Subsequently, two fractions were removed and the remains were remixed again. The procedure was repeated until the final split weighed about 500 g. This split, serving as experimental material, was pulverized to about 100 mesh using an agate mortar. A conventional oxidative alkaline fusion (OAF) was carried out in corundum crucibles; about 1 g of sodium peroxide (Na_2O_2) and 0.3 g of Na_2CO_3 were added to about 0.1 to 0.3 g of the powdered sample (amount depended on the concentration of Cr in the sample). The sample and flux materials were mixed and molten at temperatures of 700 to 800 °C in a muffle furnace during 10 min. After cooling, the fused cake was extracted from the crucible and transferred into 100 mL volumetric flasks using deionized water [2]. The solutions were finally filtered through 0.45 μm polyamide membrane filters and aliquots of these solutions were then processed for chromium isotope analysis.

2.2. Chromium Isotope Analysis

Solutions of samples in the amount which would yield about 1 μg of total chromium were pipetted into 13 mL Savillex Teflon beakers. These aliquots were spiked up with adequate amount of a ^{50}Cr–^{54}Cr double spike so that a sample to spike ratio of ~3:1 (total chromium concentrations) was achieved. The addition of a ^{50}Cr–^{54}Cr double spike of a known isotope composition to a sample before chemical purification allows accurate correction of both the chemical and the instrumental shifts in Cr isotope abundances [1,32]. The mixture was totally evaporated and 1 mL of concentrated aqua regia was subsequently added. After 3 h during which the sample was exposed to aqua regia on a hotplate at 100 °C, the sample was again dried down. The sample was then purified by passing the sample in 0.5 N HCl over an extraction column (BioRad PP columns) charged with 1 mL of 200–400 mesh BioRad AG-50W-X12 cation resin, employing a slightly modified extraction recipe published [33,34]. The Cr yield of this column extraction and purification step is usually ~70%. Samples were loaded onto Re filaments with a mixture of 3 μL silica gel, 0.3 μL 0.5 mol L^{-1} of H_3BO_3 and 0.5 μL 0.5 mol L^{-1} of H_3PO4. The samples were statically measured on a IsotopX "Phoenix" multicollector thermal ionization mass spectrometer (TIMS) at the Department of Geoscience and Natural Resource Management, University of Copenhagen, at temperatures between 1040 and 1150 °C, aiming for beam intensity at atomic mass unit (AMU) 53.9407 of 1 to 2 Volts. Every load was analyzed five times. Titanium, vanadium and iron interferences with Cr isotopes were corrected by comparing with $^{49}Ti/^{50}Ti$, $^{50}V/^{51}V$ and $^{54}Fe/^{56}Fe$ ratios. The final isotope composition of a sample was determined as the average of the repeated analyses and reported relative to the certified (standard reference material) SRM 979 standard as

$$\delta^{53}Cr(‰) = [(^{53}Cr/^{52}Cr_{sample}/^{53}Cr/^{52}Cr_{SRM979}) - 1] \times 1000$$

Repeated analysis of 1 μg loads of unprocessed double spiked SRM 979 standard during the duration of the analysis period yielded an average $\delta^{53}Cr$ value of 0.05 ± 0.06‰ (n = 12; 2σ; ^{52}Cr signal intensity at 2 V) on the "Phoenix" TIMS which we consider as a minimum external reproducibility for a sample run in this study, including separation procedure, double spike correction error, and respective internal analytical errors.

2.3. Mineral Chemistry and Whole Rock Analyses

Polished sections from chromite deposits of Greece (Table 1) were carbon or gold coated and examined by reflected light microscopy and with a SEM using EDS. SEM images and EDS analyses were carried out at the University of Athens (Department of Geology and Geoenvironment) using a JEOL JSM 5600 instrument, equipped with automated energy-dispersive analysis system ISIS 300 OXFORD, with the following operating conditions: accelerating voltage 20 kV, beam current 0.5 nA, time of measurement (dead time) 50 s and beam diameter 1–2 μm. The following X-ray lines were used: OsMα, PtMα, IrMα, AuMα, AgLα, AsLα, FeKα, NiKα, CoKα, CuKα, CrKα, AlKα, TiKα, CaKα, SiKα, MnKα, MgKα, ClKα. Standards used were pure metals for the elements Cr, Fe, Mn, Ni, Co, Ti and Si, MgO for Mg and Al_2O_3 for Al. Contents of Fe_2O_3 and FeO were calculated on the basis of the spinel stoichiometry.

Major and trace elements in massive chromitite samples were determined by inductively coupled plasma mass spectrometry (ICP-MS) analysis, at the ACME Analytical Laboratories Ltd., Vancouver, BC, Canada. The samples were dissolved using a strong multi-acid (HNO_3–$HClO_4$–HF) digestion and the residues dissolved in concentrated HCl. Platinum-group element (PGE) analyses were carried out by Ni-sulphide fire-assay pre-concentration technique, using the nickel fire assay technique from large (30 g) samples. This method allows for complete dissolution of samples. Detection limits are 5 ppb for Ru 2 ppb for Os, Ir, Pt, Pd and 1 ppb for Rh and Au. The CDN-PGMS-23 was used as standard.

3. A Brief Outline of Characteristics of Ophiolites and Hosted Chromitites

3.1. Ophiolites

The geology, petrography, mineral chemistry and geochemistry of ophiolites in the Balkan Peninsula has been a topic of extensive investigation in previous publications [10,35–48]. They are an important component of the Upper Jurassic to Lower Cretaceous Tethyan ophiolite belt, which extends through the Serbian zone of the Dinarides in the north (Mirdita zone) to the Subpelagonian zone (Pindos, Vourinos and Othrys complexes) in the south (Figure 1).

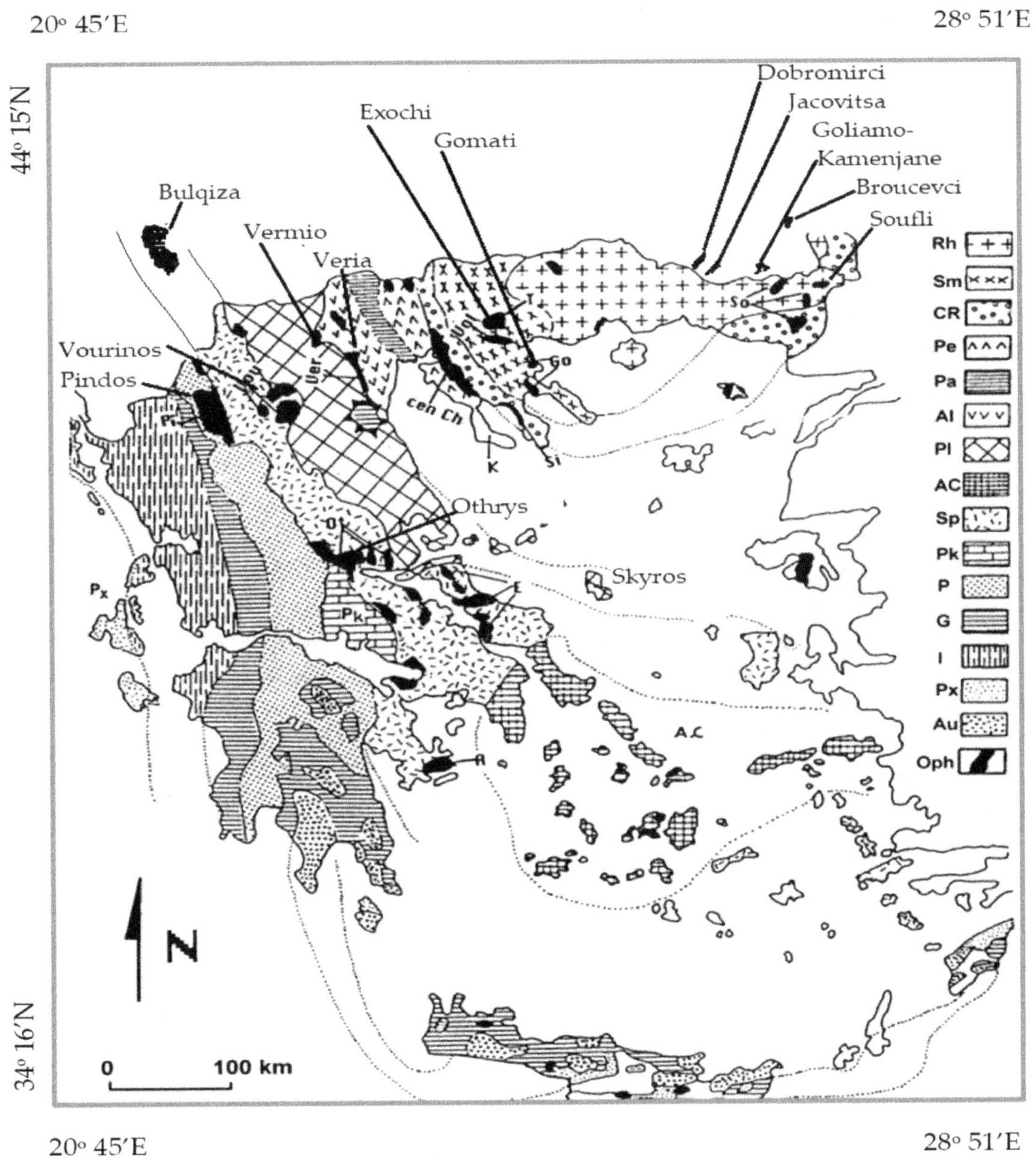

Figure 1. Simplified map of the geotectonic zones of Hellenides [35], showing the ophiolite complexes from which chromitite samples were studied. Symbols: Oph = ophiolites; Rh = Rhodope zone; Sm = Serbo–Macedonian Massif; CR = Circun–Rhodope zone; (Pe, Pa, Al) = Axios zone; Pl = Pelagonian zone; AC = Attico–Cucladic zone; Sp = Pindos zone; P = Parnassos–Giona zone; G = Gavrovou–Tripolis zone; I = Ionian zone; Px = Pakson zone.

These ophiolites are characterized by petrological and geochemical features typical of both Mid-Ocean Ridge (MOR) and Supra-Subduction Zone (SSZ)-type associations (fore-arc and back-arc ridges). They are accompanied by minor dunite bodies, and overlain by gabbroic cumulates, mafic dykes, and an extrusive sequence with a compositional basalt range between MOR and island-arc-tholeiitic to boninitic types [35–48]. Recently, ophiolite studies have focused on the importance of intra-oceanic subduction-initiation processes in ophiolite genesis [49]. Relatively small isolated ophiolite masses, of mostly serpentinized dunite and harzburgite, are located at the western margin of the Axios zone (Vermio–Veria), and the Eohellenic Pre-Cretaceous nape, including Skyros Island (Figure 1). Structural and paleomagnetic studies on those ophiolites have revealed widespread heterogeneous deformation and rotation during their original displacement and subsequent tectonic incorporation into continental margins [35–38]. The origin of major Jurassic ophiolite complexes in Greece in a hydrous SSZ environment has been evidenced by the presence of hydrous silicate inclusions, such as amphibole and phlogopite, in chromitites and by the trace and Rare Earth Elements (REE) data on separated orthopyroxenes and clinopyroxenes from harzburgites of these ophiolites [48]. Pyroxenites with variable modal contents of olivine, garnet and spinel, ranging in composition from orthopyroxenite through websterite to clinopyroxenite have been described in the Veria ophiolites [50,51].

3.2. *Characteristics of Chromitites*

Chromitites in the Balkan Peninsula are classified on the basis of the variation in the chromitite tonnage, the composition of chromite, the degree of transformation of ores and the associated ophiolites, and on the association of chromitites with sulphides.

The Othrys complex includes two tectonically separated chromite deposits, namely Eretria (Tsagli) and Domokos, in addition to several other occurrences, including Agios Stefanos, the combined tonnage being approximately 3 Mt of high-Al massive chromite ores occurring in moderately depleted harzburgite [10,35,39,44,45]. A salient feature of the Eretria (Tsagli) chromite deposits is the occurrence of Fe–Ni–Cu-sulphide mineralization with dominant minerals pyrrhotite, chalcopyrite and minor cobalt-bearing pentlandite, hosted in serpentinized harzburgites [52]. Massive sulphide mineralization occurs at the peripheral parts of podiform chromite bodies in association with chromitite and magnetite (Figure 2). Chromite is often characterized by brittle deformation and by sulphide-silicate veins (Figure 2d,e). Texture and geochemical characteristics, including PGE content, flat chondrite-normalized PGE-patterns, and very low partition coefficients for Ni and Fe between olivine and sulphides are inconsistent with sulphides having been in equilibrium with Ni-rich host rocks, at magmatic temperature [52,53]. In addition, it has been documented [54] that in the Othrys and Vourinos ophiolite complexes massive and disseminated chromitites host considerable amounts of methane in micro-fractures and in porous serpentine- or chlorite-filled veins.

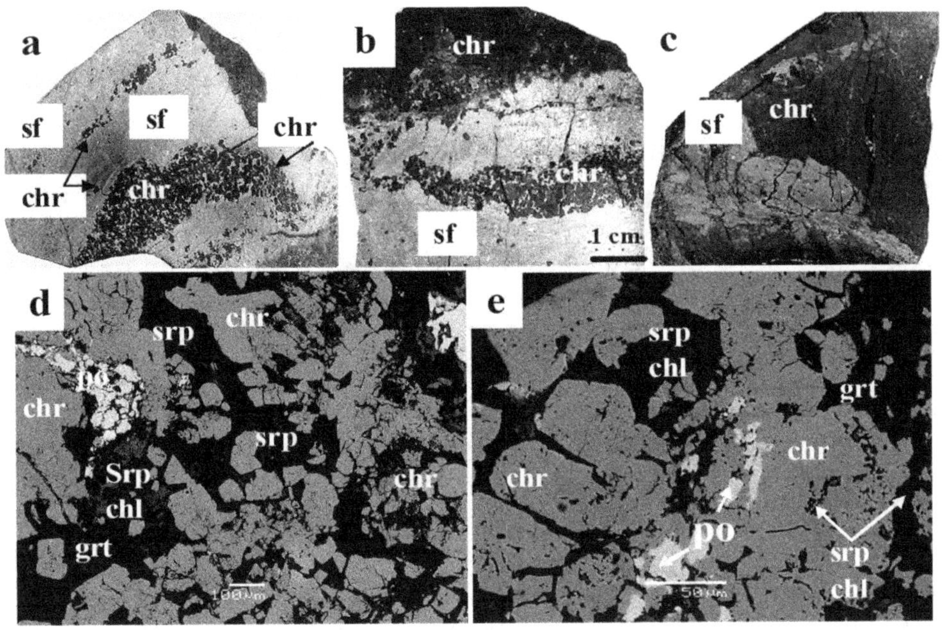

Figure 2. Photographs showing the texture of chromitite associated with sulphides and magnetite from the peripheral parts of podiform chromitite bodies at Eretria. Chromite occurs interstitially in sulphide rich portions (**a**,**b**) and sulphides occur as veins and irregular masses within zones dominated by chromite (**b**,**c**). Back-scattered electron images from massive chromitite samples from central parts of podiform bodies, showing fragmented chromite and evidence for the existence of sulphides (**d**,**e**) even distant from the peripheral parts of the chromitite bodies. Abbreviations: chr = chromite; sf = sulphides; po = pyhhrotite; srp = serpentine; chl = chlorite; grt = garnet. Scale bar: 1 cm is the same for all hand samples.

Chromite ores in the Vourinos complex occur in tectonite and cumulate sequences. Only those in the tectonites are under exploitation. All chromite ores, with a tonnage estimated to be approximately 10 Mt of high-Cr ore are found within dunite bodies or dunite envelopes in harzburgite, but there is no systematic relation between size of the dunite body and that of the ore body. The sizes of the ore bodies vary widely and contain all textural types (massive, schlieren, banded, disseminated and nodular), but usually a single type dominates. High-temperature deformation, superimposed on primary magmatic textures is very common [42,47]. The mantle sequence of the Pindos ophiolite complex resembles that of Vourinos in the presence of extensive and highly depleted harzburgite but, in contrast to Vourinos, there is only a limited number of small chromite occurrences (low potential for exploitation).

The chromitite occurrences in the Pindos ophiolite complex are small (a few tens of meters × a few tens of centimeters) and cover all textural types (massive, schlieren, banded, disseminated and nodular). Chromitites occur within completely serpentinized and weathered, intensively deformed dunite–harzburgite blocks, as a result of strong plastic and brittle deformation that was superimposed on primary magmatic textures. The chromitites are mostly fine- to medium-grained, and consist of aggregates of fractured chromite accompanied by chlorite and traces of tremolite. Primary olivine is preserved only in the form of inclusions within chromian spinel, while abundant remnants of base metal sulphides (BMS), now preserved as alloys, occur both as inclusions and interstitial phases in the silicate matrix [10,18,19]. Chromitites throughout the Pindos complex are of high-Cr and high-Al compositions, often in a spatial association with each other [45,55–57]. The most salient feature of the Pindos chromitites is the PPGE-enrichment in those bodies occurring in the area of Korydallos, with PGE_{total} concentrations reaching 7 ppm [43–45] and reportedly up to 29 ppm [18].

Relatively small massive chromitite bodies are located at the Achladones area on the island of Skyros. They are of high-Al type (average Cr/(Cr + Al) ratio of unaltered chromite is 0.56 while the $Mg/(Mg + Fe^{2+})$ ratio is 0.64) and contain elevated PGE contents, up to 3 ppm PGE_{total}. However, both high-Cr and high-Al types have been reported from Skyros [46,47,58]. Furthermore, SEM/EDS investigations of Au-coated polished sections of chromitite from Skyros revealed the presence of graphite (Figure 3).

Ophiolites associated with the Serbomacedonian massif (Gomati and Exochi) and Rhodope massif including the ophiolites Soufli–Tsoutoura (Greece), Dobromirci, Jacovitsa, Broucevci and Goliamo–Kamenjane (Bulgaria) (Figure 1) are completely serpentinized, locally sheared and metamorphosed to antigorite–tremolite and/or talc schists, and they are all small (a few thousand tons of ore). A characteristic feature of those chromitite occurrences is the common spatial association of high-Cr and high-Al ores, and the negative slope of the PGE patterns, comparable to those for podiform chromitites in ophiolites elsewhere [59–61]. The area of Ceruja is part of the western Bulqiza ophiolite complex as part of the Mirdita zone (Albanides), and includes harzburgitic to lherzolitic types [62]. Although more than 40 Mt of chromitites have been located in the uppermost part of the mantle harzburgites throughout the Bulqiza complex, in the area of Ceruja there are only small (a few thousand) chromite occurrences.

There is a wide compositional variation in chromite throughout the Bulqiza complex, with Cr/(Cr + Al) ranging from 0.46 to 0.86 and Mg/(Mg + Fe^{2+}) ratios ranging from 0.46 to 0.69. The Cr/Cr + Al) ratio in studied samples from the area of Ceruja is restricted to 0.43–0.56, their PGE content is low and the PGE-patterns show a negative slope, similar to other chromitites hosted in ophiolite complexes [62]. However, in the western Bulqiza massif, mineralization occurring as disseminations in dunite is accompanied by chromite in the transition zone and lower cumulates sequence of the complex and elevated PGE contents, up to 9000 ppb ΣPGE [62–64].

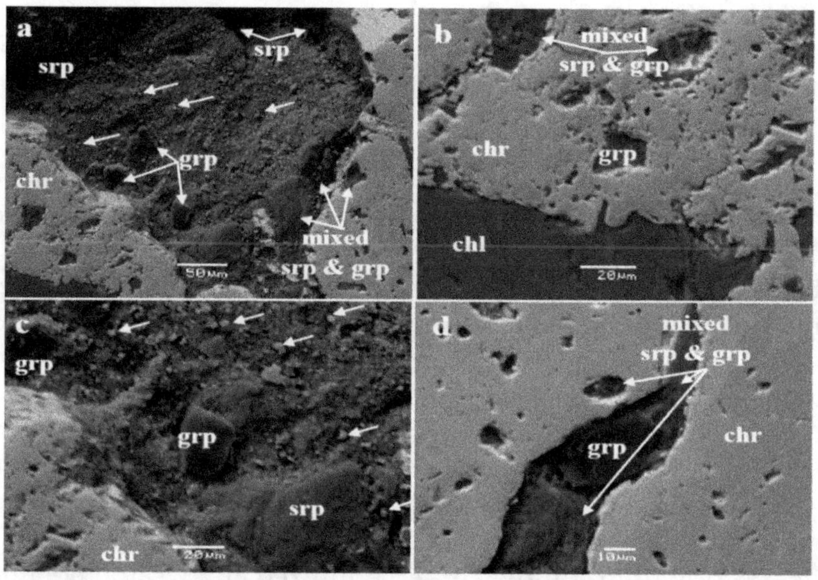

Figure 3. Back-scattered electron images (Au-coated polished sections) of strongly fragmented chromitite from Skyros Island, showing microstructural inclusions of graphite. Partial replacement of serpentine flakes by graphite (**a**,**b**,**d**); Overgrowth of fine graphite nodules (white arrows) are revealed on unpolished parts of polished sections (**a**,**c**); (**f**) Dissolution pits in silicate inclusions of chromite are filled with graphite (**b**,**d**). Abbreviations: grp = graphite-like chr = chromite; srp = serpentine; chl = chlorite.

4. Results

4.1. Distribution of Selected Trace Elements

Selected trace element contens of Ni, Co, Mn, Zn, V, Ga Ti, and platinum-group elements, which are from previous studies, are contained in Table 1.

Table 1. Chromium isotope ($\delta^{53}Cr$) and selected trace element data for chromitites from the Balkan Peninsula.

No	Location	$\delta^{53}Cr$	SEM/EDS		PGE Content (ppb)									Trace Element Content (ppm)							wt %	wt %	wt %
		(‰)	Cr≠	Mg≠	Os	Ir	Ru	Rh	Pt	Pd	ΣPGE	Pd/Ir	Pt/Pt *	Ni	Co	V	Zn	Mn	Ga	Sc	Fe	Ca	Ti
	Othrys (D), GR																						
1	Eretria	0.023	0.57	0.63	22	13	67	2	10	1	115	0.08	2.26	1600	220	900	340	1250	34	6	9.6	<0.1	0.03
2	Eretria	0.036	0.56	0.66	25	14	25	3	4	1	72	0.07	0.74	1400	210	1100	430	990	32	6	10	0.2	0.03
3	Domokos	0.025	0.53	0.74	15	11	30	5	24	2	87	0.18	2.43	1600	200	960	310	1300	32	7	9.8	<0.1	0.04
4	Domokos	0.088	0.5	0.73	20	10	35	5	8	11	89	1.1	0.34	1200	220	900	400	2500	33	6	9.6	<0.1	0.04
	Vourinos (D), GR																						
5	Kondro		0.77	0.65	26	36	55	17	4	2	140	0.06	0.22	2000	240	500	260	1560	14	6	8.1	<0.1	0.04
6	Voidolakos	−0.12	0.81	0.62	14	17	80	12	6	6	135	0.35	0.23	1580	210	560	550	1540	14	6	7.9	<0.1	0.04
7	Kissavos	−0.093	0.81	0.66	14	11	49	7	4	7	92	0.55	0.55	1900	200	400	300	1000	13	6	8.7	<0.1	0.04
8	Kissavos	−0.078	0.56	0.72	4	3	11	3	3	6	30	0.54	0.23	1800	170	780	360	850	27	6	9.2	<0.1	0.03
9	Mikrokleisoura	−0.056	0.78	0.61	30	11	40	16	8	4	109	0.63	0.26	1800	180	620	280	950	14	8	7.8	<0.1	0.05
10	Mikrokleisoura	−0.032	0.61	0.52	50	8	42	18	8	6	132	0.75	0.96	1000	230	1000	490	1760	20	12	15	0.5	0.1
	Pindos, Low PGE																						
11	P Dako's mine	−0.03	0.52	0.63	4	6	30	1	4	6	51	1.0	0.52	1300	140	600	400	760	23	5	7.8	<0.1	0.04
12	Korydallos	0.009	0.48	0.6	40	20	33	5	25	20	143	1.0	0.8	1500	260	710	410	1200	32	5	9.8	<0.1	0.08
13	Kampos Despoti	−0.096	0.81	0.58	30	18	50	3	76	4	181	0.22	7.01	1450	290	560	490	1470	17	5	10	<0.1	0.03
14	Korydallos	−0.149	0.42	0.71	8	3	11	1	75	19	117	6.33	5.5	1500	260	710	520	1300	34	5.1	7.8	<0.1	0.09
	Pindos, High-PPGE																						
15	Korydallos	−0.098	0.68	0.57	14	11	57	13	3460	1660	5215	151	7.53	1550	230	510	450	1200	30	5.7	7.8	<0.1	0.07
16	Korydallos	0.062	0.67	0.57	47	49	55	104	3020	600	3875	12.2	3.86	750	270	760	520	1280	16	6.8	9.7	<0.1	0.07
17	Korydallos	−0.102	0.69	0.49	62	47	80	112	1460	337	2098	7.2	2.4	720	240	760	620	1340	23	6.7	10	<0.1	0.08
18	Korydallos	−0.06	0.4	0.72	88	74	130	142	1720	525	2679	7.1	2.01	630	240	790	510	1300	25	6.8	9.9	<0.1	0.09
19	Korydallos	−0.086	0.4	0.72	45	38	82	78	4580	1300	6123	34.2	4.6	1630	240	580	460	1400	32	5.4	7.3	<0.1	0.07
	Pindos, High-IPGE																						
20	Milia	−0.147	0.82	0.42	150	320	350	82	150	7	1059	0.02	2.00	1600	340	980	1160	2000	15	5	12	0.1	0.01
	Skyros, high IPGE																						
21	Achladones	−0.078	0.61	0.64	140	480	1200	160	280	39	2300	0.08	1.13	1300	250	1200	540	1300	34	5	11	<0.1	0.04
22	Achladones	0.117	0.55	0.69	60	40	300	13	23	28	464	0.7	0.39	1600	200	1000	400	1650	36	5	9.4	<0.1	0.04
23	Achladones	0.159	0.58	0.51	13	31	159	7	20	21	251	0.67	0.53	1500	240	870	450	1600	40	5	9.6	<0.1	0.04
	Skyros, Low PGE																						
24	Ag. Ioannis	−0.111	0.75	0.69	30	15	85	6	4	5	145	0.33	0.23	1250	220	640	420	2150	14	6	11	<0.1	0.08
25	Parapisti	−0.084	0.79	0.69	25	20	90	5	3	2	145	0.1	0.3	1200	200	620	400	1900	11	5	10	<0.1	0.05
26	Ceruja	−0.051	0.43	0.69	3	7	18	3	25	7	63	1	1.74	2640	170	610	300	960					
27	Ceruja	−0.129	0.52	0.58	3	7	15	2	28	9	64	1.3	2.1	1640	160	1030	420	1210					
28	Ceruja	−0.047	0.56	0.65	3	7	18	3	25	7	63	1	1.74	2150	160	990	340	1100					

Table 1. *Cont.*

No	Location	$\delta^{53}Cr$	SEM/EDS		PGE Content (ppb)									Trace Element Content (ppm)							wt %	wt %	wt %
		(‰)	Cr≠	Mg≠	Os	Ir	Ru	Rh	Pt	Pd	ΣPGE	Pd/Ir	Pt/Pt *	Ni	Co	V	Zn	Mn	Ga	Sc	Fe	Ca	Ti
	Vermio-Veria																						
29	Veria	−0.075	0.75	0.62	70	15	100	9	3	2	199	0.13	0.23	1500	200	650	700	2800					
30	Vermio	−0.076	0.7	0.65	30	20	55	4	4	31	144	0.05	0.12	880	170	700	500	1400					
31	Rhodope massif, GR	−0.145	0.75	0.66	70	15	100	9	3	2	199	0.13	0.23	1770	150	550	300	1800					
	Rhodope massif, Bu																						
32	Soufli	−0.037	0.7	0.62	20	25	85	7	7	5	150	0.2	0.38	1260	210	610	360	1850					
33	Tsoutoura	−0.033	0.63	0.65	25	6	40	2	3	6	82	1.00	0.28	1430	220	670	360	1270					
34	Dobromirci	−0.184	0.65	0.68	16	32	120	6	4	14	190	0.44	0.14	520	160	590	230	2300					
35	Goliamo Kamenjane	−0.099	0.6	0.62	3	6	14	3	3	1	30			2000	780	500	3700	7500					
36	Broucevci	−0.147	0.47	0.72	5	13	23	4	2	12	60	0.92	0.09	800	180	600	370	2500					
37	Jacovitsa	−0.097	0.75	0.61	22	42	53	7	3	2	129	0.05	0.26	790	240	650	400	3500					
	SMM, GR																						
38	Exochi	0.02	0.76	0.62	23	24	73	6	5	7	140	0.29	0.25	1240	240	670	400	1850					
39	Exochi	−0.057	0.75	0.6	25	10	70	4	4	6	119	0.6	0.26	1030	150	1300	300	1270					
40	Gomati	0.084	0.49	0.67	9	20	60	6	5	4	104	0.25	0.33	1050	330	660	350	2700					

* Present data; PGE data from previous publications [24,34,43,44,50]. Symbols: Cr# = Cr/(Cr + Al); Mg# = (Mg + Fe^{2+}).

The rare earth elements La, Ce, Nd, Sm, Yb, Pr, Eu, Gd, Tb, Dy, Ho, Er, Tm and Lu as well as U, Th, Cs, Sb, Y, Zr, Ge, W, Mo, Sc, Pb, Li, Sr and K were lower than the detection limits in detail specified in the section of the analytical methods, except for La and Ce contents in the chromitite samples from Othrys and Skyros island, which are slightly higher than the respective detection limit. The best defined inverse correlations are those between the Cr/(Cr + Al) ratio and V (r = –0.80 to –0.85) for chromitites from Vourinos, Pindos (low PGE) and Skyros, and between the Cr/(Cr + Al) ratio (Figure 4a), and Ga (r = –0.90 to –0.96) for chromitites from Vourinos, Pindos (low PGE) and Skyros and for high-PGE chromitites (r = –0.65; Figure 4b).

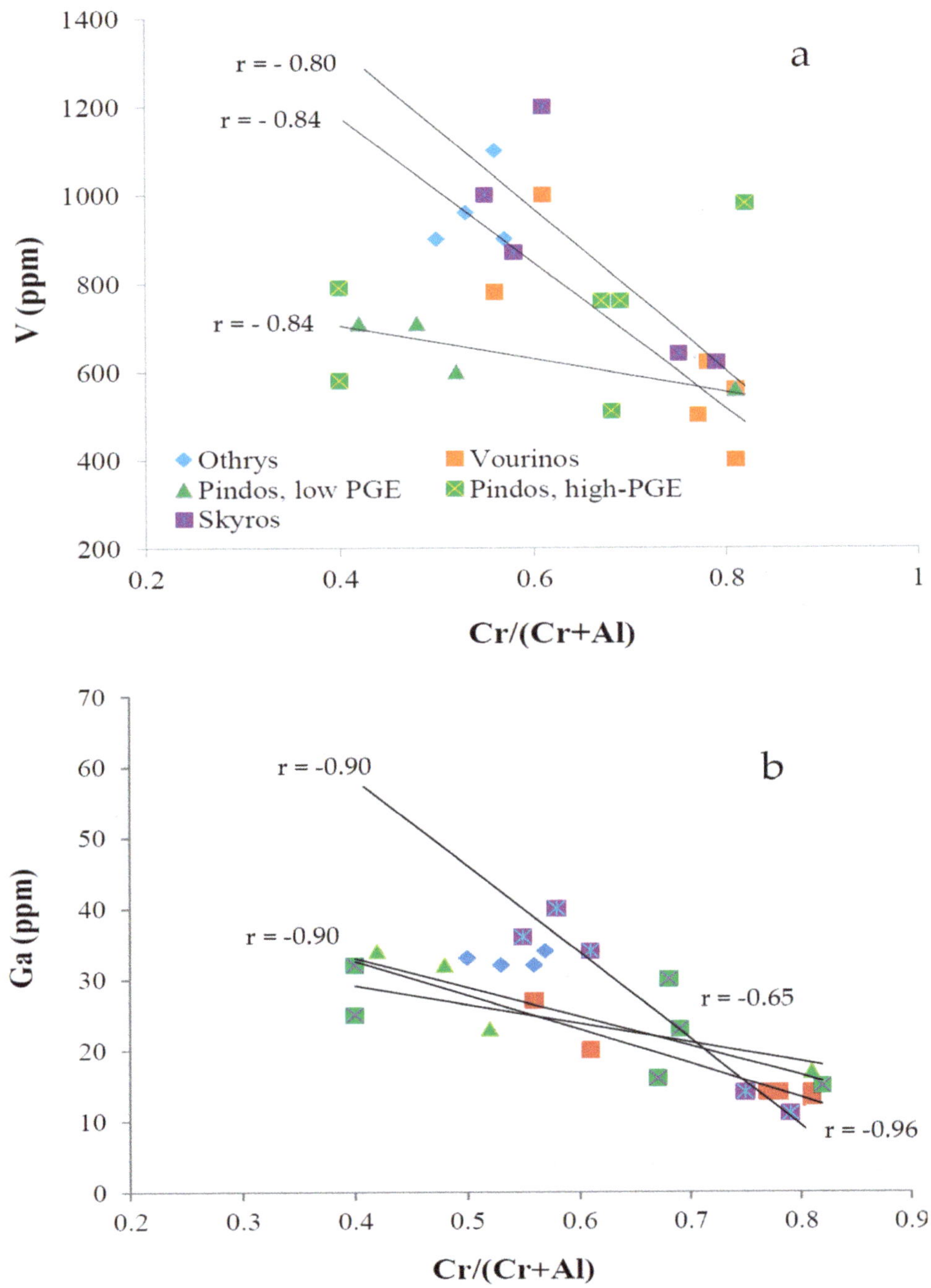

Figure 4. Plots of the Cr/(Cr + Al) ratio versus V and Ga for chromites from the Balkan Peninsula. The best pronounced negative correlations are those between Cr/(Cr + Al) ratio and V for chromitites from Vourinos, Pindos (low PGE) and Skyros (**a**), and Ga for chromitites from Vourinos, Pindos (low PGE) and Skyros (**b**). Data from Table 1.

4.2. Chromium Isotope Results

The analysed chromitites, representative of a wide range in their major element (high-Cr and high-Al) and PGE composition, show a relatively wide range of chromium isotope compositions, expressed as $\delta^{53}Cr$ values, even in spatially related occurrences, throughout the entire metallogenic belt. The $\delta^{53}Cr$ values range from −0.184‰ (in the Bulgarian Rhodope massif) to +0.159‰ on Skyros Island (Figure 5; Table 1). The lower values (−0.246‰) are comparable to those having been reported by [2]. $\delta^{53}Cr$ as low as −0.21‰ are compatible with the $\delta^{53}Cr$ isotope range for mantle-derived rocks [1], while the more positively fractionated values of up to +0.159‰ exceed those values around +0.068‰ reported for global chromitites from layered intrusions and podiform-type occurrences [2].

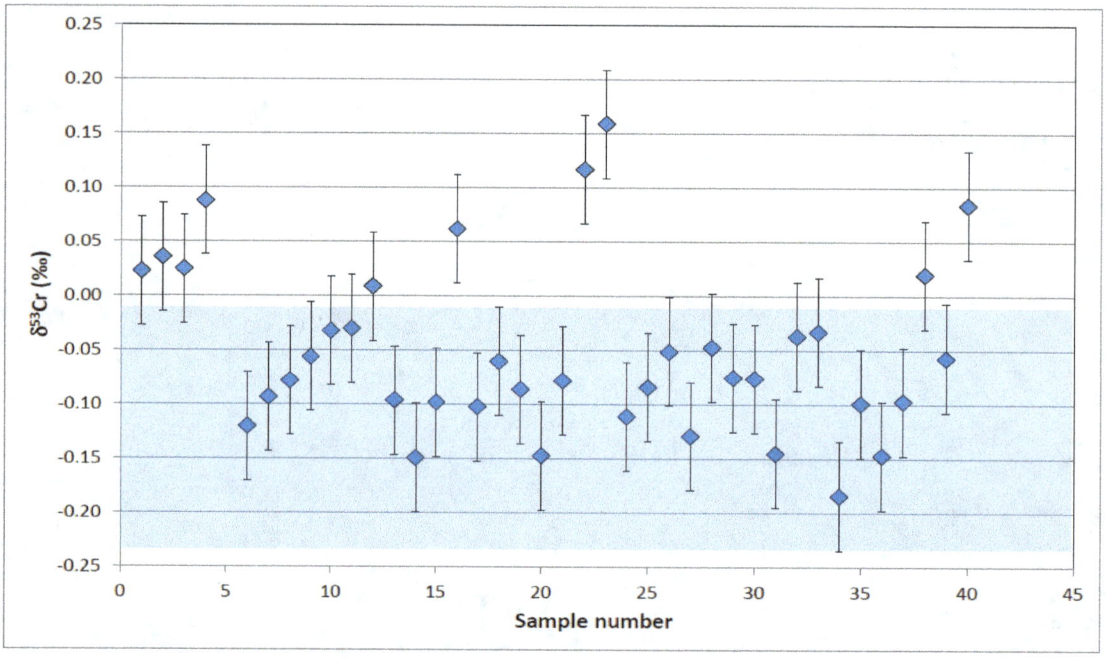

Figure 5. $\delta^{53}Cr$ values of the chromitites (n = 40) from the Balkan Peninsula (Table 1) studied herein. The blue coloured band displays the range of high-temperature magmatic signatures as has been defined by [1]. Errors are 2sd. The reproducibility of the (standard reference material) SRM 979 standard under similar measuring conditions is at +/−0.06‰. The horizontal axis corresponds to the number of analyzed samples 1–40 (Table 1).

Along with the $\delta^{53}Cr$ values of the chromitites, selected major and trace element compositions reported in previous publications are given (Table 1) in order to delineate potential geochemical constraints that control their composition and the Cr isotope compositions. It is remarkable that all chromite samples from Othrys and certain ores from Skyros exhibit positive $\delta^{53}Cr$ values, while chromitites from the Serbo-macedonian-Rhodope massifs and Pindos show a wide range from positive, slightly negative to the most negative $\delta^{53}Cr$ values (Figures 5 and 6; Table 1). Also, the results do not show any correlation trend either between total PGE content or Mg/(Mg + Fe^{2+}) ratio and $\delta^{53}Cr$ values. There is a slightly negative trend between $\delta^{53}Cr$ values and Cr/(Cr + Al) ratio, which is better pronounced for the chromitite samples from Othrys and Vourinos (Figure 6a) and a positive trend between $\delta^{53}Cr$ values and Ga content (Figure 6b).

The Pd/Ir values for chromitites from the Balkan Peninsula are plotted *versus* normalized values (Pt/Pt* = (Pt/8.3)/(Rh/1.6)(Pd/4.4) (Figure 7, [65])). The negative Pt/Pt* values (Pt/Pt* < 1) and low Pd/Ir ratios are a characteristic feature of IPGE-elevated chromitites from Skyros and of low-PGE chromitites, in contrast to PPGE-elevated ones from the Pindos complex (Figure 7; Table 1).

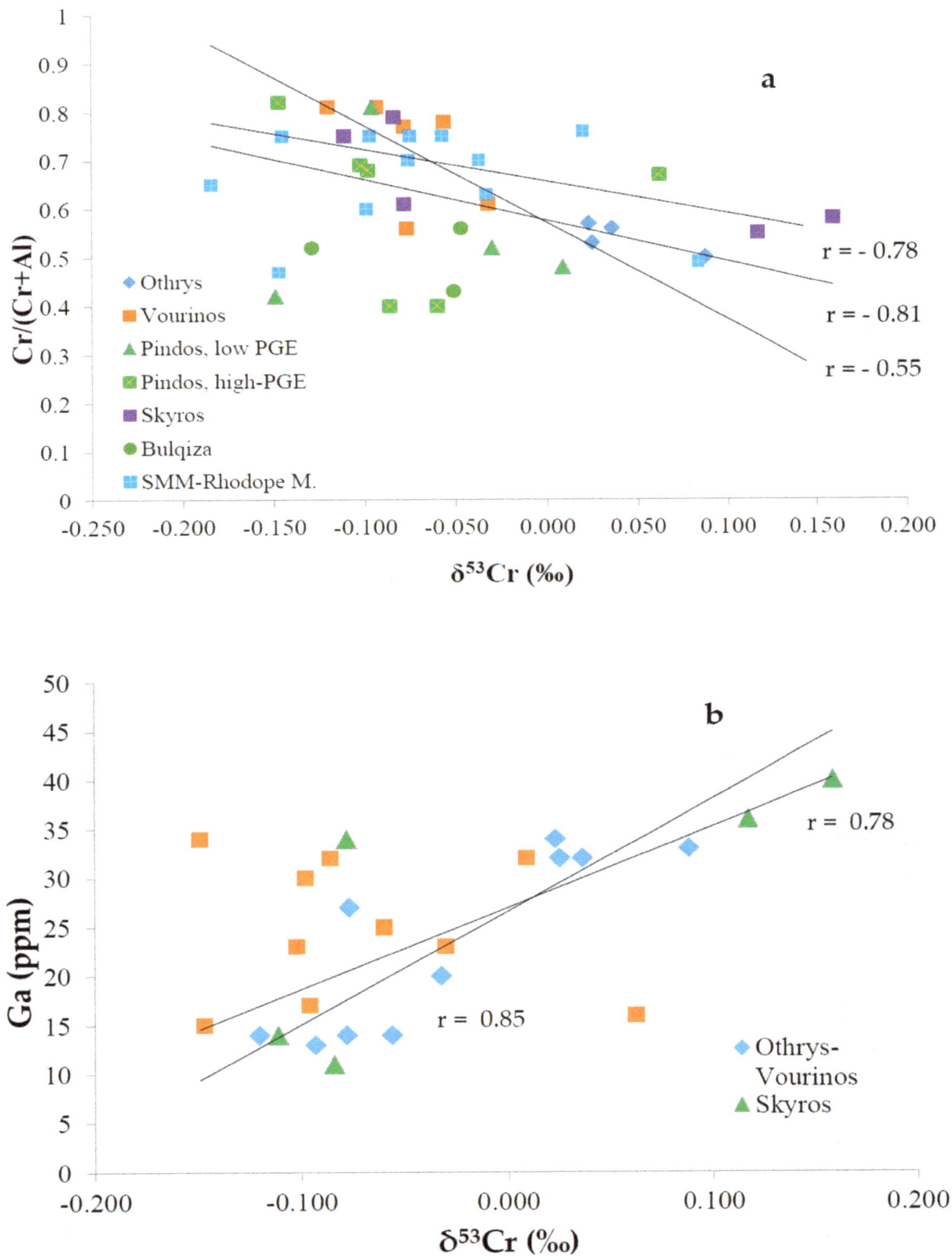

Figure 6. Plots of δ^{53}Cr values *versus* Cr/(Cr + Al) ratios and Ga for certain chromitites from the Balkan Peninsula. There is a good negative trend between δ^{53}Cr values and Cr/(Cr + Al) ratios for chromitites from Othrys and Skyros (**a**), and a positive trend between δ^{53}Cr values and Ga contents for chromitites from Skyros and Vourinos–Othrys (**b**). Data from Table 1.

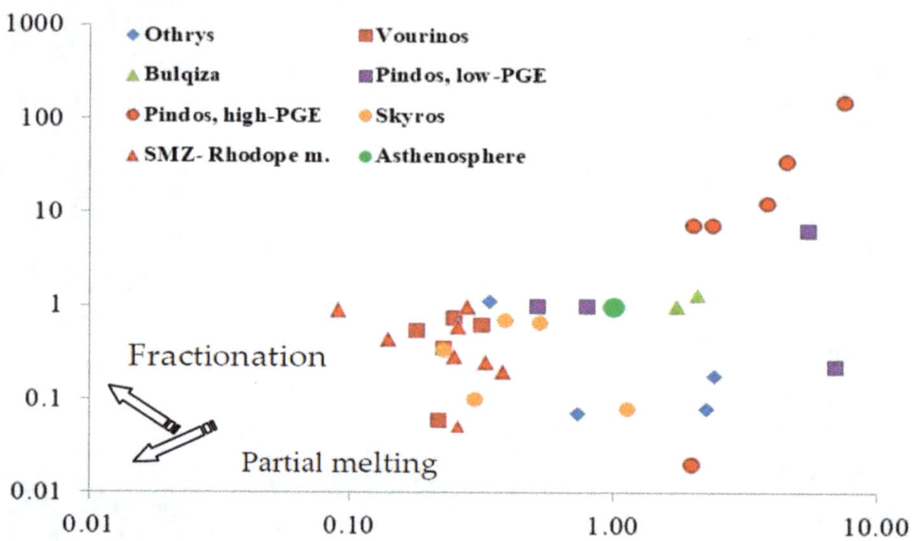

Figure 7. Plots of Pd/Ir *versus* Pt/Pt* normalized values [(Pt/Pt* = (Pt/8.3)/(Rh/1.6) × (Pd/4.4)] after [65] of chromitities studied herein, grouped by locality. Green circle = Asthenosphere mantle, calculated by [66]. Data from Table 1.

5. Discussion

Geochemical compositions and mineral chemistry characteristics of peridotites and chromitites from Greece have shown that the partial melting degree of primitive mantle and the hydrous nature of parent magma are major factors that control the compositional variations in them [10,18,19,42,45,56,57]. In addition, the known widespread geochemical heterogeneities in the bulk-rock compositions of chromitites from the Balkan Peninsula (Table 1) would be inconsistent with their evolution through simple partial melting processes, and suggest that their original geochemical compositions may have been modified by subsequent, post-magmatic processes taking place at shallow mantle depths [8,66–69].

5.1. Potential Pathways Contributing to Chromium Isotope Fractionation

Chromium occurs in different oxidation states in nature, with relatively immobile Cr(III) and very soluble oxidized Cr(VI) being the most abundant species [70]. Redox processes are accompanied by significant isotope fractionation, also in systems where Cr(VI) is transformed to Cr(III) [71]. During reduction, the lighter isotopes are preferentially reduced, resulting in an enrichment of ^{53}Cr relative to ^{52}Cr values in the remaining Cr(VI) pool. This enrichment is measured as the change in the ratio of $^{53}Cr/^{52}Cr$, and is expressed as $\delta^{53}Cr$ values in units permil (‰) relative to a standard [32]. In the following we consider potential pathways for changes in redox conditions that might affect the Cr isotope signatures and their changes in magmatic to post-magmatic processes: (1) measured $\delta^{53}Cr$ signatures in mantle chromitites reflect primary magmatic compositional variations, (2) oxidative fluids may affect solubility of Cr-bearing minerals and release of Cr(VI) during the formation of ferrian-chromite, (3) reductive fluids may cause reduction of Cr(VI) back to Cr(III), deposition of graphite and Fe–Cr alloys and (4) weathering of Cr-bearing minerals [19–21,25,60–64], and related dissolution/re-precipitation processes potentially having a control on the bulk $\delta^{53}Cr$ signatures of chromitites. Based on the observation that bulk silicate earth (BSE) carries a slightly lighter $\delta^{53}Cr$ signature compared to that of mantle-derived chromitites, it has been suggested that small but discernible high-temperature Cr isotope fractionations might occur within the Earth's mantle, potentially imparted during fractional crystallization and/or partial melting of mantle sources [1].

5.1.1. Magmatic Processes

It has been established that high-Cr and high-Al podiform chromitites are derived from melts formed by high and low degrees of partial melting in the upper mantle, respectively [3,13,14,72].

Considering the data distribution in the scatter-diagram of Pd/Ir versus the Pt/Pt* ratios for chromitites from the Balkan Peninsula, it appears that chromitites from the large deposits (Vourinos and Othrys) exhibit Pd/Ir < 1 and Pt/Pt* < 1, suggesting partial melting of already depleted mantle [65], and an increasing partial melting trend from high-Al toward high-Cr deposits (Table 1; Figure 7). In addition, the relatively high Pd/Ir ratio, that is considered to indicate the fractionation degree of the PGE [66], provides evidence for fractionation to some extent of parent magmas for the PPGE-elevated chromitites from Pindos compared to the IPGE-elevated chromitites from Skyros and the Milia area of Pindos (Table 1; Figure 7).

The closer inspection of the Cr isotope data from chromitites studied herein, ranging from negative (−0.184‰) to positive (+0.159‰) δ^{53}Cr values (Table 1; Figures 5 and 6), may provide an opportunity for a better understanding of the role of magmatic or/and post-magmatic processes on the Cr isotope signatures of chromitites [2,21–23]. A comparison of the Cr isotope data for chromitites from our study (Table 1) with published values from other chromitites reveals a significant overlap in the δ^{53}Cr data. Our lowest δ^{53}Cr signature (−0.184‰) is slightly higher than the most negatively fractionated δ^{53}Cr value of −0.246‰ reported by [19], and the most negatively fractionated value of −0.21 ‰ in the data set published [1,2], while the isotopically most positively fractionated value in our data set (+0.159‰) is somewhat higher than the hitherto highest reported δ^{53}Cr value (+0.068‰) for global chromitites from ophiolite complexes and layered intrusions [1,2].

The diagrams of δ^{53}Cr values versus Cr/(Cr + Al) ratios depict better defined negative trends for chromitites from large deposits such as those from Othrys–Vourinos and Skyros (Figure 6a), than for chromitites from smaller occurrences. Given that the former chromitites are characterized by small degrees of fractionation of the respective parental melts, as exemplified by the low Pd/Ir ratios (Table 1), the observed correlation trends may point to the lowering of δ^{53}Cr values in the melts alongside with the partial melting degree of the respective magma sources. In addition, the most evolved samples from the lower cumulate sequence of the Vourinos ophiolite complex (sample 10) has higher Cr isotope compositions, along with higher Fe contents, compared to the mantle chromitites (represented by Voidolakkos, Kondro and Kissavos (Table 1), that may suggest a coupling of increasing trends of δ^{53}Cr values with the degree of fractionation of the parent magma. Also, with respect to trace elements, including Ni, Co, Zn, V, Mn, Ti and Ga, defined correlations are only depicted between Ga and V concentrations, and Cr/(Cr + Al) ratios. Negative trends are revealed by the Othrys–Vourinos and Skyros chromitites, both of which including high-Cr and high-Al ores (Figure 4). Considering the low Pd/Ir ratios shown by those chromitites (Table 1), these negative trends could be attributed to compositional variations in the parent magmas, rather than to changes inherent during fractionation of these melts.

5.1.2. Post-Magmatic Processes

In general, the lack of any clear correlation between major and trace elements, and δ^{53}Cr values may be due to post-magmatic processes (dehydration, serpentinization, metamorphism), which may modify the primary magmatic compositional trends recorded by chromite and silicates (depending on the chromite/silicate ratio), during an extended period of deformation (ductile asthenospheric mantle flow to shallow crustal brittle deformation) [68,72–80]. With respect to the Rhodope massif, on the basis of major-, minor- and trace-element compositions of chromite and Re–Os isotopic compositions of magmatic PGM on the Dobromirtsi chromitites, a long-term history involving reworking of the ancient Archean mantle, during the Phanerozoic, has been suggested [78–82]. Geochemical and mineral chemistry data demonstrated a complex interplay of substitutions, related to the ability of fluids to infiltrate the chromite and the extent of the interaction between pre-existing cores and rims, and that during metamorphism minor and trace elements in chromite can be strongly modified [68]. Also, a sequence of petrological, deformation and redox events has been described for the upper-mantle rocks of the Othrys ophiolite complex. These events are seen to have affected the chromitite bodies during their obduction, emplacement and subsequent alteration [36]. The compositional variations in the Skyros chromitites in particular [83], may have been caused by redox changes during multistage transformations in the Othrys chromites. Cr oxidation and reduction processes accompanying these deformational events may have caused the variations of Cr isotope signatures measured in those chromitites. (Table 1; Figures 4 and 6).

The variations of $\delta^{53}Cr$ signatures in the high-temperature magmatic environment in which the chromitites from the Balkan peninsula were formed, seems to be in a good agreement with the variations measured in some metamorphic Cr-rich minerals such as uvarovites, Cr-tremolite, Cr-diopside, Cr-pyrope. Some of the positively fractionated $\delta^{53}Cr$ signatures of these phases have been interpreted as a consequence of heavy Cr isotope enrichments caused by metamorphism [2,21]. Moreover, the investigation of Cr isotopic composition of different types of mantle xenoliths from diverse geological settings (fertile to refractory off-craton spinel and garnet peridotites, pyroxenite veins, metasomatised spinel lherzolites and associated basalts) revealed $\delta^{53}Cr$ values of peridotites which also span a range from negatively to positively fractioned values of −0.51 to +0.75‰, and a slightly negative correlation between $\delta^{53}Cr$ and Al_2O_3 and CaO contents for most mantle peridotites [23].

5.2. The Role of Sulphides on the $\delta^{53}Cr$ Signatures

The most positively fractionated $\delta^{53}Cr$ values are depicted by all massive chromitite samples from the large Othrys complex, and by a few samples from small chromite occurrences (Table 1). Chromitites in certain ophiolite complexes, such as Othrys, Lemesos, Cyprus, Shetland, Bulqiza, Oregon, Moa-Baracoa and elsewhere are associated with Fe–Ni–Cu-sulphide mineralization with dominant minerals pyrrhotite, chalcopyrite and pentlandite, hosted in mantle serpentinized peridotites and ultramafic rocks of the cumulate sequence of the ophiolite complexes [9,17,52,53,62,63]. Although an originally primary magmatic origin for these sulphides is not precluded, characteristics of the highly transformed ore at the Eretria chromite deposit (Greece) may indicate that the original magmatic features have been overprinted and that metals were released from the host rocks by a low-level hydrothermal circulation process [52]. Therefore, in addition to a potential release of Cr from the chromite ore affected by post-magmatic processes, such as brittle deformation and fluid circulation, the less alkaline environment which would be controlled by the dissolution of sulphides could be potentially important for the control of redox-mediated Cr mobilizations and accompanying changes in $\delta^{53}Cr$ signatures.

5.3. The Potential Role of Abiotic Methane (CH_4)

A salient feature of the presented data that potentially could explain the rather large compositional and isotope variations, could be the presence of graphite (Figure 3), and the observation of methane [54] occurring in fluid inclusions of fracture-filling minerals, along with the presence of chlorite and serpentine. These features are especially present in the Othrys chromitites, which also show the most positively fractionated $\delta^{53}Cr$ values (Table 1; Figure 5). Recently, the investigation of microdiamonds and graphite in chromite from ophiolite-type chromitites hosted in the Tehuitzingo serpentinite (southern Mexico), have been attributed a secondary origin and process related to the retrograde evolution of the respective chromitite. These processes apparently took place at relatively low temperatures (520–670 °C), at low pressures and in shallow depths [69]. The presence of minerals indicative of super-reducing phases, such as graphite-like, Fe–Ni–Cr alloys, awaruite (Ni_3Fe), and heazlewoodite (Ni_3S_2) [44,45,82,83], in the chromitite ores from the islands of Othrys and Skyros may be suggestive of the introduction of continental crust derived, reduced C–O–H fluids, during post-magmatic alteration processes. These were probably facilitated during brittle deformation stages in the shallow crust and in the subducted oceanic slab [69,84,85]. The association of graphite with sperrylite and sulphides in magnetite ore from the Skyros ophiolite complex has also been attributed to its deposition from a CO_2^- and CH_4^-dominated reducing fluid at low oxygen fugacity at relatively low temperatures (500–300 °C) during serpentinization [58]. In addition, methane is widespread in microfractures and porous serpentine- or chlorite-filled veins [54]. The widespread occurrence of methane and Ni–V–Co phosphides in the Othrys chromitites [54,84,85] and graphite in the Skyros island ophiolite (Figure 3) seems to be compatible with the positively fractionated δ^{53}Cr values (Table 1), features pointing to the circulation of reducing fluids, probably during serpentinization, which facilitated the back-reduction of mobilized Cr(VI) fractions to secondary immobile Cr(III)-bearing mineral phases [32,86].

Therefore, the relatively elevated, slightly heavy Cr isotope signatures recorded in chromitites may be the result of a combination of (a) fractionation during magmatic processes, i.e., degree of partial melting and differentiation of the parent magma, and (b) multi-stage post-magmatic overprinting, which seem to depend on the size of the respective chromitite occurrence (in terms of tonnage of chromite) and on a probable relationship between the size and degree of shielding towards alteration of the chromitite bodies.

5.4. *Weathering of Cr-Bearing Minerals and Environmental Significance*

Contamination of groundwater and soil by heavy metals is becoming a serious threat to our environments worldwide. The presence of harmful Cr(VI) in soils and in turn in groundwater may be related to natural processes or/and human activities, such as transfer of weathered material from rocks and primary raw materials, wastes and/or application of large amounts of fertilizers and pesticides for a long time in cultivated areas. The European Soil Data Centre has established a link between the effects of metal bio-availability and metal bio-accumulation, and human health and negative impacts on ecosystems [25–30,87–96]. Dose-dependent differences in toxicities of elements, the particle size, and the oxidation state require serious consideration in health-risk assessments [90]. A very common source of salts in irrigated soils is the irrigation water itself, affected by sea water in low-lying areas along the coast. The salinization of shallow aquifers may have a major effect on plant/crop growth and toxicity. The use of multidisciplinary methods in the study of ecosystem processes in response to groundwater and soil system has great potentials for sustainable developments [87–96].

The application of Cr isotope measurements to evaluate Cr(VI) contamination in groundwater and rock leachates from the central Evia and Assopos basin in Greece has revealed positively fractionated δ^{53}Cr values ranging from +0.56 to +0.96‰ in water leachates of ultramafic rocks, which are comparable to those from Central Europe [26–28]. These results imply oxidative mobilization of Cr(VI) from the ultramafic host rocks, and successive back-reduction of so mobilized Cr(VI) fractions [29]. The widespread distribution of ophiolite complexes in Europe and in orogenic zones elsewhere around the globe, emphasizes the need for further more detailed studies addressing the variation of δ^{53}Cr values in Cr-bearing rocks and ores.

In summary, the original geochemical compositions of chromitites may have been modified by subsequent, post-magmatic processes taking place at shallow mantle depths [35–38,56,57,68,69,72–89,97]. The wide range of δ^{53}Cr values, from positive, slightly negative to the most negatively fractionated signatures, even present in individual relatively large chromitite deposits, together with negative trends observed between δ^{53}Cr values and Cr/(Cr + Al) ratios in chromitites, may reflect a control of Cr isotope compositions in chromitites by degrees of partial melting and by magma fractionation. This is best exemplified by the high-Al chromitites from the cumulate sequence of the Vourinos complex. In addition, post-magmatic redox cycling of Cr-isotopes may occur in response to brittle deformation of chromitites and the subsequent interaction of the Cr-bearing rocks with oxidative fluids, also potentially leading to the formation of Fe-chromite through substitution of Cr, Al and Mg by Fe(III) and Fe(II). The formation of Cr-bearing minerals, such as serpentine, chlorite, and some garnets, and the concomitant release of Cr(VI)-bearing fluids, under alkaline conditions, may take place during serpentinization of respective ore bodies. Mobilized Cr(VI) bearing fluids then are prone to back-reduction promoted by the reduced environment in serpentinized ultramafic rocks, causing the formation of graphite, secondary Fe–Cr-bearing silicates, and other secondary alteration phases that potentially can carry a positively fractionated Cr isotope signal.

6. Conclusions

The $\delta^{53}Cr$ values presented herein for chromitites from the Balkan Peninsula offer an opportunity to study the effects of Cr isotope fractionations that potentially could be related to primary magmatic processes and to compare these with compositional trends eventually depicted by the respective chromite deposits and chromite occurrences.

A modification of the magmatic control on the geochemical characteristics of chromitites is revealed by positively fractionated $\delta^{53}Cr$ values in ophiolite zones affected by brittle deformation/metamorphism and characterized by the presence of secondary epigenetic Cr-bearing minerals, formed by the circulation of reducing fluids carrying abiotic methane.

Chromitites from the Balkan Peninsula depict a wide range in $\delta^{53}Cr$ values. Signatures range from positively-, to slightly negatively fractionated $\delta^{53}Cr$ values, even in individual relatively large deposits. Positively fractionated $\delta^{53}Cr$ values of all chromitite samples from Othrys and of high-Al chromitites from Skyros, and a negatively correlated trend between $\delta^{53}Cr$ and Cr/(Cr + Al), may reflect a control of $\delta^{53}Cr$ by degree of partial melting and by magma fractionation. This is best exemplified by high-Al chromitites from the cumulate sequence of the Vourinos complex.

The application of Cr isotopes to the evaluation of Cr(VI) contamination in groundwater and rock leachates relies on oxidative mobilization of Cr(VI) and successive back-reduction to C(III). The widespread distribution of significant natural Cr-sources in the Balkan peninsula, enhanced by industrial activities involving Cr in their processes, present a latently dangerous framework which is prone to have negative effects on the environment, particularly with respect to hexavalent Cr toxicity. Detailed studies addressing the variation of $\delta^{53}Cr$ values in Cr-bearing rocks, ores, soils, surface run-off and groundwater are required to understand the complicated mutual interplay between original natural Cr sources and the weathering-induced release of Cr into the environment.

Author Contributions: Conceptualization, R.F. and M.E.-E.; Methodology, M.E.-E., R.F. and I.M.; R.F. provided the chromium isotope analyses and their validation. M.E.-E. and I.M. provided the field information and performed the other analyses. All authors (M.E.-E., R.F. and I.M.) contributed to the elaboration, interpretation of the data and to the writing of the manuscript. All authors have read and agreed to the published version of the manuscript.

Acknowledgments: Many thanks are expressed to the two anonymous reviewers for their constructive criticism and suggestions on an earlier draft of the manuscript. Vassilios Skounakis in thanked for his assistance with SEM/EDS analysis of the samples. RF thanks Toby Leeper for always keeping the mass spectrometers at IGN in perfect running condition, and Toni Larsen for help in the ion chromatographic separation of the samples. This paper is dedicated to the memory of Maria Zhelyaskova-Panayiotova, University of Sofia, who provided the chromitite samples from Bulgaria, and passed away in May 2018.

References

1. Schoenberg, R.; Zink, S.; Staubwasser, M.; Von Blanckenburg, F. The stable Cr isotope inventory of solid Earth reservoirs determined by double spike MC-ICP-MS. *Chem. Geol.* **2008**, *249*, 294–306. [CrossRef]
2. Farkaš, J.; Chrastny, V.; Novak, M.; Cadkova, E.; Pasava, J.; Chakrabarti, R.; Jacobsen, S.B.; Ackerman, L.; Bullen, T.D. Chromium isotope variations ($\delta^{53/52}Cr$) in mantle derived sources and their weathering products: Implications for environmental studies and the evolution of $\delta^{53/52}Cr$ in the Earth's mantle over geologic time. *Geochim. Cosmochim. Acta* **2013**, *123*, 74–92. [CrossRef]
3. Melcher, F.; Grum, W.; Simon, G.; Thalhammer, T.V.; Stumpfl, E. Petrogenesis of the ophiolitic giant chromite deposit of Kempirsai, Kazakhstan: A study of solid and fluid inclusions in chromite. *J. Petrol.* **1997**, *10*, 1419–1458. [CrossRef]
4. Arai, S. Conversion of low-pressure chromitites to ultrahigh-pressure chromitites by deep recycling: A good inference. Earth Planet. *Sci. Lett.* **2013**, *379*, 81–87. [CrossRef]
5. Robinson, P.T.; Trumbull, R.B.; Schmitt, A.; Yang, J.S.; Li, J.W.; Zhou, M.F.; Erzinger, J.; Dare, S.; Xiong, F. The origin and significance of crustal minerals in ophiolitic chromitites and peridotites. *Gondwana Res.* **2015**, *27*, 486–506. [CrossRef]

6. Griffin, W.L.; Afonso, J.C.; Belousova, E.A.; Gain, S.E.; Gong, X.H.; González-Jiménez, J.M.; Howell, D.; Huang, J.X.; McGowan, N.; Pearson, N.J.; et al. Mantle recycling: Transition zone metamorphism of Tibetan ophiolitic peridotites and its tectonit implications. *J. Petrol.* **2016**, *57*, 655–684. [CrossRef]
7. Ballhaus, C.; Wirth, R.; Fonseca, R.O.C.; Blanchard, H.; Pröll, W.; Bragagni, A.; Nagel, T.; Schreiber, A.; Dittrich, S.; Thome, V.; et al. Ultra-high pressure and ultra-reduced minerals in ophiolites may form by lightning strikes. *Geochem. Perspect. Lett.* **2017**, *5*, 42–46. [CrossRef]
8. González-Jiménez, G.M.; Camprubì, A.; Colás, V.; Griffin, W.L.; Proenza, J.A.; O'Reilly, S.Y.; Centeno-García, E.; Garcia-Casco, A.; Belousova, E.; Talavera, C.; et al. The recycling of chromitites in ophiolites from southwestern North America. *Lithos* **2017**, *294–295*, 53–72. [CrossRef]
9. Prichard, H.M.; Neary, C.R.; Potts, P.J. Platinum-group minerals in the Shetland ophiolite complex. In *Metallogeny of Basic and Ultrabasic Rocks*; Gallagher, M.J., Ixer, R.A., Neary, C.R., Prichard, H.M., Eds.; Institution of Mining and Metallurgy: Edinburgh, UK, 1986; pp. 395–414.
10. Economou-Eliopoulos, M. Platinum-group element distribution in chromite ores from ophiolite complexes: Implications for their exploration. *Ore Geol. Rev.* **1996**, *11*, 363–381. [CrossRef]
11. Tarkian, M.; Economou-Eliopoulos, M.; Sambanis, G. Platinum-group minerals in chromitites from the Pindos ophiolite complex, Greece. *N. J. Mineral. Mon.* **1996**, *4*, 145–160.
12. Ohnenstetter, M.; Johan, Z.; Cocherie, A.; Fouillac, A.M.; Guerrot, C.; Ohnenstetter, D.; Chaussidon, M.; Rouer, O.; Makovicky, E.; Makovicky, M.; et al. New exploration methods for platinum and rhodium deposits poor in base-metal sulphides-NEXTRIM. *Tran. Inst. Min. Metall. Sect. B Appl. Earth Sci.* **1999**, *108*, B119–B150.
13. Garuti, G.; Zaccarini, F. In situ alteration of platinum-group minerals at low temperature evidence from serpentinized and weathered chromitites of the Vourinos complex (Greece). *Can. Mineral.* **1997**, *35*, 611–626.
14. Proenza, J.A.; Gervilla, F.; Melgarejo, J.C.; Bodinier, J.L. Al- and Cr-rich chromitites from the Mayarí-Baracoa ophiolitic belt (eastern Cuba); consequence of interaction between volatile-rich melts and peridotites in suprasubduction mantle. *Econ. Geol.* **1999**, *94*, 547–566. [CrossRef]
15. Proenza, J.A.; Zaccarini, F.; Escayola, M.; Cábana, C.; Shalamuk, A.; Garuti, G. Composition and textures of chromite and platinum-group minerals in chromitites of the western ophiolitic belt from Córdoba Pampeans Ranges, Argentine. *Ore Geol. Rev.* **2008**, *33*, 32–48. [CrossRef]
16. Uysal, I.; Kapsiotis, A.; Akmaz, R.M.; Saka, S.; Seitz, H.M. The Guleman ophiolitic chromitites (SE Turkey) and their link to a compositionally evolving mantle source during subduction initiation. *Ore Geol. Rev.* **2018**, *93*, 98–113. [CrossRef]
17. Prichard, H.M.; Brough, C. Potential of ophiolite complexes to host PGE deposits. In *New Developments in Magmatic Ni-Cu and PGE Deposits*; Li, C., Ripley, E.M., Eds.; Geological Publishing House: Beijing, China, 2009; pp. 277–290.
18. Kapsiotis, A.; Grammatikopoulos, T.A.; Tsikouras, B.; Hatzipanagiotou, K. Platinum-group mineral characterization in concentrates from high-grade PGE Al-rich chromitites of Korydallos area in the Pindos ophiolite complex (NW Greece). *Resour. Geol.* **2010**, *60*, 178–191. [CrossRef]
19. Kapsiotis, A.; Grammatikopoulos, T.; Tsikouras, B.; Hatzipanagiotou, K.; Zaccarini, F.; Garuti, G. Chromian spinel composition and Platinum-group element mineralogy of chromitites from the Milia area, Pindos ophiolite complex, Greece. *Can. Mineral.* **2009**, *47*, 1037–1056. [CrossRef]
20. O'Driscoll, B.; González-Jiménez, J.M. Petrogenesis of the Platinum-group minerals. *Rev. Mineral. Geochem.* **2016**, *81*, 489–578. [CrossRef]
21. Shen, J.; Liu, J.; Qin, L.; Wang, S.J.; Li, S.; Xia, J.; Zhang, Q.; Yang, J. Chromium isotope signature during continental crust subduction recorded in metamorphic rocks. *Geochem. Geophys. Geosyst.* **2015**, *16*, 3840–3854. [CrossRef]
22. Xia, C.; Guodong, L.; Yue, H.; Yuchuan, M. Precipitation Stable Isotope Variability in Tropical Monsoon Climatic Zone of Asia. *IOP Conf. Ser. Mater. Sci. Eng.* **2018**, *392*, 042028. [CrossRef]
23. Xia, J.; Qina, L.; Shena, J.; Carlsonb, R.W.; Ionovd, D.A.; Mockbet, T.D. Chromium isotope heterogeneity in the mantle. *Earth Planet Sci. Lett.* **2017**, *464*, 103–115. [CrossRef]
24. Bonnand, P.; Williams, H.M.; Parkinson, I.J.; Wood, B.J.; Halliday, A.N. Stable chromium isotopic composition of meteorites and metal-silicate experiments: Implications for fractionation during core formation. *Earth Planet. Sci. Lett.* **2016**, *3*, 4–7. [CrossRef]

25. Slejko, F.F.; Petrini, R.; Lutman, A.; Forte, C.; Ghezzi, L. Chromium isotopes tracking the resurgence of hexavalent chromium contamination in a past-contaminated area in the Friuli Venezia Giulia Region, northern Italy. *Isot. Environ. Health Stud.* **2019**, *55*, 56–69. [CrossRef] [PubMed]
26. Novak, M.; Chrastny, V.; Cadkova, E.; Naldras, J.; Bullen, T.D.; Tylcer, J.; Szurmanova, Z.; Cron, M.; Prechova, E.; Curik, J.; et al. Common occurrence of a positive δ^{53}Cr shift in Central European waters contaminated by geogenic/industrial chromium relative to source values. *Environ. Sci. Technol.* **2014**, *48*, 6089–6096. [CrossRef] [PubMed]
27. Novak, M.; Kram, P.; Sebek, O.; Kurik, J.; Andronikov, A.; Veselovsky, F.; Chrastny, V.; Martinkova, E.; Stevanova, M.; Prechova, E.; et al. Temporal changes in Cr fluxes and δ^{53}Cr values in runoff from a small serpentinite catchment (Slavkov Forest, Czech Republic). *Chem. Geol.* **2017**, *472*, 22–30. [CrossRef]
28. Economou-Eliopoulos, M.; Frei, R.; Atsarou, C. Application of chromium stable isotopes to the evaluation of Cr (VI) contamination in groundwater and rock leachates from central Euboea and the Assopos basin (Greece). *Catena* **2014**, *122*, 216–228. [CrossRef]
29. Novak, M.; Martinkova, E.; Chrastny, V.; Stepanova, M.; Sebek, O.; Andronikov, A.; Curik, J.; Veselovsky, F.; Prechova, E.; Houskova, M. The fate of Cr(VI) in contaminated aquifers 65 years after the first spillage of plating solutions: A δ^{53}Cr study at four Central European sites. *Catena* **2017**, *158*, 371–380. [CrossRef]
30. Qin, L.; Wang, X. Chromium isotope geochemistry. *Rev. Mineral. Geochem.* **2017**, *82*, 379–414. [CrossRef]
31. Economou-Eliopoulos, M.; Eliopoulos, D.G.; Tsoupas, G. On the diversity of the PGE content in chromitites hosted in ophiolites and in porphyry-Cu systems: Controlling factors. *Ore Geol. Rev.* **2017**, *88*, 156–173. [CrossRef]
32. Ellis, A.S.; Johnson, T.M.; Bullen, T.D. Chromium isotopes and the fate of hexavalent chromium in the environment. *Science* **2002**, *295*, 2060–2062. [CrossRef]
33. Trinquier, A.; Elliott, T.; Ulfbeck, D.; Coath, C.; Krot, A.N.; Bizzarro, M. Origin of nucleosynthetic isotope heterogeneity in the solar protoplanetary disk. *Science* **2009**, *324*, 295–424. [CrossRef]
34. Bonnand, P.; Parkinson, I.J.; James, R.H.; Karjalainen, A.M.; Fehr, M.A. Accurate and precise determination of stable Cr isotope compositions in carbonates by double spike MC-ICP-MS. *J. Anal. At. Spectrom.* **2011**, *26*, 528–535. [CrossRef]
35. Moudrakis, D. Introduction to the geology and Macedonia and Trace: Aspects of the geotectonic evolution of the Hellenides. *Bull. Geol. Soc. Greece* **1992**, *30*, 31–46.
36. Jones, G.; Robertson, A. Tectono-stratigraphy and evolution of the Pindos ophiolite and associated units. *J. Geol. Soc.* **1991**, *148*, 267–288. [CrossRef]
37. Rassios, A.; Dilek, Y. Rotational deformation in the Jurassic Mesohellenic Ophiolites, Greece, and its tectonic significance. *Lithos* **2009**, *108*, 207–223. [CrossRef]
38. Rassios, A.H.E.; Moores, E.M. Heterogenous mantle complex, crustal processes, and obduction kinematics in a unified Pindos–Vourinos ophiolitic slab (northern Greece). In *Tectonic Development of the Eastern Mediterranean Region*; Robertson, A.H.F., Mountrakis, D., Eds.; Geological Society of London, Special Publications: London, UK, 2006; Volume 260, pp. 237–266.
39. Rassios, A.; Konstantopoulou, G. Emplacement tectonism and the position of chrome ores in the Mega Isoma peridotites, SW Othris, Greece. *Bull. Geol. Soc. Greece* **1993**, *28*, 463–474.
40. Saccani, E.; Photiades, A.; Santato, A.; Zeda, O. New evidence for supra-subduction zone ophiolites in the Vardar Zone from the Vermion Massif (northern Greece): Implication for the tectono-magmatic evolution of the Vardar oceanic basin. *Ofioliti* **2008**, *33*, 17–37.
41. Pearce, J.A.; Lippard, S.J.; Roberts, S. Characteristics and Tectonic Significance of Supra-Subduction Zone Ophiolites. *Geol. Soc. Lond.* **1984**, *16*, 77–94. [CrossRef]
42. Konstantopoulou, G.; Economou-Eliopoulos, M. Distribution of Platinum-group Elements and Gold in the Vourinos Chromitite Ores, Greece. *Econ. Geol.* **1991**, *86*, 1672–1682. [CrossRef]
43. Beccaluva, L.; Coltorti, M.; Saccani, E.; Siena, F. Magma generation and crustal accretion as evidenced by supra-subduction ophiolites of the Albanide-Hellenide Sub Pelagonian zone. *Isl. Arc* **2005**, *14*, 551–563. [CrossRef]
44. Kapsiotis, A.; Rassios, A.; Antonelou, A.; Tzamos, E. Genesis and Multi-episodic Alteration of Zircon-bearing Chromitites from Ayios Stefanos, Othris Massif, Greece: Assessment of an Unconventional Hypothesis on the Origin of Zircon in Ophiolitic Chromitites. *Minerals* **2016**, *6*, 124. [CrossRef]

45. Kapsiotis, A.; Economou-Eliopoulos, M.; Zhenga, H.; Sud, B.X.; Lenaz, D.; Jing, J.J.; Antoneloug, A.; Velicogna, M.; Xia, B. Refractory chromitites recovered from the Eretria mine, East Othris massif (Greece): Implications for metallogeny and deformation of chromitites within the lithospheric mantle portion of a forearc-type ophiolite. *Chemie der Erde* **2019**, *79*, 139–152. [CrossRef]
46. Barth, M.G.; Mason, P.R.D.; Davies, G.R.; Drury, M.R. The Othris Ophiolite, Greece: A snapshot of subduction initiation at a mid-ocean ridge. *Lithos* **2008**, *100*, 234–254. [CrossRef]
47. Economou, M.; Dimou, E.; Economou, G.; Migiros, G.; Vacondios, I.; Grivas, E.; Rassios, A.; Dabitzias, S. *Chromite Deposits of Greece: Athens*; Theophrastus Publications: Athens, Greece, 1986; pp. 129–159.
48. Bizimis, M.; Salters, V.J.M.; Bonatti, E. Trace and REE content of clinopyroxenes from supra-subduction zone peridotites. Implications for melting and enrichment processes in island arcs. *Chem. Geol.* **2000**, *165*, 67–85. [CrossRef]
49. Stern, R.J. The anatomy and ontogeny of modern intra-oceanic arc systems. In *the Evolving Continents: Understanding Processes of Continental Growth*; Kusky, T.M., Zhai, M.G., Xiao, W., Eds.; Geological Society of London: London, UK, 2010; Volume 338, pp. 7–34.
50. Tsoupas, G.; Economou-Eliopoulos, M. High PGE contents and extremely abundant PGE-minerals hosted in chromitites from the Veria ophiolite complex, northern Greece. *Ore Geol. Rev.* **2008**, *33*, 3–19. [CrossRef]
51. Rogkala, A.; Petrounias, P.; Tsikouras, B.; Hatzipanagiotou, K. New occurrence of pyroxenites in the Veria-Naousa ophiolite (North Greece): Implications on their origin and petrogenetic evolution. *Geosciences* **2017**, *7*, 92. [CrossRef]
52. Economou, M.; Naldrett, A.J. Sulfides associated with podiform bodies of chromite at Tsangli, Eretria, Greece. *Mineral. Depos.* **1984**, *19*, 289–297. [CrossRef]
53. Foose, M.P.; Economou, M.; Panayotou, A. Compositional and mineralogic constraints in the Limassol Forest portion of the Troodos ophiolite complex, Cyprus. *Mineral. Depos.* **1985**, *20*, 234–240. [CrossRef]
54. Etiope, G.; Ifandi, E.; Nazzari, M.; Procesi, M.; Tsikouras, B.; Ventura, G.; Steele, A.; Tardini, R.; Szatmari, P. Widespread abiotic methane in chromitites. *Scient. Rep.* **2018**, *8*, 8728. [CrossRef]
55. Economou-Eliopoulos, M.; Vacondios, I. Geochemistry of chromitites and host rocks from the Pindos ophiolite complex, northwestern Greece. *Chem. Geol.* **1995**, *122*, 99–108. [CrossRef]
56. Economou-Eliopoulos, M.; Sambanis, G.; Karkanas, P. Trace element distribution in chromitites from the Pindos ophiolite complex, Greece: Implications for the chromite exploration. In *Mineral Deposits*; Stanley, C.J., Ed.; Balkema: Rotterdam, The Netherlands, 1999; pp. 713–716.
57. Prichard, H.; Economou-Eliopoulos, M.; Fisher, P.C. Contrasting Platinum-group mineral assemblages from two different podiform chromitite localities in the Pindos ophiolite complex, Greece. *Can. Mineral.* **2008**, *46*, 329–341. [CrossRef]
58. Tarkian, M.; Economou-Eliopoulos, M.; Eliopoulos, D. Platinum-group minerals and tetraauricuprite in ophiolitic rocks of the Skyros Island, Greece. *Mineral. Petrol.* **1992**, *47*, 55–66. [CrossRef]
59. Scarpelis, N.; Economou, M. Genesis and metasomatism of chromite ore from the Gomati area, Chalkidiki, Greece. *Ann. Geol. Pays Hell.* **1978**, *29*, 716–728.
60. Magganas, A.; Economou, M. On the chemical composition of chromite ores from the ophiolitic complex of Soufli, NE Greece. *Ofioliti* **1988**, *13*, 15–27.
61. Zhelyaskova-Panayiotova, M.; Economou-Eliopoulos, M. Platinum-group element and gold concentrations in oxide and sulfide mineralizations from ultramafic rocks of Bulgaria. *Ann. Univ. Sofia Geol. Geogr.* **1994**, *86*, 196–218.
62. Cina, A. Pentlandite Mineralization Related to Albanian Ophiolites. In Proceedings of the XIX CBGA Congress Proceedings, Thessaloniki, Greece, 23–26 September 2010; Volume 100, pp. 317–323.
63. Tashko, A.; Economou-Eliopoulos, M. An overview of the PGE distribution in the Bulqiza ophiolite complex, Albania. *Bull. Geol. Soc. Greece* **1994**, *32*, 193–201.
64. Karaj, N. Reportition des platinoides chromites et sulphures dans le massif de Bulqiza, Albania. In *Incidence sur le Processus Metallogeniques Dans les Ophiolites (These)*; Universite de Orleans: Orleans, France, 1992; p. 379.
65. Garuti, G.; Fershtater, G.; Bea, F.; Montero, P.; Pushkarev, E.V.; Zaccarini, F. Platinum-group elements as petrological indicators in mafic-ultramafic complexes of central and southern Urals: Preliminary results. *Tectonophysics* **1997**, *276*, 181–194. [CrossRef]

66. Barnes, S.-L.; Naldrett, A.J.; Gorton, M.P. The origin of the fractionation of the platinum-group elements in terrestrial magmas. *Chem. Geol.* **1985**, *53*, 303–323. [CrossRef]
67. Garuti, G.; Zaccarini, F.; Economou-Eliopoulos, M. Paragenesis and composition of laurite from the chromitites of Othrys (Greece): Implications for Os-Ru fractionation in ophiolitic upper mantle of the Balkan Peninsula. *Miner. Depos.* **1999**, *34*, 312–319. [CrossRef]
68. Colás, V.; González-Jiménez, J.M.; Griffin, W.L.; Fanlo, I.; Gervilla, F.; O'Reilly, S.Y.; Pearson, N.J.; Kerestedjian, T.; Proenza, J.A. Fingerprints of metamorphism in chromite: New insights from minor and trace elements. *Chem. Geol.* **2014**, *389*, 137–152. [CrossRef]
69. Pujol-Solà, N.; Proenza, I.A.; Garcia-Casco, A.; González-Jiméne, J.M.; Andreazini, A.; Melgarejo, J.C.; Gervilla, F. An Alternative Scenario on the Origin of Ultra-High Pressure (UHP) and Super-Reduced (SuR) Minerals in Ophiolitic Chromitites: A Case Study from the Mercedita Deposit (Eastern Cuba). *Minerals* **2018**, *8*, 433. [CrossRef]
70. Losi, M.E.; Amrhein, C.; Frankenberger, W.T. Environmental biochemistry of chromium. *Rev. Environ. Contam. Toxicol.* **1994**, *136*, 91–121. [PubMed]
71. Schauble, E.; Rossman, G.R.; Taylor, H.P., Jr. Theoretical estimates of equilibrium chromium-isotope fractionations. *Chem. Geol.* **2004**, *205*, 99–114. [CrossRef]
72. Zhou, M.F.; Robinson, P.T.; Su, B.X.; Gao, J.F.; Li, J.W.; Yang, J.S.; Malpas, J. compositions of chromite, associated minerals, and parental magmas of podiform chromite deposits: The role of slab contamination of asthenospheric melts in suprasubduction zone environments. *Gondwana Res.* **2014**, *26*, 262–283. [CrossRef]
73. Tzamos, E.; Filippidis, A.; Rassios, A.; Grieco, G.; Michailidis, K.; Koroneos, A.; Gamaletsos, P.N. Major and minor element geochemistry of chromite from the Xerolivado–Skoumtsa mine, Southern Vourinos: Implications for chrome ore exploration. *J. Geochem. Explor.* **2016**, *165*, 81–93. [CrossRef]
74. Tzamos, E.; Kapsiotis, A.; Filippidis, A.; Koroneos, A.; Grieco, G.; Rassios, A.E.; Godelitsas, A. Metallogeny of the Chrome Ores of the Xerolivado–Skoumtsa Mine, Vourinos Ophiolite, Greece: Implications on the genesis of IPGE-bearing high-Cr chromitites within a heterogeneously depleted mantle section. *Ore Geol. Rev.* **2017**, *90*, 226–242. [CrossRef]
75. Grieco, G.; Bussolesi, M.; Tzamos, E.; Rassios, A.E.; Kapsiotis, A. Processes of primary and re-equilibration mineralization affecting chromitite ore geochemistry within the Vourinos ultramafic sequence, Vourinos ophiolite (West Macedonia, Greece). *Ore Geol. Rev.* **2018**, *95*, 537–551. [CrossRef]
76. Kamenetsky, V.S.; Crawford, A.J.; Meffre, S. Factors controlling chemistry of magmatic spinel: An empirical study of associated olivine, Cr-spinel and melt inclusions from primitive rocks. *J. Petrol.* **2001**, *42*, 655–671. [CrossRef]
77. Zaccarini, F.; Proenza, A.J.; Ortega-Gutierrez, F.; Garuti, G. Platinum group minerals in ophiolitic chromitites from Tehuitzingo (Acatlan complex, southern Mexico): Implications for post-magmatic modification. *Mineral. Petrol.* **2005**, *84*, 147–168. [CrossRef]
78. Gervilla, F.; Padrón-Navarta, J.A.; Kerestedjian, T.; Sergeeva, I.; González-Jiménez, J.M.; Fanlo, I. Formation of ferrian chromite in podiform chromitites from the Golyamo Kamenyane serpentinte, Eastern Rhodopes, SE Bulgaria: A two-stage process. *Contr. Mineral. Petrol.* **2012**, *164*, 643–657. [CrossRef]
79. González-Jiménez, J.M.; Locmelis, M.; Belousova, E.; Griffin, W.L.; Gervilla, F.; Kerestedjian, T.N.; Pearson, N.J.; Sergeeva, I. Genesis and tectonic implications of podiform chromitites in the metamorphosed Ultramafic Massif of Dobromirtsi (Bulgaria). *Gondwana Res.* **2015**, *27*, 555–574. [CrossRef]
80. Grammatikopoulos, T.A.; Kapsiotis, A.; Tsikouras, B.; Hatzipanagiotou, K.; Zaccarini, F.; Garuti, G. Spinel composition, PGE geochemistry and mineralogy of the chromitites from the Vourinos ophiolite complex, northwestern Greece. *Can. Mineral.* **2011**, *49*, 1571–1598. [CrossRef]
81. Mposkos, E.; Krohe, A. Pressure–temperature–deformation paths of closely associated ultra-high-pressure (diamond-bearing) crustal and mantle rocks of the Kimi complex: Implications for the tectonic history of the Rhodope Mountains, northern Greece. *Can. J. Earth Sci.* **2006**, *43*, 1755–1776. [CrossRef]
82. Mposkos, E.; Baziotis, I.; Proyer, A. Pressure–temperature evolution of eclogites from the Kechros complex in the Eastern Rhodope (NE Greece). *Int. J. Earth Sci.* **2012**, *101*, 973–996. [CrossRef]
83. Economou-Eliopoulos, M. On the origin of the PGE-enrichment in chromitites associated with ophiolite complexes: The case of Skyros Island, Greece. In *Digging Deeper Proceedings of the Ninth Biennial SGA Meeting, Dublin 2007*; Andrew, C.J., Ed.; Irish Association of Economic Geology: Dublin, Ireland, 2007; Volume 2, pp. 1611–1614.

84. Ifandi, E.; Zaccarini, Z.; Tsikouras, B.; Grammatikopoulos, T.; Garuti, G.; Karipi, S. First occurrences of Ni-V-Co phosphides in chromitite from the Agios Stefanos Mine, Othrys Ophiolite, Greece. *Ofioliti* **2018**, *43*, 131–145.
85. Tsikouras, B.; Etiope, G.; Ifandi, E.; Kordella, S.; Papatheodorou, G.; Hatzipanagiotou, K. Petrological Implications for the Production of Methane and Hydrogen in Hyperalkaline Springs from the Othrys Ophiolite, Greece. In Proceedings of the 13th International Congress, Chania, Greece, 5–8 September 2013; Bulletin of the Geological Society of Greece: Athens, Greece, 2013; Volume XLVII, pp. 449–457.
86. Economou-Eliopoulos, M.; Tsoupas, G.; Skounakis, V. Occurrence of Graphite-Like Carbon in Podiform Chromitites of Greece and Its Genetic Significance. *Minerals* **2019**, *9*, 152. [CrossRef]
87. Xu, X.-Z.; Cartigny, P.; Yang, J.-S.; Dilek, Y.; Xiong, F.; Guo, G. Fourier trans-form infrared spectroscopy data and carbon isotope characteristics of the ophiolite-hosted diamonds from the Luobusa ophiolite, Tibet, and Ray-Iz ophiolite, Polar Urals. *Lithosphere* **2018**, *10*, 156–169. [CrossRef]
88. Apollaro, C.; Marini, L.; Critelli, T.; Barca, D.; Bloise, A.; De Rosa, R.; Liberi, F.; Miriello, D. Investigation of rock-to-water release and fate of major, minor, and trace elements in the metabasaltserpentinite shallow aquifer of Mt. Reventino (CZ, Italy) by reaction path modeling. *Appl. Geochem.* **2011**, *26*, 1722–1740. [CrossRef]
89. Apollaro, C.; Fuoco, I.; Brozzo, G.; De Rosa, R. Release and fate of Cr(VI) in the ophiolitic aquifers of Italy: The role of Fe(III) as a potential oxidant of Cr(III) supported by reaction path modelling. *Sci. Total Environ.* **2019**, *660*, 1459–1471. [CrossRef]
90. Van der Putten, W.; Ramirez, K.; Poesen, J.; Lenka, L.; Šimek, M.; Anne, W.; Mari, M.; Philippe, L.; Heikki, S.; Andrey, Z.; et al. *Opportunities for Soil Sustainability in Europe*; European Academies' Science Advisory Council: Halle, Germany, 2018; 41p.
91. Vithanage, M.; Kumarathilaka, P.; Oze, C.; Karunatilake, S.; Seneviratne, M.; Hseu, Z.Y.; Gunarathne, V.; Dassanayake, M.; Ok, Y.S.; Rinklebe, J. Occurrence and cycling of trace elements in ultramafic soils and their impacts on human health: A critical review. *Environ. Int.* **2019**, *131*, 104974. [CrossRef]
92. Izbicki, J.A.; Ball, J.W.; Bullen, T.D.; Sutley, S.J. Chromium, chromiumisotopes and selected trace elements, western Mojave Desert, USA. *Appl. Geochem.* **2008**, *23*, 1325–1352. [CrossRef]
93. Kazakis, N.; Kantiranis, N.; Kalaitzidou, K.; Kaprara, M.; Mitrakas, M.; Frei, R.; Vargemezis, G.; Tsourlos, P.; Zouboulis, A.; Filippidis, A. Origin of hexavalent chromium in groundwater: The example of the Sarigkiol Basin, Northern Greece. *Sci. Total Environ.* **2017**, *593–594*, 552–566. [CrossRef] [PubMed]
94. Sun, Z.; Wang, X.; Planavsky, N. Cr isotope systematics in the Connecticut River estuary. *Chem.Geol.* **2019**, *506*, 29–39. [CrossRef]
95. Frei, R.; Gaucher, C.; Døssing, L.N.; Sial, A.N. Chromium isotopes in carbonates—A tracer for climate change and for reconstructing the redox state of ancient seawater. *Earth Planet. Sci. Lett.* **2011**, *312*, 114–125. [CrossRef]
96. Frei, R.; Poire, D.; Frei, K.M. Weathering on land and transport of chromium to the ocean in a subtropical region (Misiones, NW Argentina): A chromium stable isotope perspective. *Chem. Geol.* **2014**, *381*, 110–124. [CrossRef]
97. Filippidis, A. Chemical variation of chromite in the central sector of xerolivado chrome mine of Vourinos, Western Macedonia, Greece. *Neues Jahrb. Mineral. Mon.* **1997**, *8*, 354–370. [CrossRef]

PERMISSIONS

All chapters in this book were first published in MDPI; hereby published with permission under the Creative Commons Attribution License or equivalent. Every chapter published in this book has been scrutinized by our experts. Their significance has been extensively debated. The topics covered herein carry significant findings which will fuel the growth of the discipline. They may even be implemented as practical applications or may be referred to as a beginning point for another development.

The contributors of this book come from diverse backgrounds, making this book a truly international effort. This book will bring forth new frontiers with its revolutionizing research information and detailed analysis of the nascent developments around the world.

We would like to thank all the contributing authors for lending their expertise to make the book truly unique. They have played a crucial role in the development of this book. Without their invaluable contributions this book wouldn't have been possible. They have made vital efforts to compile up to date information on the varied aspects of this subject to make this book a valuable addition to the collection of many professionals and students.

This book was conceptualized with the vision of imparting up-to-date information and advanced data in this field. To ensure the same, a matchless editorial board was set up. Every individual on the board went through rigorous rounds of assessment to prove their worth. After which they invested a large part of their time researching and compiling the most relevant data for our readers.

The editorial board has been involved in producing this book since its inception. They have spent rigorous hours researching and exploring the diverse topics which have resulted in the successful publishing of this book. They have passed on their knowledge of decades through this book. To expedite this challenging task, the publisher supported the team at every step. A small team of assistant editors was also appointed to further simplify the editing procedure and attain best results for the readers.

Apart from the editorial board, the designing team has also invested a significant amount of their time in understanding the subject and creating the most relevant covers. They scrutinized every image to scout for the most suitable representation of the subject and create an appropriate cover for the book.

The publishing team has been an ardent support to the editorial, designing and production team. Their endless efforts to recruit the best for this project, has resulted in the accomplishment of this book. They are a veteran in the field of academics and their pool of knowledge is as vast as their experience in printing. Their expertise and guidance has proved useful at every step. Their uncompromising quality standards have made this book an exceptional effort. Their encouragement from time to time has been an inspiration for everyone.

The publisher and the editorial board hope that this book will prove to be a valuable piece of knowledge for researchers, students, practitioners and scholars across the globe.

LIST OF CONTRIBUTORS

So Jeong Park and Choong Kyun Rhee
Department of Chemistry, Chungnam National University, Daejeon 34134, Korea

Min Hee Joo, Sung Min Hong and Youngku Sohn
Department of Chemistry, Chungnam National University, Daejeon 34134, Korea
Department of Chemical Engineering and Applied Chemistry, Chungnam National University, Daejeon 34134, Korea

Anja Breuker, Simon F. Ritter and Axel Schippers
Federal Institute for Geosciences and Natural Resources (BGR), Stilleweg 2, 30655 Hannover, Germany

Martiya Sadeghi, Nikolaos Arvanitidis and Anna Ladenberger
Department of Mineral Resources, Geological Survey of Sweden, Uppsala, Sweden

Leonid Chaikin, Andrei Shoppert, Irina Loginova and Julia Napol'skikh
Department of Non-Ferrous Metals Metallurgy Ural Federal University Named after the First President of Russia B.N. Yeltsin, 620002 Yekaterinburg, Russia

Dmitry Valeev
Laboratory of New Metallurgical Processes (#24), A.A. Baikov Institute of Metallurgy and Materials Science, Russian Academy of Sciences, 49, Leninsky Prospect, 119334 Moscow, Russia
Scientific Research Centre "Thermochemistry of Materials", National University of Science & Technology "MISIS", 4, Leninsky Prospect, 119049 Moscow, Russia

Demetrios G. Eliopoulos
Institute of Geology and Mineral Exploration (IGME), Sp. Loui 1, Olympic Village, GR-13677 Acharnai, Greece

Maria Economou-Eliopoulos
Department of Geology and Geoenvironment, University of Athens, 15784 Athens, Greece

Kenneth N. Han
South Dakota School of Mines and Technology, Rapid City, SD 57701-3995, USA

Connor A. Gray and Adrian D. Van Rythoven
Department of Environmental, Geographical, and Geological Sciences, Bloomsburg University of Pennsylvania, Bloomsburg, PA 17815, USA

Yee-Kwong Leong
Department of Chemical Engineering, University of Western Australia, Crawley, WA 6009, Australia

Harry Watts
Department of Chemical Engineering, University of Western Australia, Crawley, WA 6009, Australia
Watts & Fisher Pty Ltd., Crawley, WA 6009, Australia

Raisa G. Kravtsova, Vladimir L. Tauson, Artem S. Makshakov, Nikolay V. Bryansky and Nikolay V. Smagunov
A.P. Vinogradov Institute of Geochemistry, Siberian Branch of Russian Academy of Sciences, Favorskiy str 1A, 664033 Irkutsk, Russia

Xinbo Yang and Rick Q. Honaker
Department of Mining Engineering, University of Kentucky, Lexington, KY 40506, USA

Luca Bindi
Dipartimento di Scienze della Terra, Università degli Studi di Firenze, I-50121 Florence, Italy

Federica Zaccarini and Giorgio Garuti
Department of Applied Geological Sciences and Geophysics, University of Leoben, A-8700 Leoben, Austria

Basilios Tsikouras
Faculty of Science, Physical and Geological Sciences, Universiti Brunei Darussalam, BE 1410 Gadong, Brunei Darussalam

Elena Ifandi
Faculty of Science, Physical and Geological Sciences, Universiti Brunei Darussalam, BE 1410 Gadong, Brunei Darussalam
Department of Geology, Section of Earth Materials, University of Patras, 265 00 Patras, Greece

Chris Stanley
Department of Earth Sciences, Natural History Museum, London SW7 5BD, UK

Daniela Mauro
Dipartimento di Scienze della Terra, Università degli Studi di Pisa, I-56126 Pisa, Italy

Rina Kim, Kyeong Woo Chung, Ho-Sung Yoon and Chul-Joo Kim
Resources Recovery Research Center, Mineral Resources Division, Korea Institute of Geoscience and Mineral Resources (KIGAM), Daejeon 34132, Korea

Heechan Cho and Jinan Jeong
Department of Energy Resources Engineering, Seoul National University, Seoul 08826, Korea

Jihye Kim
Department of Chemical Engineering and Applied Chemistry, University of Toronto, Toronto, ON M5S 3E5, Canada

Sugyeong Lee
Robert M. Buchan Department of Mining, Queen's University, Kingston, ON K7L 3N6, Canada

Ioannis Mitsis
Department of Geology and Geoenvironment, University of Athens, 15784 Athens, Greece

Robert Frei
Department of Geosciences and Natural Resource Management, University of Copenhagen, Øster Voldgade 10, 1350 Copenhagen K, Denmark

Index

www.ingramcontent.com/pod-product-compliance
Lightning Source LLC
LaVergne TN
LVHW081021210926
848485LV00004B/916
* 9 7 8 1 6 4 7 4 0 4 0 2 4 *